Visible Light Active Photocatalysts for Environmental Remediation and Organic Synthesis

Visible Light Active Photocatalysts for Environmental Remediation and Organic Synthesis

Editor

Vincenzo Vaiano

MDPI • Basel • Beijing • Wuhan • Barcelona • Belgrade • Manchester • Tokyo • Cluj • Tianjin

Editor
Vincenzo Vaiano
Industrial Engineering
Salerno
Fisciano
Italy

Editorial Office
MDPI
St. Alban-Anlage 66
4052 Basel, Switzerland

This is a reprint of articles from the Special Issue published online in the open access journal *Photochem* (ISSN 2673-7256) (available at: www.mdpi.com/journal/photochem/special_issues/Visible_Light_Photo).

For citation purposes, cite each article independently as indicated on the article page online and as indicated below:

LastName, A.A.; LastName, B.B.; LastName, C.C. Article Title. *Journal Name* **Year**, *Volume Number*, Page Range.

ISBN 978-3-0365-3648-4 (Hbk)
ISBN 978-3-0365-3647-7 (PDF)

Contents

About the Editor

Vincenzo Vaiano

Vincenzo Vaiano graduated in Chemical Engineering at the University of Naples II. In 2005, he conducted research activities at the University of Bradford (UK) under the supervision of Prof. Roger I. Bickley (pioneer in the field of heterogeneous photocatalysis and author of several scientific papers on the study of photocatalysts). In March 2006, he was awarded the PhD in Chemical Engineering (Thesis: Photocatalytic Selective Oxidation of Cyclohexane). Actually, he is Associate Professor of Industrial Chemistry at Department of Industrial Engineering of University of Salerno. From 22 December 2016 to 2019, he was Assistant Professor at the Department of Industrial Engineering of the University of Salerno. From 22 January 2007 to 2016, he was a Research Fellow at the University of Salerno (Projects: applications of photocatalysis for the selective partial oxidation of hydrocarbons under mild conditionsand of selective partial oxidation products by photocatalytic reactions). The main research lines of Vincenzo Vaiano regard photocatalysis for sustainable chemistry, photocatalytic and photo-Fenton processes for pollutant removal in wastewater, catalytic combustion of sewage sludge, decomposition and oxidative decomposition of H_2S, hydrolysis of COS in the liquid phase.

MDPI

Article

Study of the Response Surface in the Photocatalytic Degradation of Acetaminophen Using TiO_2

Adriana Marizcal-Barba [1], Jorge Alberto Sanchez-Burgos [2], Victor Zamora-Gasga [2] and Alejandro Perez Larios [1,*]

[1] Materials, Water and Energy Research Laboratory, Department of Engineering, Los Altos University Center, University of Guadalajara, Tepatitlán de Morelos 47600, Mexico; adriana.marizcal7736@alumnos.udg.mx

[2] Tecnológico Nacional de México/Instituto Tecnológico de Tepic, Av. Tecnológico 2595, Lagos del Country, Tepic 63175, Mexico; jsanchezb@ittepic.edu.mx (J.A.S.-B.); vzamora@ittepic.edu.mx (V.Z.-G.)

* Correspondence: alarios@cualtos.udg.mx

Abstract: An effective way to obtain the optimal parameters of a process or experiment is the response surface method. Using the Box–Behnken design further decreases the number of experiments needed to obtain sufficient data to obtain a reliable equation. From the equation, it is possible to predict the behavior of the response with respect to the combination of variables involved. In this study we evaluated the photocatalytic activity of the synthesized TiO_2 for the degradation of acetaminophen, a frequently used and uncontrolled drug that has been detected with increasing frequency in wastewater effluents. The variables used for this study were pH, contaminant concentration (acetaminophen) and catalyst dose. We found, with a 95% confidence level, that 99% of the contaminant can be degraded to pH 10, contaminant to 35 mg/L and a catalyst dose of 0.15 g TiO_2.

Keywords: acetaminophen; photocatalytic activity; TiO_2; response surface method; Box–Behnken design

Citation: Marizcal-Barba, A.; Sanchez-Burgos, J.A.; Zamora-Gasga, V.; Perez Larios, A. Study of the Response Surface in the Photocatalytic Degradation of Acetaminophen Using TiO_2. *Photochem* **2022**, *2*, 225–236. https://doi.org/10.3390/photochem2010017

Academic Editor: Vincenzo Vaiano

Received: 5 February 2022
Accepted: 7 March 2022
Published: 10 March 2022

1. Introduction

The response surface method (RSM) is one of the experimental statistical design techniques that is applied to building models and investigating the effects and interactions of all selected operating conditions on the response of a given experiment [1,2]. This method is very effective for the optimization of complex processes, and allows researchers to obtain the optimal conditions of an operation [3]; this results in more convenience, in that it saves time, labor and costs [4,5]. Among the most important RSM methodologies, are the 2^k and 3^k factorial designs, where 2 and 3 are the number of levels to test, and k is the number of controllable factors [3]. However, a full factorial design (FFD) is impractical because of the large number of experiments required to predict the outcome [6]. Due to the above, in a statistical design, what is sought is to accurately predict all the positions of the factorial space that are equidistant from the center [6]–the most common being Artificial Neural Networks (ANN), the Central Composite Design (CCD) and the Box–Benkhen Design (BBD)–n order to optimize the response. The Box–Behnken Design (BBD) provides us with a second-order response model and can maximize the amount of complex information with minimum experimentation time [7,8]; even more importantly, it can avoid the need for analyses of their extreme combinations. With this technique, it is possible to investigate the effects from three to seven factors, each of them with three levels of experimental condition (level: low, medium and high). This is proven to be more efficient than other response surface designs [2,9]. The data obtained from the BBD have been evaluated using the analysis of variance (ANOVA) technique, to determine which of the controllable factors are statistically significant, since it compares the quadratic sum of the sources of variation and provides a confidence interval above 90% [3,6,9]. Among the most investigated variables in photocatalysis in an RSM are the dose of the catalyst, the concentration of the contaminant and the pH of the solution [2,5,8–10]. On a smaller scale, the following have been studied:

the wavelength of the irradiation source [11], the dose of O_2 [10], and the size of the TiO_2 particles [6].

Titanium dioxide (TiO_2) has been a subject of study in recent decades due to its wide range of applications in areas such as pigments [12], lubricants [7], optical sensors [13], food and cosmetics [14], either in thin films as two-dimensional material [15], or in nanoparticles—as well as nanocomposites of TiO_2-Fe_2O_3 [16], Ti-Zr [17] and graphene-TiO_2 [18]—for photocatalytic applications in hydrogen production [19,20], and water treatment [21]. In this sense, the elimination of mainly phenolic compounds [6], organic pollutants [10], colorants [2,22], pesticides and herbicides [23,24], and drugs [3,8,25–27] has been studied. In addition, its ability to be used for prolonged periods of time and the reproducibility of the results of photocatalytic activity have been studied [19,26]. The photocatalytic process is an advanced oxidation method based on the generation of hydroxyl radicals, when UV light is irradiated on a semiconductor catalyst [2]. When TiO_2 nanoparticles are exposed to a light source that has an energy higher than its bandgap (TiO_2 = 3.2 eV), it produces the most powerful intermediate oxidative radicals [10]. Irradiation causes electrons in the valence band to migrate to the conduction band, creating the electron-hole pair (e^-–h^+) [14]. On the surface of TiO_2, the holes (h^+) of the valence band can react with hydroxide ions (OH) or adsorbed H_2O; therefore, the adsorption properties of the substrate affect the reaction rate [5,6]. Meanwhile, the electrons (e^-) of the conduction band can be captured or interact with oxygen molecules; therefore, the availability of oxygen in the aqueous phase has an effect on the photocatalytic activity to generate hydroxyl radicals [20]. Due to their high oxidation potential, they react with most organic compounds to form simpler species such as CO_2 and H_2O, achieving mineralization [3]. On the other hand, statistical methods have been reported to investigate the effects of catalyst dose, pH, contaminant concentration, light source wavelength, etc., albeit studying one parameter at a time [22,23]. However, these methods are inefficient to estimate the effects between the interaction of the factors, so a reliable prediction cannot be made in a photocatalytic system [9,28].

In this work, the efficiency of the photocatalytic degradation of acetaminophen with TiO_2 is investigated, using the analysis of the influence of three factors and their interactions on the determination of the optimal conditions of the experiment. The factors to be studied are catalyst dose, reaction pH and contaminant concentration, using a BBD as an RSM optimization method.

2. Materials and Methods

2.1. Chemical Reagents

Ethyl alcohol (Aldrich 99.4%), distilled H_2O and titanium (IV) butoxide ($C_{16}H_{36}O_4Ti$, Sigma-Aldrich 97%) were used as precursors.

2.2. Synthesis of TiO_2

TiO_2 was synthesized by the sol-gel method. Ethyl alcohol and H_2O with a molar ratio of 8:1 were placed in a three-necked flask, and the solution was heated to 70 °C to add titanium butoxide dropwise. After 24 h, the material was dried at 100 °C and ground in an agate mortar. Finally, the material was calcined at 500 °C for 5 h with a heating ramp of 2 °C/min.

2.3. Characterization of TiO_2

For the micrographs, a TESCAN brand scanning electron microscope, model MIRA3 (LMU, London, UK), with a power of 20.0 KV, was used.

X-ray diffraction studies were performed in Panalytical equipment, empyrean model (Empyrean, Almelo, The Netherland), with Cu Kα radiation (λ = 0.154 nm) and with a diffraction angle (2θ) of 10 to 90°, using a step of 0.03° and a time of 3 s per step.

UV-vis diffuse reflectance spectra were obtained with a Shimadzu UV-Vis spectrophotometer, model UV-2600 (Shimadzu UV-2600, Tokyo, Japan) coupled with an integrating sphere for diffuse reflectance studies. The diffuse reflectance spectrum was obtained and transformed to a magnitude proportional to the extinction coefficient (α) through the Kubelka–Munk function, using wavelengths in the range of 900 to 190 nm.

X-ray photoelectron spectroscopy was analyzed in SPECS model III equipment (Thermo Scientific K Alpha, Tokyo, Japan), with monochromatic Al K radiation (1486 eV) and a scanning resolution of 0.1 eV. The survey and high-resolution spectra of the sample were recorded in a constant step energy mode at 60 eV, using a spot size of 400 mm. The vacuum in the analysis chamber was maintained at 1×10^{-9} Torr during the analysis. Sample loading effects were corrected for using the O1s offset from 531.0 eV. The system was calibrated by determining the position of the Au 4f peak at 84.00 ± 0.05 eV.

2.4. Photocatalytic Activity of TiO_2

The photocatalytic activity of TiO_2 for the degradation of acetaminophen (ACTP) was investigated using a 400 mL pyrex reactor under UV light, where a 1 mW cm^{-2} UV lamp with a wavelength of 256 nm was placed, covered with a quartz tube for immersion. 350 mL of aqueous solution was prepared for each reactor, according to the experimental design developed, to evaluate the variables of contaminant concentration (20, 30 and 40 mg L^{-1}), pH (4, 7 and 10) and catalyst dose (50, 100 and 150 mg) under constant stirring. Before proceeding to the photoactivity test, the solution was kept in a dark environment to create an adsorption–desorption equilibrium [29]. The samples were analyzed in a Shimadzu 2600 model UV-Vis spectrophotometer (Shimadzu UV-2600, Tokyo, Japan), from 500 to 190 nm. The degradation curves were obtained by measuring the absorbance at the wavelength (243–246 nm) corresponding to acetaminophen [11,27], as a function of time (every 30 min for 3 h). To calculate the percentage efficiency of acetaminophen degradation, Equation (1) was used:

$$\text{Yield } (\%) = \frac{C_A - C_t}{C_A} * 100 \tag{1}$$

where C_A is the initial concentration of the contaminant and C_t is the final concentration of the contaminant.

The total organic carbon of the samples was analyzed using the Shimadzu model TOC-LCSN equipment (Shimadzu, Tokyo, Japan), and applying Equation (2)

$$TOC = TC - IC, \tag{2}$$

where TOC is the amount of total organic carbon (mg L^{-1}), TC is the total amount of carbon (mg L^{-1}), and IC is the amount of inorganic carbon (mg L^{-1}) present in the aqueous solution.

To calculate the percentage yield of TOC conversion, the Equation (3) was used:

$$\text{Yield } (\%) = \frac{TOC_0 - TOC_{residual}}{TOC_0} * 100 \tag{3}$$

where TOC_0 is the initial concentration and $TOC_{residual}$ is the final concentration.

2.5. Response Surface Method (RSM)

The RSM establishes a mathematical relationship between the variables evaluated and the results obtained to adapt a second-order polynomial model according to Equation (4):

$$y = \beta_0 + \sum_{i=1}^{k} \beta_i x_i + \sum_{i=1}^{k} \beta_{ii} x_i^2 + \sum_{i=1}^{k} \sum_{j=i+j}^{k} \beta_{ij} x_i x_j + \varepsilon \tag{4}$$

where y is the response variable, x_i is the value for the parameters X_A, X_D and X_{pH}, β_0 is a constant, β_i is the value of the regression coefficient, k is the number of independent variables and ε is the effect of the experimental error.

For the statistical analysis, the analysis of variance method (ANOVA) was applied, establishing the probability value (p) of 95% ($p < 0.05$), as the statistical significance level parameter for the proposed model. On the other hand, the efficiency of the model was evaluated by means of the correlation coefficient (R^2) and the adjusted correlation coefficient (R^2 adjusted) and the reproducibility was determined by the experimental error.

Data were expressed as mean standard deviation (n). The analyzes of variance used to assess whether a term has a significant effect ($p < 0.05$) were performed using one-factor ANOVA and the Tukey test, both analyzed using the STATISTICA v software. 10 (Statsoft, Tulsa, OK, USA).

To determine the optimal conditions for degradation and mineralization of acetaminophen in a photocatalytic reactor, the 3-factor Box–Benkhen design with 3 central points and 3 repetitions was applied to determine the optimal conditions to maximize degradation efficiency of the acetaminophen contaminant. The method consisted of defining a minimum or low level (denoted as −1), a central or medium level (denoted as 0), and a high or maximum level (denoted as 1) for each experimental factor, as shown in Table 1. The effects of the interactions between the experimental factors and their influence on the response were quantified to optimize: the initial concentration of the acetaminophen contaminant (X_A mg L^{-1}), the dose of the catalyst (X_D mg) and the pH of the solution (X_{pH} pH).

Table 1. Variables and levels established for statistical analysis.

Variables	Factor	Range and Established Levels		
		−1	**0**	**1**
XA	ACTP concentration [1]	20	30	40
XD	Catalyst dosageTiO_2 [2]	50	100	150
XpH	pH of solution	4	7	10

[1] mg∗L^{-1}, [2] mg.

Experimental factors and levels were selected for each factor based on literature values, available resources and results of preliminary experiments.

3. Results and Discussion

3.1. Characterization

3.1.1. Scanning Electron Microscopy (SEM)

In Figure 1, the micrographs of TiO_2 SEM, which presents irregularly shaped hemispherical agglomerated morphology, can be seen. It is suggested that this type of agglomeration is due to electrostatic attractions and/or ionization energy [9].

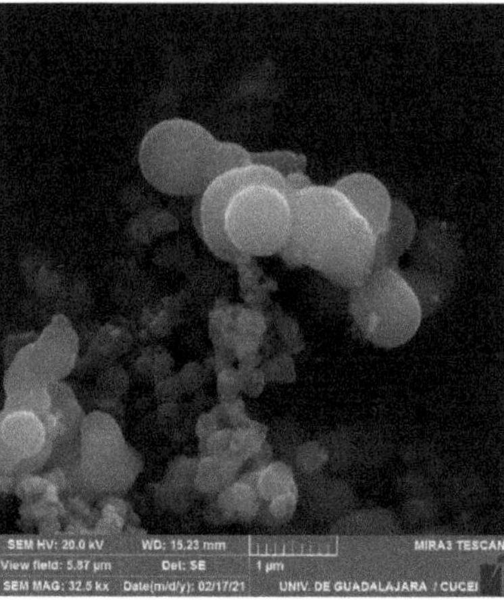

Figure 1. Micrograph of the TiO_2 NPs.

3.1.2. X-ray Diffraction (XRD)

Figure 2 shows the XRD patterns of the sample (calcination at 500 °C). The TiO_2 pattern is consistent with diffractions at 2θ = 25.4°, 36°, 46°, 53, 53°, 65°, which are assigned to the anatase crystal phase (JCPDD: 21 1272) and correspond to index Miller (101), (004), (200), (105), (211) and (204). This is corroborated with previous studies where the anatase phase is obtained after anneal at 500 °C [17,30]. It also suggests that the bonds of the original structure are maintained after calcination.

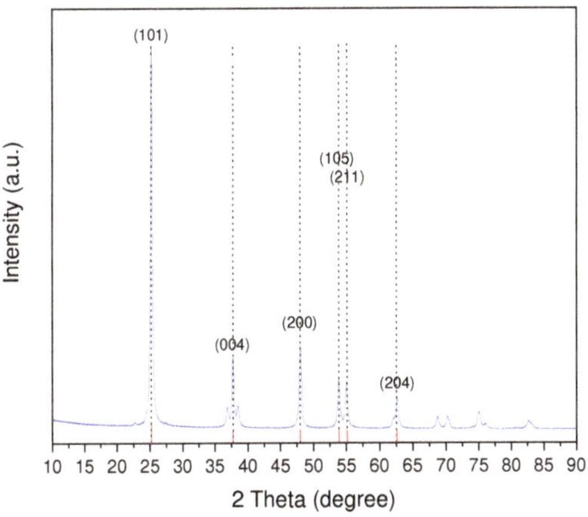

Figure 2. X-ray diffractogram of TiO_2.

3.1.3. UV-Vis Spectroscopy

Figure 3 shows the UV-Vis spectrum of TiO_2 as a photocatalyst. The spectrum shows an important absorption within the UV range (<400 nm), where two peaks are shown in the region from 400 to 190 nm; this suggests that the peak at 210 nm (green line) could be associated with charge transfer of the ligand–metal between the tetrahedral Ti^{4+} and an an oxygenated ligand such as O^{2-}, and the second peak at 350 nm (red line) is attributed to the Ti^{4+} cations in the octahedral environment, as well as to the anatase phase [5]. On the other hand, the calculated E_g value was 3.20 eV (Figure 4), which coincides with those reported in the literature [31].

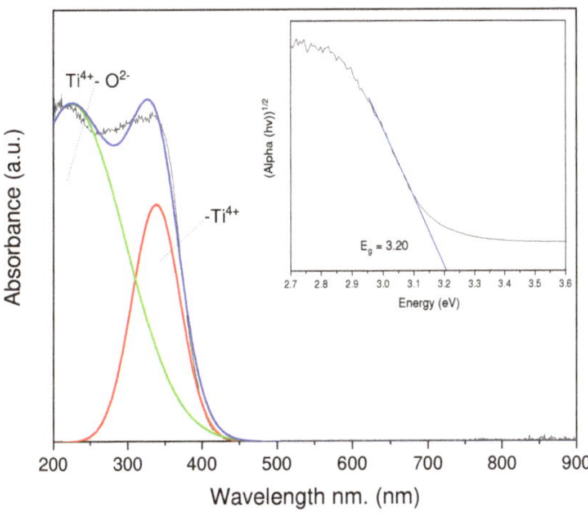

Figure 3. UV−Vis spectrum of TiO_2.

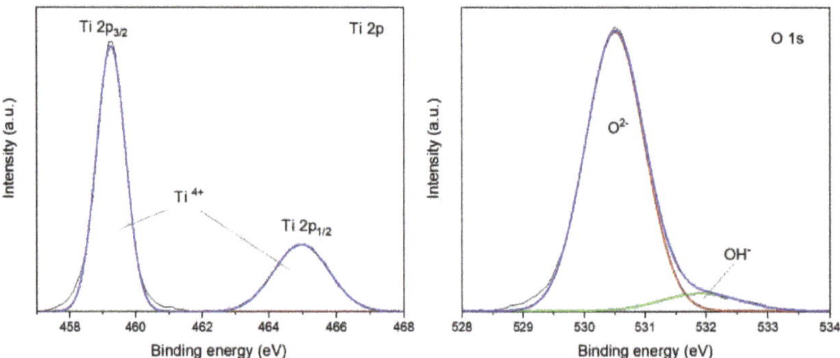

Figure 4. XPS spectra of TiO_2.

3.1.4. X-ray Photoelectron Spectroscopy

The superficial chemical composition of TiO_2 was analyzed by X-ray photoelectron spectroscopy. Figure 4 shows the spectra for Ti 2p, O1s and C1s. In Figure 4 (Ti 2p), the components of the split spin-orbit of Ti 2p, related to the components Ti $2p_{1/2}$ and Ti $2p_{3/2}$, can be seen; these are found in the band position of 455 and 468 eV, respectively; these peaks are associated with the presence of Ti^{4+} species, suggesting that there are no defects associated with the TiO_2 lattice. On the other hand, the behavior of oxygen (O1s), is characteristic of the material located at 530.5 eV, where oxygen tends to attract electrons, making the nucleus more electronegative [17]. Furthermore, the bands located at 532 eV, are attributable to the presence of weakly adsorbed species associated with surface hydroxyl (OH) groups (Figure 4 (O 1s)) [30].

3.2. Photocatalytic Activity

3.2.1. Model Validation

For the response surface optimization study, the photocatalytic degradation of acetaminophen was performed at each design point of the three factors (catalyst weight, contaminant concentration, and pH solution) at three levels each. Considering this design, three replicates of 37 experiments were performed.

In order to avoid any systematic bias in the results, the experiments were performed randomly and the responses of other process factors, not selected for the object of study of the experimental design, are considered an error for the experimental design. The coefficients of the quadratic model, which describes the percentage of degradation (efficiency) as a function of the reaction condition (independent variable), were calculated by means of a multiple regression analysis on the experimental data.

The analysis of the BBD was carried out considering a quadratic model for its prediction. The results of the coefficients obtained from the effect of the following factors are presented individually in Table 2, in addition to the linear (L) and quadratic (Q) interactions of the model: X_A—concentration of acetaminophen (ACTP); X_{pH}—pH; and X_D—dose of catalyst (Wt of catalyst). Those factors and interactions that present a significant effect ($p < 0.05$) are shown, and are that of the ACTP factor that had the greatest effect (linear and quadratic coefficient, respectively), followed by the linear coefficient of the pH factor, the linear coefficient of the "Wt of catalyst" factor, and the interaction of the linear coefficient of the ACTP and pH factors. The obtained model has an R^2 fit of 0.85 and R^2 adjusted of 0.81, for which the model is considered to have a good prediction.

Table 2. Effect estimates for the efficiency quadratic model.

Factor	Effect	Std.Err.	t (27)	*p*
Mean/Interc.	87.117	0.977	89.178	0.000
(1) [ACTP] (L)	−20.361	2.398	−8.492	0.000
[ACTP] (Q)	11.037	1.718	6.423	0.000
(2) pH (L)	12.128	2.41	5.032	0.000
pH (Q)	3.03	1.727	1.755	0.091
(3) Wt of catalyst (L)	4.934	2.346	2.103	0.045
Wt of catalyst (Q)	−0.309	1.718	−0.18	0.859
1L by 2L	16.57	3.481	4.761	0.000
1L by 3L	4.832	3.334	1.449	0.159
2L by 3L	−1.133	3.3	−0.343	0.734

According to the mathematical model, the optimal conditions of the process are those presented in Table 3. It is observed that the optimal conditions are an ACTP level of 35, with a pH of 10 and Wt of catalyst of 150 mg. Table 4 shows the prediction under optimal conditions, breaking down the values that would be obtained in the model coefficients and that would allow an efficiency value of 99.03%.

Table 3. Optimal process conditions.

Factor	Level Factor	Predicted Efficiency	Desirability Value	−95% CI Efficiency	+95% CI Efficiency
ACTP concentration [1]	35	99.034	1.000	93.062	105.006
Catalyst dosage TiO_2 [2]	150				
pH of solution	10				

[1] $mg{*}L^{-1}$, [2] mg.

Table 4. Breakdown of model coefficients under optimal conditions.

Factor	Regression Coefficients	Value	Value
Constant	63.005	0	0
(1) [ACTP] (L)	3.188	35	111.569
[ACTP] (Q)	−0.11	1225	−135.202
(2) pH (L)	−1.174	10	−11.736
pH (Q)	−0.337	100	−33.662
(3) Wt of catalyst (L)	−0.094	150	−14.084
Wt of catalyst (Q)	0	22,500	2.781
1L by 2L	0.276	350	96.661
1L by 3L	0.005	5250	25.366
2L by 3L	−0.004	1500	−5.664
Predicted			99.034
−95% Conf.			93.062
+95% Conf.			105.006

Acetaminophen concentration (ACTP); Catalyst dosage TiO_2 (Wt of catalyst); pH of solution (pH).

The t-value of effects is set on the Pareto chart (Figure 5). The Pareto diagram was obtained to observe the hierarchy of the effects in each term on the efficiency. If the t-value of effects set on the Pareto charts is less than or equal to the significance level ($p < 0.05$), this reveals that there is a statistically significant association between the response variable and the term in the model. Significant effects on efficiency were found in the following order: ACTP Linear > ACTP Quadratic > pH Linear > ACTP Linear * pH Linear > Wt of Catalyst Linear. The regression coefficients of the quadratic model (Equation (5) and Table 4) show the effects of each term on the efficiency. Positive coefficients indicated an increase in efficiency, while negative coefficients indicated a decrease in efficiency. In this sense, it was observed that ACTP Linear and ACTP Linear * pH Linear favored greater efficiency. However, the quadratic ACTP, linear pH and quadratic Wt of catalyst terms decreased the efficiency, and the other terms in the model were not statistically significant ($p > 0.05$); therefore, they do not influence efficiency.

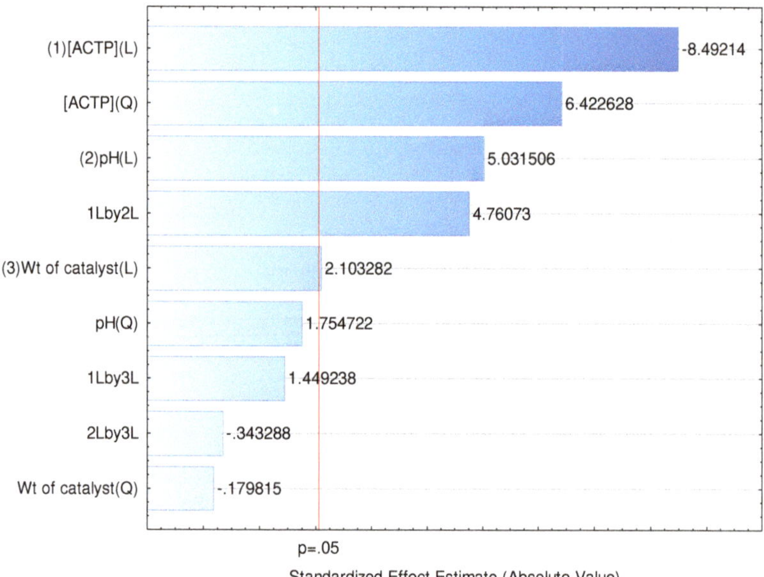

Figure 5. Pareto chart.

Corresponding to the experimental dates, the general mathematical model is the one presented in Equation (5):

$$\begin{aligned}\text{Efficiency} = \;& 63.00 + 3.18 * \text{ACTP} - 0.11 * \text{ACTP}^2 - 1.17 * \text{pH} - 0.33 * \text{pH}^2 \\ & -0.09\text{Wt of Catalyst} + 0.00 * \text{Wt of Catalyst}^2 + 0.27 * \text{ACTP} \\ & *\text{pH} + 0.00 * \text{ACTP} * \text{Wt of Catalyst} - 0.00 * \text{pH} \\ & *\text{Wt of Catalyst}\end{aligned} \quad (5)$$

This model describes the regression coefficients (Table 4, column of regression coefficients), corresponding to the quadratic and linear parts of the equation as a function of the factors (ACTP, pH, Wt of catalyst) and their interactions, with which the behavior of the response variable (efficiency) can be predicted.

In Figure 6a, the relationship between the pH and ACTP response surface is presented; when the Wt of Catalyst value is set to 100, there is an efficiency increase with a decrease in ACTP concentration, but it remains stable throughout the basic pH. In Figure 6b, the response surface graph of Wt of Catalyst and ACTP at pH 7.24 is shown; it can be noted that higher efficiency can be obtained when Wt of Catalyst increases, depending on the increase in the initial ACTP concentration. On the other hand, Figure 6c shows the response surface plot of pH and Wt of Catalyst at a fixed ACTP value of 29.72. It can be seen that the highest efficiency values are found in pH values from 8 to 11, and at any Wt of Catalyst; however, due to process conditions, the appropriate pH is 10. In addition, the initial ACTP concentration also has an effect, with an optimized dose of 35 mg/L for maximum degradation efficiency.

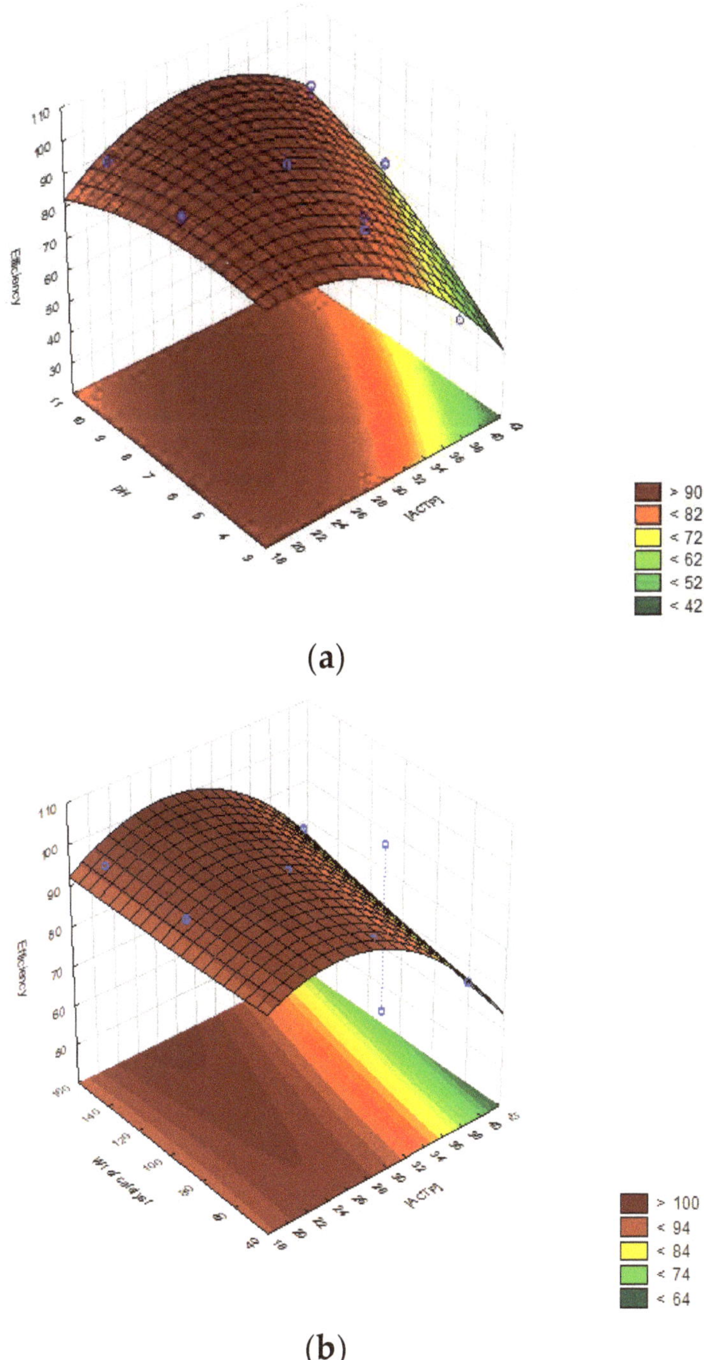

Figure 6. *Cont.*

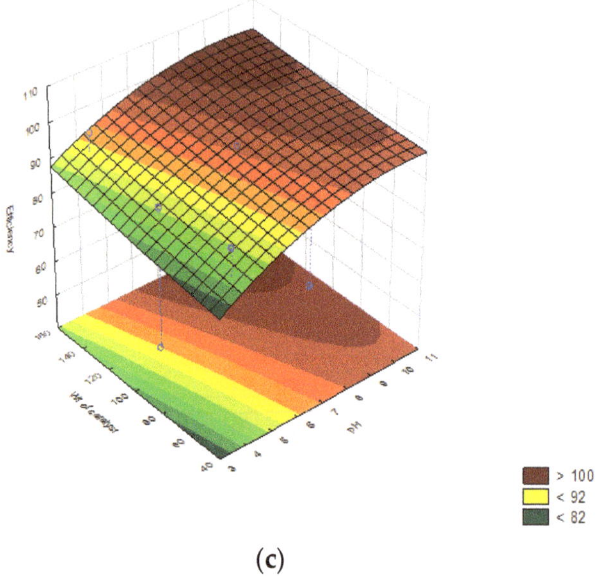

(c)

Figure 6. Response surface plot of: (**a**) pH and ACTP at a fixed Wt of Catalyst value of 100; (**b**) Wt of Catalyst and ACTP at a fixed pH value of 7.24; and (**c**) pH and Wt of Catalyst at a fixed ACTP value of 29.72.

Additional experiments were performed to confirm the validity and accuracy of the response surface model within the design variables considered. It was compared with the degradation percentage calculated using experimental data at a 0.15 g TiO_2 dose, a 35 mg/L contaminant concentration and 10 pH in the solution, obtaining a degradation percentage of 97.19%, with an S.D. of 0.04%. The experimental response is 1.86% lower than the expected maximum response.

3.2.2. TOC Analysis

Furthermore, in the additional experiments, the samples used for the TOC analysis were also obtained. The percentage of mineralization was obtained using Equation (3), obtaining 60% during the first 3 h of irradiation and continuing up to 95% at 6 h of irradiation.

3.2.3. Proposed Photocatalytic Mechanism

TiO_2 is capable of absorbing photons from UV-light-generating holes (h^+) in the valence band and e^- in the conduction band, as shown in reaction 1 (r1). Those h^+ can then form hydroxyl radicals ($OH^\bullet$) through oxidative reactions 2 to 4 ((r2) to (r4)).

$$\text{Light irradiation} + TiO_2 \rightarrow TiO_2\ (h^+ + e^-) \quad \text{(r1)}$$

$$h^+ + H_2O \rightarrow H^+ + OH^\bullet \quad \text{(r2)}$$

$$2h^+ + 2H_2O \rightarrow 2H^+ + H_2O_2 \quad \text{(r3)}$$

$$H_2O_2 \rightarrow OH^\bullet + OH^\bullet \quad \text{(r4)}$$

Furthermore, the electrons in the conduction band form super oxide radical anions ($O_2^{\bullet-}$) via reductive reactions, as shown in the follow reactions 5 to 7 ((r5) to (r7))

$$2e^- + O_2 \rightarrow O_2^{\bullet-} \quad \text{(r5)}$$

$$O_2^{\bullet-} + 2H^+ \rightarrow H_2O_2 + O_2 \quad \text{(r6)}$$

$$H_2O_2 \rightarrow OH^{\bullet} + OH^{\bullet} \quad (r7)$$

Radicals $O_2^{\bullet-}$ and $OH^{\bullet}$ are strong oxidants that can oxidize recalcitrant compounds such as acetaminophen, because $OH^{\bullet}$ tends to remove hydrogen or attack the C-C unsaturated bonds, and the $O_2^{\bullet-}$ can lead to total mineralization [5].

4. Conclusions

TiO_2 was successfully prepared by the sol-gel method. The TiO_2 nanoparticles possess a spherical agglomeration in cylindrical form. The photocatalyst band gap energy is 3.2 eV and exhibits photocatalytic activity in the degradation of ACTP under UV-Vis light irradiation. The regression coefficients of the equation obtained show the effects of each variable and their interactions on the efficiency response of our degradation experiments. A catalyst loading of 0.15 g is the optimum dosage to enhance the removal rate. The amount of ACTP removed increases with the initial concentration of the photocatalyst. Increasing the pH value does not improve removal efficiency due to the surface charge. A multivariate experimental design was used to develop a quadratic model as the functional relationship between the percentage removal of ACTP and the three independent variables. Response surface methodology with a BBD was successfully employed to investigate the significance of the factors at different levels during the ACTP removal. A satisfactory goodness-of-fit was observed between the predictive and experimental results. This indicates that response surface methodology is applicable in optimization of the removal of ACTP by TiO_2.

Author Contributions: A.M.-B., writing—original draft; J.A.S.-B. and V.Z.-G., formal analysis; A.P.L., writing—review and editing. All authors have read and agreed to the published version of the manuscript.

Funding: This research received no external funding.

Institutional Review Board Statement: Not applicable.

Informed Consent Statement: Not applicable.

Data Availability Statement: Not applicable.

Acknowledgments: A. Marizcal-Barba thanks CONACYT for the scholarship (799894) and appreciates the support in the characterization of photocatalysts from Martín Flores and Milton Vázquez. Thanks are also extended to the technicians of the equipment, Ing. Sergio Oliva and José Rivera, for the characterization of the SEM and XPS analysis (project 270660, Support for the Strengthening and Development of the Scientific and Technological Infrastructure).

Conflicts of Interest: The authors declare no conflict of interest.

References

1. Hafeez, A.; Taqvi, S.A.A.; Fazal, T.; Javed, F.; Khan, Z.; Amjad, U.S.; Bokhari, A.; Shehzad, N.; Rashid, N.; Rehman, S.; et al. Optimization on cleaner intensification of ozone production using Artificial Neural Network and Response Surface Methodology: Parametric and comparative study. *J. Clean. Prod.* **2020**, *252*, 119833. [CrossRef]
2. Mortazavian, S.; Saber, A.; James, D.E. Optimization of Photocatalytic Degradation of Acid Blue 113 and Acid Red 88 Textile Dyes in a UV-C/TiO_2 Suspension System: Application of Response Surface Methodology (RSM). *Catalysts* **2019**, *9*, 360. [CrossRef]
3. Vebber, M.C.; da Silva Crespo, J.; Giovanela, M. Self-assembled thin films of PAA/PAH/TiO_2 for the photooxidation of ibuprofen. Part I: Optimization of photoactivity using design of experiments and surface response methodology. *Chem. Eng. J.* **2019**, *360*, 1447–1458. [CrossRef]
4. Peng, T.; Ray, S.; Veeravalli, S.S.; Lalman, J.A.; Arefi-Khonsari, F. The role of hydrothermal conditions in determining 1D TiO_2 nanomaterials bandgap energies and crystal phases. *Mater. Res. Bull.* **2018**, *105*, 104–113. [CrossRef]
5. Chaker, H.; Ameur, N.; Saidi-Bendahou, K.; Djennas, M.; Fourmentin, S. Modeling and Box-Behnken design optimization of photocatalytic parameters for efficient removal of dye by lanthanum-doped mesoporous TiO_2. *J. Environ. Chem. Eng.* **2021**, *9*, 104584. [CrossRef]
6. Ray, S.; Lalman, J.A.; Biswas, N. Using the Box-Benkhen technique to statistically model phenol photocatalytic degradation by titanium dioxide nanoparticles. *Chem. Eng. J.* **2009**, *150*, 15–24. [CrossRef]
7. Mohamad, N.S.; Kasolang, S. Optimized characterization of response surface methodology on lubricant with titanium oxide nanoparticles. *Ind. Lubr. Tribol.* **2017**, *69*, 387–392. [CrossRef]

8. Li, L.; Ma, Q.; Wang, S.; Song, S.; Li, B.; Guo, R.; Cheng, X.; Cheng, Q. Photocatalytic performance and degradation mechanism of aspirin by TiO_2 through response surface methodology. *Catalysts* **2018**, *8*, 118. [CrossRef]
9. KC, C.; Prabhu, T.N.; Kiran, R.R.S.; Krishna, R.H. Applications of artificial neural network and Box-Behnken Design for modelling malachite green dye degradation from textile effluents using TiO_2 photocatalyst. *Environ. Eng. Res.* **2021**, *27*, 200553. [CrossRef]
10. Tetteh, E.K.; Rathilal, S.; Naidoo, D.B. Photocatalytic degradation of oily waste and phenol from a local South Africa oil refinery wastewater using response methodology. *Sci. Rep.* **2020**, *10*, 8850. [CrossRef]
11. Yap, H.C.; Pang, Y.L.; Lim, S.; Lai, C.W.; Abdullah, A.Z. Enhanced sonophotocatalytic degradation of paracetamol in the presence of Fe-doped TiO_2 nanoparticles and H_2O_2. *Environ. Earth Sci.* **2020**, *79*, 457. [CrossRef]
12. Sun, S.; Ding, H.; Hou, X. Preparation of $CaCO_3$-TiO_2 composite particles and their pigment properties. *Materials* **2018**, *11*, 1131. [CrossRef] [PubMed]
13. Kim, W.T.; Choi, W.Y. Fabrication of TiO_2 photonic crystal by anodic oxidation and their optical sensing properties. *Sens. Actuators A Phys.* **2017**, *260*, 178–184. [CrossRef]
14. Musial, J.; Krakowiak, R.; Mlynarczyk, D.T.; Goslinski, T.; Stanisz, B.J. Titanium dioxide nanoparticles in food and personal care products—What do we know about their safety? *Nanomaterials* **2020**, *10*, 1110. [CrossRef]
15. Tareen, A.K.; Khan, K.; Aslam, M.; Liu, X.; Zhang, H. Confinement in two-dimensional materials: Major advances and challenges in the emerging renewable energy conversion and other applications. *Prog. Solid State Chem.* **2021**, *61*, 100294. [CrossRef]
16. Ameur, N.; Bachir, R.; Bedrane, S.; Choukchou-Braham, A. A Green Route to Produce Adipic Acid on TiO_2–Fe_2O_3 Nanocomposites. *J. Chin. Chem. Soc.* **2017**, *64*, 1096–1103. [CrossRef]
17. Pérez-Larios, A.; Rico, J.L.; Anaya-Esparza, L.M.; Vargas, O.A.G.; González-Silva, N.; Gómez, R. Hydrogen production from aqueous methanol solutions using Ti–Zr mixed oxides as photocatalysts under UV irradiation. *Catalysts* **2019**, *9*, 938. [CrossRef]
18. Lin, L.; Wang, H.; Xu, P. Immobilized TiO2-reduced graphene oxide nanocomposites on optical fibers as high performance photocatalysts for degradation of pharmaceuticals. *Chem. Eng. J.* **2017**, *310*, 389–398. [CrossRef]
19. Li, F.; Huang, Y.; Peng, H.; Cao, Y.; Niu, Y. Preparation and Photocatalytic Water Splitting Hydrogen Production of Titanium Dioxide Nanosheets. *Int. J. Photoenergy* **2020**, *2020*, 3617312. [CrossRef]
20. Eidsvåg, H.; Bentouba, S.; Vajeeston, P.; Yohi, S.; Velauthapillai, D. TiO_2 as a photocatalyst for water splitting—An experimental and theoretical review. *Molecules* **2021**, *26*, 1687. [CrossRef]
21. Abel, S.; Jule, L.T.; Belay, F.; Shanmugam, R.; Dwarampudi, L.P.; Nagaprasad, N.; Krishnaraj, R. Application of Titanium Dioxide Nanoparticles Synthesized by Sol-Gel Methods in Wastewater Treatment. *J. Nanomater.* **2021**, *2021*, 3039761. [CrossRef]
22. Lee, J.; Seong, S.; Jin, S.; Jeong, Y.; Noh, J. Synergetic photocatalytic-activity enhancement of lanthanum doped TiO_2 on halloysite nanocomposites for degradation of organic dye. *J. Ind. Eng. Chem.* **2021**, *100*, 126–133. [CrossRef]
23. Rao, T.N.; Hussain, I.; Anwar, M.S.; Koo, H. Photocatalytic degradation kinetics of pesticide residues in different pH waters using metal-doped metal oxide nanoparticles. *EQA Int. J. Environ. Qual.* **2020**, *36*, 37–44. [CrossRef]
24. Martínez, C.P.C.; Ortega, I.A.A.; Sarmiento, H.A.R.; Morales, F.J.T.; Romero, J.R.G. Photocatalytic degradation of the 2,4-dichlorophenoxyacetic acid herbicide using supported iridium materials. *Cienc. Desarro.* **2021**, *12*, 125–134. [CrossRef]
25. da Cunha, R.; do Carmo Batista, W.V.F.; de Oliveira, H.L.; dos Santos, A.C.; dos Reis, P.M.; Borges, K.B.; Martelli, P.B.; Furtado, C.A.; de Fátima Gorgulho, H. Carbon Xerogel/TiO_2 composites as photocatalysts for acetaminophen degradation. *J. Photochem. Photobiol. A Chem.* **2021**, *412*, 113248. [CrossRef]
26. Jallouli, N.; Elghniji, K.; Trabelsi, H.; Ksibi, M. Photocatalytic degradation of paracetamol on TiO_2 nanoparticles and TiO_2/cellulosic fiber under UV and sunlight irradiation. *Arab. J. Chem.* **2017**, *10*, S3640–S3645. [CrossRef]
27. Rimoldi, L.; Meroni, D.; Falletta, E.; Ferretti, A.M.; Gervasini, A.; Cappelletti, G.; Ardizzone, S. The role played by different TiO_2 features on the photocatalytic degradation of paracetamol. *Appl. Surf. Sci.* **2017**, *424*, 198–205. [CrossRef]
28. Inger, M.; Dobrzyńska-Inger, A.; Rajewski, J.; Wilk, M. Optimization of ammonia oxidation using response surface methodology. *Catalysts* **2019**, *9*, 249. [CrossRef]
29. Hegyi, A.; Szilagyi, H.; Grebenișan, E.; Sandu, A.V.; Lăzărescu, A.-V.; Romila, C. Influence of TiO_2 Nanoparticles Addition on the Hydrophilicity of Cementitious Composites Surfaces. *Appl. Sci.* **2020**, *10*, 4501. [CrossRef]
30. Pérez-Larios, A.; Torres-Ramos, I.; Zanella, R.; Rico, J.L. Ti-Co mixed oxide as photocatalysts in the generation of hydrogen from water. *Int. J. Chem. React. Eng.* **2022**, *20*, 129–140. [CrossRef]
31. Li, S.; Yang, Y.; Su, Q.; Liu, X.; Zhao, H.; Zhao, Z.; Li, J.; Jin, C. Synthesis and photocatalytic activity of transition metal and rare earth element co-doped TiO_2 nano particles. *Mater. Lett.* **2019**, *252*, 123–125. [CrossRef]

Article

Enhanced Photocatalytic Activity of TiO_2 Thin Film Deposited by Reactive RF Sputtering under Oxygen-Rich Conditions

Takaya Ogawa *, Yuekai Zhao, Hideyuki Okumura and Keiichi N. Ishihara

Graduate School of Energy Science, Kyoto University, Kyoto 606-8501, Japan; yuekai1994@163.com (Y.Z.); okumura.hideyuki.4e@kyoto-u.ac.jp (H.O.); ishihara.keiichi.6w@kyoto-u.ac.jp (K.N.I.)
* Correspondence: ogawa.takaya.8s@kyoto-u.ac.jp

Abstract: TiO_2 thin films are promising as photocatalysts to decompose organic compounds. In this study, TiO_2 thin films were deposited by reactive radio-frequency (RF) magnetron sputtering under various flow rates of oxygen and argon gas. The results show that the photocatalytic activity decreases as the oxygen-gas ratio is increased to 30% or less, while the activity increases under oxygen-rich conditions. It was observed that the crystal structure changed from anatase to a composite of anatase and rutile, where the oxygen-gas ratio during RF sputtering is more than 40%. Interestingly, the oxygen vacancy concentration increased under oxygen-rich conditions, where the oxygen-gas ratio is more than 40%. The sample prepared under the most enriched oxygen condition, 70%, among our experiments exhibited the highest concentration of oxygen vacancy and the highest photocatalytic activity. Both the oxygen vacancies and the composite of anatase and rutile structure in the TiO_2 films deposited under oxygen-rich conditions are considered responsible for the enhanced photocatalysis.

Keywords: TiO_2 thin film; photocatalyst; oxygen and argon gas flow rates

Citation: Ogawa, T.; Zhao, Y.; Okumura, H.; Ishihara, K.N. Enhanced Photocatalytic Activity of TiO_2 Thin Film Deposited by Reactive RF Sputtering under Oxygen-Rich Conditions. *Photochem* **2022**, *2*, 138–149. https://doi.org/10.3390/photochem2010011

Academic Editor: Vincenzo Vaiano

Received: 15 January 2022
Accepted: 11 February 2022
Published: 18 February 2022

1. Introduction

Thin films of photocatalysts have been intensively studied as clean technology [1–3]. Photocatalytic thin films can decompose various organic compounds in a solution using energy from sunlight, and it is relatively easy to separate the catalysts from the solution. Among many photocatalysts, TiO_2 is promising due to its low cost, stable availability, high durability, and less risk to the human body and environment [4–7]. There are three major TiO_2 crystalline forms: anatase, rutile, and brookite. The rutile phase is the most stable in bulk at room temperature and ambient pressure.

Several techniques can be used to deposit TiO_2 thin films [8], such as the sol-gel method [3,9], spray pyrolysis [10], pulsed laser deposition [11], e-beam evaporation [12], chemical vapor deposition [13], and sputtering [14]. Notably, reactive radio-frequency (RF) magnetron sputtering is a promising technique with desirable features such as excellent film adhesion and uniform distribution of thickness [15]. It is possible to form films with various electronic states and microstructures via changing sputtering conditions, such as RF power, deposition pressure, substrate temperature, and gas flow rate. Especially, it is the characteristic of reactive sputtering that can control oxygen contents. Then, the oxygen gas ratio (R_{O2}: [O_2 gas flow rate)]/[(Ar + O_2) gas flow rate]) is one of the most critical parameters for photocatalytic activity [16], although the effect of the R_{O2} is still controversial. Zhang et al. reported that TiO_2 film deposited under a high argon flow rate absorbs more light irradiation, which results in more electron–hole pairs generated in the TiO_2 film and thus enhanced photocatalytic activity [17]. Huang et al. and Chiou et al. demonstrated that the photocatalytic activity of TiO_2 films is raised with increasing oxygen flow rate to some extent but finally dropped at the high values of R_{O2} [18,19]. Furthermore, the wide range of R_{O2} conditions was not tested without heating substrate (see Table S1 in Supporting Information).

In this study, we synthesized TiO_2 thin films via reactive RF sputtering at the condition with the R_{O2} values ranging from 0 to 70%. In this study, we synthesized TiO_2 thin films via reactive RF sputtering with changing R_{O2} and examined the parameters that influence photocatalytic activity. As the R_{O2} was increased, the photocatalytic activity decreased with the R_{O2} values up to 30% and increased with the R_{O2} values over 40%. We investigated the oxygen vacancy concentration, and it decreased with the R_{O2} values up to 30–40%, which is a general behavior under the oxygen-rich conditions [15,20]. On the other hand, interestingly, the vacancy concentration was increased with the R_{O2} values of more than 40%. In addition, other film properties also showed a change in their tendency at around R_{O2} = 40%. These are considered to involve enhanced photocatalytic activity to decompose an organic dye. A possible explanation of the phenomena will be provided to suggest suitable deposition conditions for highly active thin-film photocatalysts.

2. Experimental Details

2.1. Synthesis of TiO_2 Thin Films

TiO_2 thin films were deposited on a glass slide substrate using a reactive RF magnetron sputtering method with pure titanium (Ti) target. The deposition was performed without a substrate heater, while the temperature should be slightly raised by the sputtering energy. The distance between the substrate and the target was 70 mm. The glass slide substrates with the size of 24 × 48 × 1.2 mm were purchased from Toshinriko Co. Ltd (Tokyo, Japan). The sputtering equipment is RSC-MG2 produced by CryoVac Inc. (Osaka, Japan). The target is 99.9% pure titanium provided from Kojundo Chemical Laboratory Co., Ltd., (Sakado, Japan). Before deposition, pre-sputtering with a sputtering power of 300 W was conducted to clean the surface of the Ti target under Ar gas (100 sccm). The deposition time for a sample was 2 h with a sputtering power of 300 W. During the sputtering, the total flow rate of Ar and O_2 gases was 100 sccm under a pressure of 4.0×10^{-3} Pa. Eight samples were prepared with different ratios of O_2 gas (R_{O2}: [O_2 gas flow rate]/[(Ar + O_2) gas flow rate)] = 5%, 10%, 20%, 30%, 40%, 50%, 60%, and 70%). The following characterization methods were conducted on these samples as deposited.

2.2. Characterization

The crystalline structures were characterized by X-ray diffractometry (XRD) (RINT2100CMJ, produced by Rigaku Co., Ltd., Tokyo, Japan) using Cu Kα radiation (λ = 1.54184 Å), operated at 40 kV and 30 mA.

The transmittance measurements of thin-film samples were performed using UV-vis spectroscopy (Lambda750S produced by PerkinElmer Co., Ltd., Waltham, MA, USA) in the wavelength range between 250 nm and 800 nm. The average transparency was determined in the visible light range from 380 nm to 740 nm. The band gaps of thin film samples were estimated using the Tauc plot method. The TiO_2 was reported to have both a direct forbidden and an indirect allowed transition, where the latter transition dominates the optical absorption due to the weak strength of the former transition. Thus, we assumed only the indirect allowed transitions.

A field emission scanning electron microscope (FE-SEM, SU6600 produced by Hitachi High-Technologies Co., Ltd., Tokyo, Japan) was used to observe the surface morphology of the thin-film samples at an accelerating voltage of 20.0 kV. The average diameter was estimated via ImageJ.

The chemical composition and chemical bonding states of TiO_2 thin films were investigated by X-ray photoelectron spectroscopy (XPS) analysis (JPS-9030 produced by JEOL Co., Ltd., Tokyo, Japan), operated at 10.0 kV and 20.0 mA. A standard MgKα X-ray source was employed, and the measurement area had a diameter of 10 mm. A high-etching-rate ion gun with Ar ions was employed to remove the contaminants and hydroxyl (OH) groups on the specimen surface. The approximate etched depth was 0.2 nm. The obtained data were analyzed using the software SPECSURF Analysis, which is built-in for JEOL XPS. The Ti^{3+} and Ti^{4+} peak areas were utilized to estimate oxygen vacancy concentrations (Ti^{3+} ratio:

[Ti^{3+} peak area]/[(Ti^{3+} + Ti^{4+}) peak area]) because one Ti^{3+} ion in TiO_2 requires one oxygen vacancy to accommodate charge neutrality [21].

2.3. *Measurement of Photocatalytic Activity*

The photocatalytic activity was evaluated via the decomposition rate of methyl orange (MO). The MO granular solid (purity: 99.9%; produced by Nacalai Tesque Inc., Kyoto, Japan) was used to prepare a MO solution with a concentration of 0.01 mmol/L, and 16 mL was utilized for photocatalytic measurements.

A xenon lamp (UXL-500D-O produced by Ushio Co., Ltd., Tokyo, Japan) located above the solution was used as the light source, which included both ultraviolet light and visible light (the spectra of the lamp were shown in Figure S1 in Supporting Information). The power of the lamp was 500 W. A tube with 15 cm (height) of water was placed between the Xenon lamp and the MO solution to filter the infrared light of the Xenon lamp irradiation and maintain the temperature. During the photoactivity measurement, the solution was stirred at 1200 rpm, and the concentration changes in MO were monitored every 20 min via spectrophotometry.

3. Results

3.1. *XRD Measurements*

Figure 1 shows XRD patterns of the TiO_2 thin films deposited under various oxygen flow rates. The patterns indicate that the anatase structure is dominant except for R_{O2} = 70%. As the R_{O2} value increases (from 5%), the diffraction peak corresponding to the (110) plane of the rutile phase at around 2θ = 27° was not prominent until R_{O2} = 30–40%. However, at R_{O2} greater than 40%, the rutile structure was distinguished and became comparable with the anatase structure at R_{O2} = 70%. The crystallite sizes of the films estimated via Scherrer's equation are shown in Table S1 in the Supporting Information. The crystallite sizes did not indicate the correlation with R_{O2}.

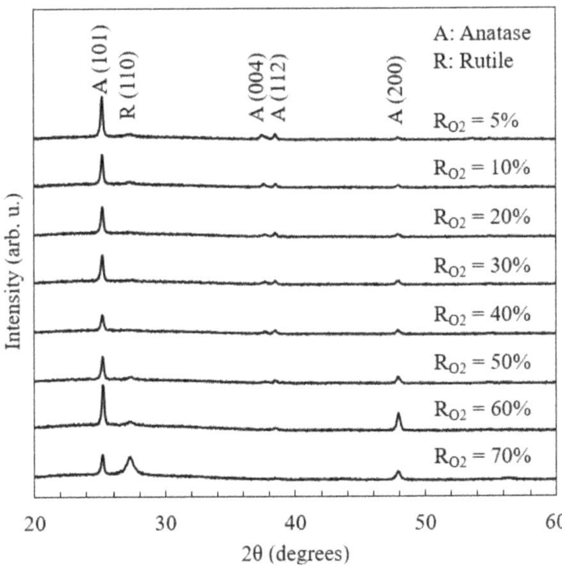

Figure 1. XRD patterns of TiO_2 thin films.

3.2. *Optical Transmittance Spectra*

Figures 2–4 show the optical transmittance spectra of TiO_2 thin films, an example of a Tauc plot based on the spectra, and the band gaps based on Tauc plots, respectively. As the R_{O2} value was increased, the band gap was also increased until R_{O2} = 20%, remained constant until 30–40%, and decreased over R_{O2} = 40%. The reported band gaps of the anatase and rutile structures are 3.2 eV and 3.0 eV, respectively, for bulk TiO_2 [22]. The constant value, around 3.32 eV, at R_{O2} = 20–40% could be derived from the anatase structure, where the difference of 0.12 eV larger than 3.2 eV is explained by the tendency that the band gap of thin films is generally wider than that of bulk [23,24]. The relatively smaller values of band gap at R_{O2} = 5–10% should originate from the rutile structure, which was slightly detected in XRD results. The rutile structure has a narrower band gap than the anatase structure. The narrower band gaps in the other R_{O2} region should be due to the composite of the rutile and the anatase structure, which is consistent with XRD results. In fact, the rutile structure mostly disappeared in XRD patterns at R_{O2} = 20–40%. Therefore, the change in the band gap is mainly attributed to the crystal structure, though the film strain could modify the band gap value.

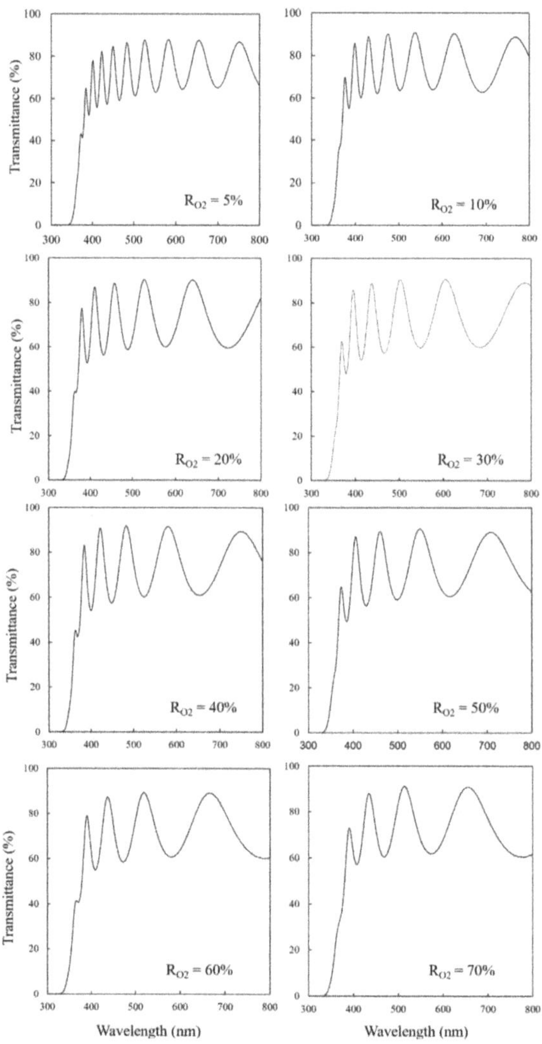

Figure 2. Optical transmittance spectra of TiO_2 thin films.

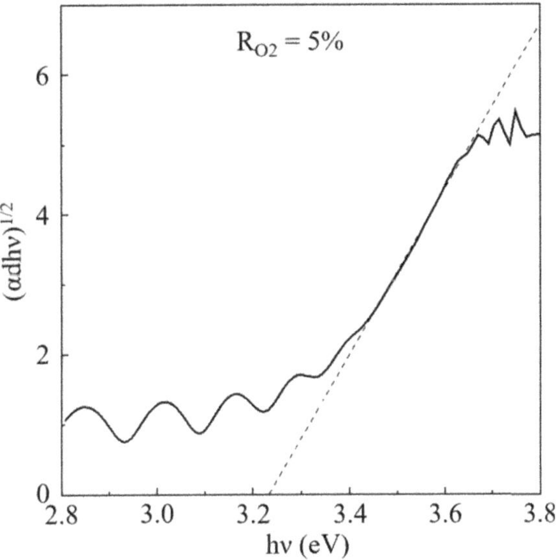

Figure 3. An example of Tauc plot (the sample deposited under R_{O2} = 5%).

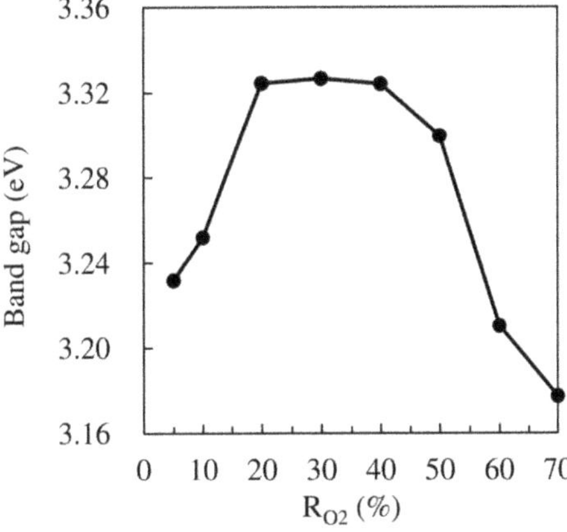

Figure 4. Band gaps of TiO2 thin.

3.3. SEM Images

The surface morphologies were investigated by FE-SEM measurements (Figure 5). The average diameters were 23 and 25 nm for the film deposited under R_{O2} = 5% and 50%, respectively. Thus, it is suggested that the morphology does not change at the value of R_{O2} = 50%.

The thickness of TiO_2 films was measured by the cross-sectional FE-SEM images and the interference of transmitted light. Figure 6 shows that the films gradually became thinner as the R_{O2} increased, indicating that the growth rate of the films became slower with increasing R_{O2}. At high R_{O2} (low Ar gas ratio), the plasma generation generally becomes more difficult because Ar can more easily be dissociated than O_2; the dissociation energy of the Ar was about 15.76 eV, and O_2 was approximately 48.77 eV [19]. Therefore, a possible explanation of the thinner film at high R_{O2} is that the plasma sputtering rate becomes less intensive.

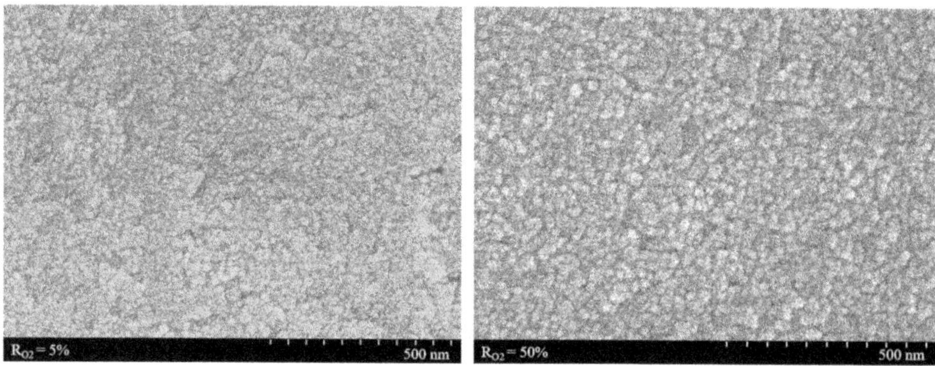

Figure 5. Surface morphologies of TiO_2 thin films prepared at R_{O2} = 5% (**left**) and 50% (**right**).

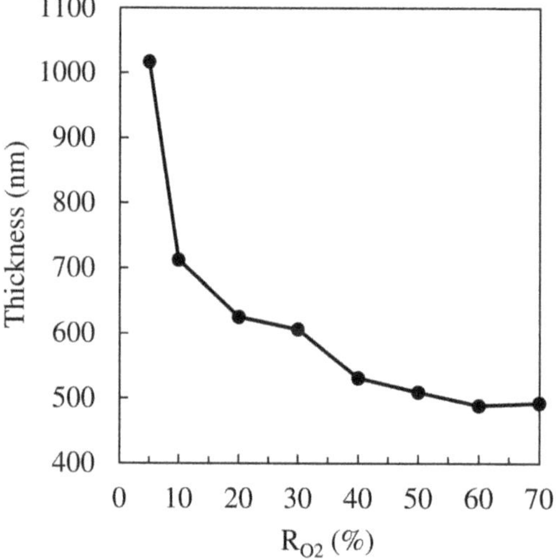

Figure 6. Thickness of TiO_2 thin films.

3.4. XPS Measurements

The XPS measurements were performed to examine the chemical states of Ti in proximity to surfaces of TiO_2 thin films. Figure 7 shows the high-resolution narrow scan of XPS spectra around the Ti $2p^{3/2}$ spin orbital. The main peak (458.0 eV) is ascribed to the Ti^{4+} state, and the shoulder peak in the lower binding energy region (456.5 eV) is assigned to the Ti^{3+} state. The ratio of Ti^{3+} to Ti^{4+} corresponds to the oxygen vacancy ratio in TiO_2 thin films because Ti^{3+} is generated when Ti^{4+} in TiO_2 is reduced and releases oxygen [25]. Figure 8 summarizes the Ti^{3+} ratio changes as a function of R_{O2}. It is noteworthy that the TiO_2 film deposited under the most oxygen-rich condition (R_{O2} = 70%) has the highest amount of oxygen vacancies.

3.5. Photocatalytic Activity

Figure 9a shows the photocatalytic activity of the TiO_2 thin films for MO degradation. As indicated by the control test (blank concentration) fluctuation, the error in the concentration measurement is at least 4.3%. To evaluate the photocatalytic activities, we employed the MO concentrations at 80 min (Figure 9b), where the decomposed MO amount is significant even considering the measurement error. We then found similar behavior in the degradation rate as well as other properties: that is, with the R_{O2} increase, the degradation rate is decreased until R_{O2} = 30%, while the rate is increased over R_{O2} = 40%. The TiO_2 film synthesized under R_{O2} = 70% showed the highest photocatalytic activity. The correlation between catalytic activity and the other properties will be discussed in the next section.

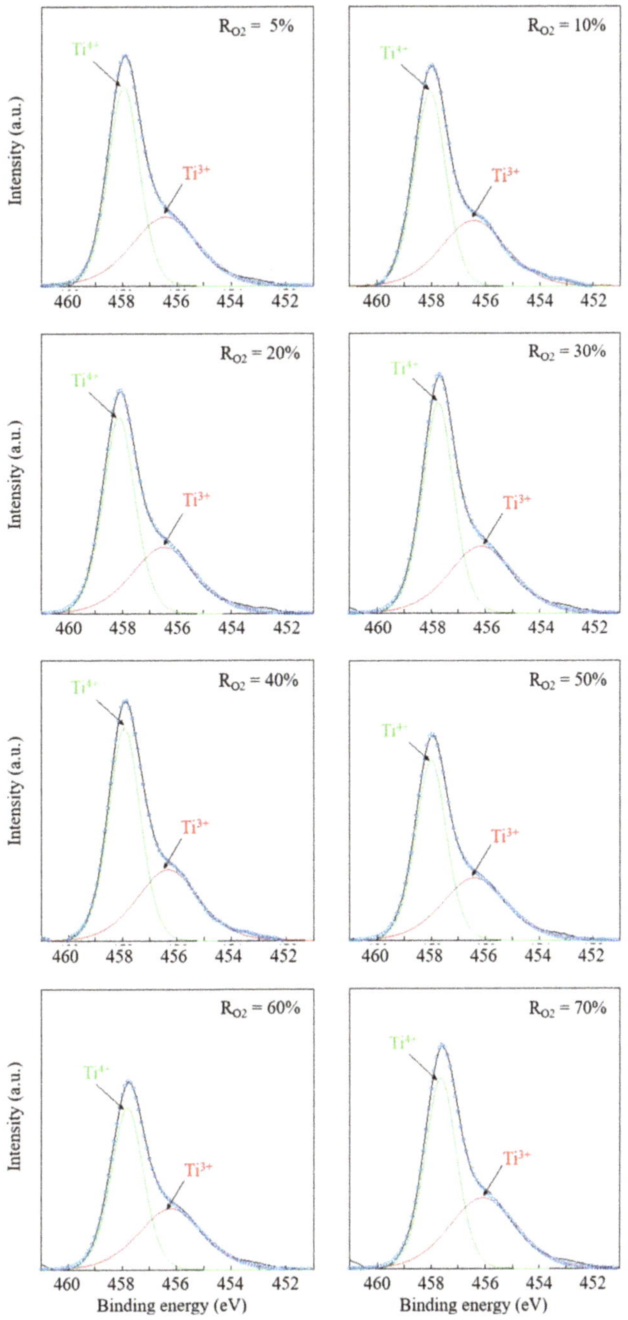

Figure 7. XPS results for the binding energy of Ti^{3+} versus Ti^{4+} in TiO_2 thin films.

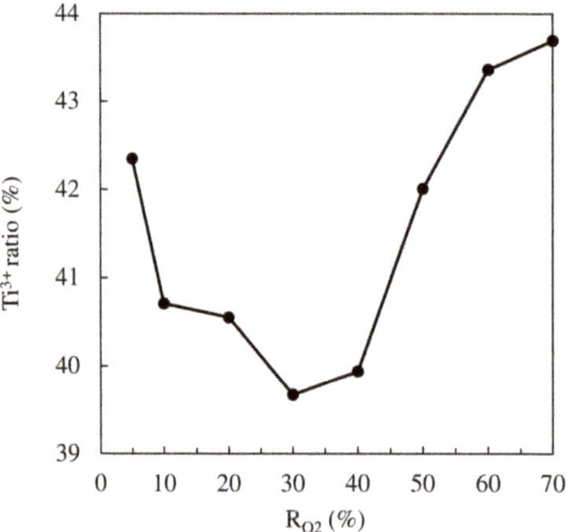

Figure 8. The ratio of Ti^{3+} to Ti^{4+} in TiO_2 thin films.

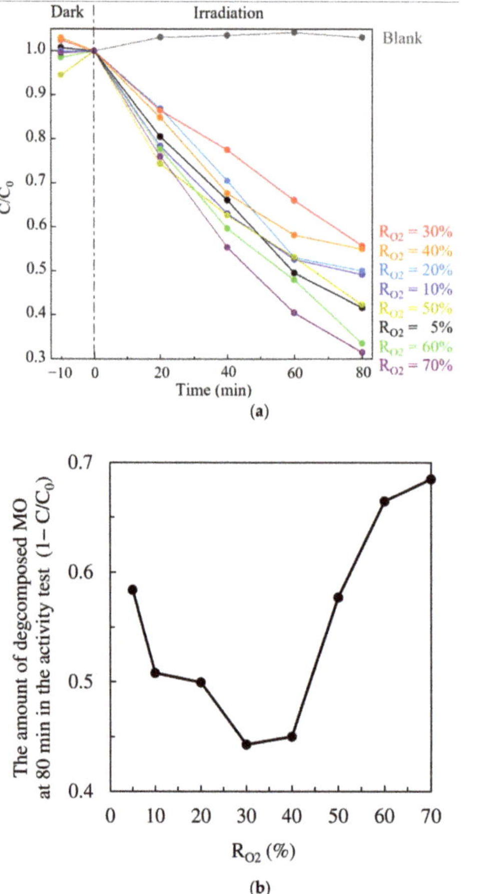

Figure 9. The catalytic activity of TiO_2 thin films for MO degradation. C_0 and C are the concentrations of MO solution at the initial and at each minute, respectively. (**a**) The concentration changes over time and (**b**) the amount of decomposed MO at 80 min during the activity test ($1-C/C_0$).

To check whether Ti^{3+} reacts with MO to become Ti^{4+}, we estimated the amount of Ti^{3+} using the thickness of the films and compared them with the amount of decomposed MO after 80 min irradiation. The amount of decomposed MO is about six orders of magnitude higher than that of Ti^{3+} in TiO_2 thin film. Therefore, the contribution of Ti^{3+} to the decomposition of MO, via sacrificing itself to become Ti^{4+}, was negligible.

4. Discussion

4.1. Plausible Explanation of a Less Oxidized State Induced under the Oxygen-Rich Condition

In this section, we suggest a possible explanation for why the less oxidized state appeared in the TiO_2 thin films prepared under the oxygen-rich condition, R_{O2} = 40–70%. According to a previous study [26], oxygen-rich conditions facilitate the nucleation of the rutile structure. Similar behavior regarding crystal structure changes was also reported in another study; as R_{O2} was increased, the rutile structure disappeared first and appeared again when the synthesis condition became oxygen-rich, and they discussed these phenomena from reactive mechanics in the deposition [27,28]. Hence, the increase in the rutile structure in the oxygen-rich condition is reasonable. Furthermore, the ab initio calculation indicates that the rutile structure with (110) facets favors the oxygen vacancies on the surface, rather than the anatase structure with its most stable (101) facet, since their formation energy at the rutile (110) surface is smaller than that at the subsurface and much smaller than that in bulk rutile [29,30], which is consistent with our XPS results. Therefore, a higher concentration of oxygen vacancies is mainly observed due to an increased amount of the rutile structure, which is produced under oxygen-rich conditions.

4.2. Origin of High Photocatalytic Activity

The photocatalytic activity is generally related to the crystal structure [31–33], band gap [4], surface morphology [3], and the amount of oxygen vacancy [34–36]. However, the FE-SEM images show that the surface morphology did not change when comparing the samples prepared under R_{O2} = 5% and 50%. Hence, the surface morphology is unlikely related to the change in the degradation rate. The other three parameters (crystal structure, band gap, and the amount of oxygen vacancy) are then considered to further examine the origin of enhanced photocatalytic activity.

Figure 10 shows the relation between the photocatalytic activity and the band gap (a) or the Ti^{3+} ratio (b). Based on the XRD patterns, the TiO_2 thin films synthesized under R_{O2} = 20–30% have only the anatase structure, whereas the samples prepared under R_{O2} = 5–10 and 40–70% were composites of the anatase and the rutile structure. Although the reference [27] indicated that only one of the two structures was formed dominantly at R_{O2} = 70%, our sample showed both peaks are distinguished in XRD results, which seems derived from some other conditions. Generally, the anatase structure is more active than the rutile structure [37]. However, rutile/anatase composites could show higher activity than the anatase structure alone [31–33], because the composite interface (heterojunction) may disturb the recombination of electrons and holes. In addition, the narrow band gap is preferable for absorbing more light, though a weak correlation between band gap and decomposition rate is seen, as shown in Figure 10a. Therefore, higher activity is likely influenced by the existence of composites.

Oxygen vacancies are also involved in promoting charge separation and disturbing the electron–hole recombination process, resulting in high photocatalytic activity [34–36]. Then, a relatively large amount of oxygen vacancy in the TiO_2 samples could be responsible for high photocatalytic activity. Although the pre-sputtering in XPS measurement removes the outermost surface layer (~0.2 nm) and detects the chemical composition in the subsurface, the oxygen vacancies in the subsurface layer could have a strong potential to enhance the photocatalytic activity [38]. Thus, the seemingly strong correlation shown in Figure 10b can be significant and explained in a consistent manner. It is noted that, though the discussed Ti^{3+} ratio may not be exactly the same with the as-sputtered surface due to some reduction of Ti^{4+} to Ti^{3+} via Ar ion etching, the difference of the reduction among samples could be negligible, and our XPS results on the tendency of Ti^{3+} ratio are inconsistent with the change in XRD patterns, band gap, and photocatalytic activity. It is then plausible that our XPS results are correlated to the oxygen vacancies in the original states. Therefore, the excellent correlation between Ti^{3+} ratio and degradation rate, as shown in Figure 10b, strongly suggests that the oxygen vacancies have a solid potential to enhance photocatalytic activity.

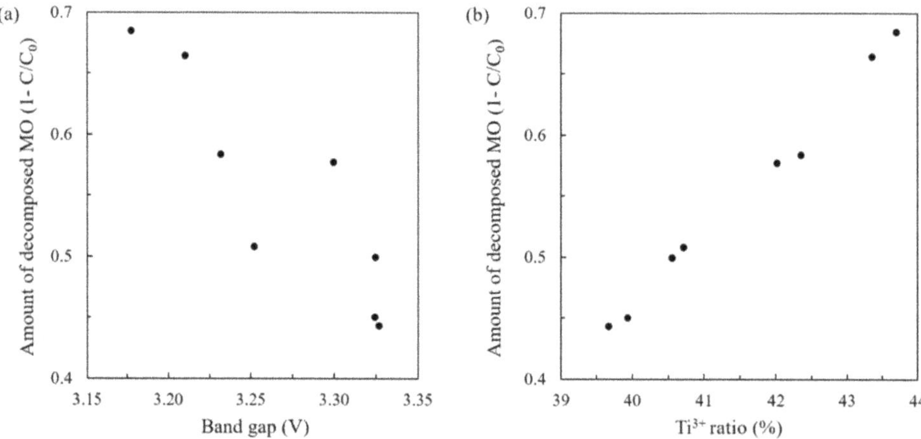

Figure 10. The relationship between the amount of decomposed MO at 80 min during the activity test ($1-C/C_0$) and (**a**) Ti^{3+} ratio and (**b**) band gap. C_0 and C are the concentration of MO solution at the initial and at each minute, respectively.

Hence, we can conclude that a large amount of oxygen vacancy, the narrower band gap, and the existence of rutile/anatase composites play an essential role in enhancing photocatalytic activity. It could be easy to prepare the thin films using reactive RF sputtering under the less oxygen flow condition to attain these three factors in the satisfactory range. However, this study shows that it is also possible or preferable in higher oxygen flow conditions.

5. Conclusions

The TiO_2 thin films were prepared via reactive RF sputtering under various flow rates of oxygen and argon gas. We observed the highest photocatalytic activity for the films prepared under the most oxygen-enriched condition among the prepared samples, 70%. The origin of the enhanced activity is attributed to the composite of anatase/rutile structure and the increased oxygen vacancy despite the oxygen-rich sputtering condition. The reason why a less oxidized state appeared under the oxygen-enriched condition is as follows: the rutile structure prefers to grow under oxygen-rich conditions at around room temperature and tends to generate a higher concentration of oxygen vacancies. Therefore, this research presents the insight that the less oxidized state of TiO_2 films can be prepared even in an oxygen-rich condition. The conditions could be useful for preparing highly photoactive TiO_2 thin films.

Supplementary Materials: The following supporting information can be downloaded at: https://www.mdpi.com/article/10.3390/photochem2010011/s1, Figure S1: Spectral irradiance of the Xenon lamp; Table S1: The comparison of sputtering condition in several pieces of literature; Table S2: The crystallite size of TiO_2 thin films estimated via Scherrer's equation based on anatase (101) plane.

Author Contributions: Writing—original draft: T.O.; Formal analysis, Writing—review & editing, Validation: Y.Z.; K.N.I.; Data curation: T.O.; Y.Z.; H.O.; Investigation: T.O.; Y.Z.; All authors have read and agreed to the published version of the manuscript.

Funding: This research received no external funding.

Institutional Review Board Statement: Not applicable.

Informed Consent Statement: Not applicable.

Conflicts of Interest: The authors declare no conflict of interest.

References

1. Patel, N.; Fernandes, R.; Guella, G.; Kale, A.; Miotello, A.; Patton, B.; Zanchetta, C. Structured and nanoparticle assembled Co−B thin films prepared by pulsed laser deposition: A very efficient catalyst for hydrogen production. *J. Phys. Chem. C* **2008**, *112*, 6968–6976. [CrossRef]
2. Anpo, M.; Takeuchi, M. The design and development of highly reactive titanium oxide photocatalysts operating under visible light irradiation. *J. Catal.* **2003**, *216*, 505–516. [CrossRef]
3. Yu, J.G.; Zhao, X.J.; Zhao, Q.N. Effect of surface structure on photocatalytic activity of TiO2 thin films prepared by sol-gel method. *Thin Solid Films* **2000**, *379*, 7–14. [CrossRef]
4. Hoffmann, M.R.; Martin, S.T.; Choi, W.Y.; Bahnemann, D.W. Environmental applications of semiconductor photocatalysis. *Chem. Rev.* **1995**, *95*, 69–96. [CrossRef]
5. Carp, O.; Huisman, C.L.; Reller, A. Photoinduced reactivity of titanium dioxide. *Prog. Solid State Chem.* **2004**, *32*, 33–177. [CrossRef]
6. Hieu, V.Q.; Lam, T.C.; Khan, A.; Thi Vo, T.-T.; Nguyen, T.-Q.; Doan, V.D.; Tran, D.L.; Le, V.T.; Tran, V.A. TiO2/Ti3C2/g-C3N4 ternary heterojunction for photocatalytic hydrogen evolution. *Chemosphere* **2021**, *285*, 131429. [CrossRef] [PubMed]
7. Hieu, V.Q.; Phung, T.K.; Nguyen, T.-Q.; Khan, A.; Doan, V.D.; Tran, V.A.; Le, V.T. Photocatalytic degradation of methyl orange dye by Ti3C2–TiO2 heterojunction under solar light. *Chemosphere* **2021**, *276*, 130154. [CrossRef] [PubMed]
8. Chung, C.K.; Liao, M.W.; Lai, C.W. Effects of oxygen flow ratios and annealing temperatures on raman and photoluminescence of titanium oxide thin films deposited by reactive magnetron sputtering. *Thin Solid Films* **2009**, *518*, 1415–1418. [CrossRef]
9. Pomoni, K.; Vomvas, A.; Trapalis, C. Electrical conductivity and photoconductivity studies of TiO2 sol–gel thin films and the effect of N-doping. *J. Non-Cryst. Solids* **2008**, *354*, 4448–4457. [CrossRef]
10. Natarajan, C.; Fukunaga, N.; Nogami, G. Titanium dioxide thin film deposited by spray pyrolysis of aqueous solution. *Thin Solid Films* **1998**, *322*, 6–8. [CrossRef]
11. Terashima, M.; Inoue, N.; Kashiwabara, S.; Fujimoto, R. Photocatalytic TiO2 thin-films deposited by pulsed laser deposition technique. *IEEJ Trans. Fundam. Mater.* **2001**, *121*, 59–64. [CrossRef]
12. Sun, L.; Hou, P. Spectroscopic ellipsometry study on e-beam deposited titanium dioxide films. *Thin Solid Films* **2004**, *455–456*, 52–529. [CrossRef]
13. Sun, H.; Wang, C.; Pang, S.; Li, X.; Tao, Y.; Tang, H.; Liu, M. Photocatalytic TiO2 films prepared by chemical vapor deposition at atmosphere pressure. *J. Non-Cryst. Solids* **2008**, *354*, 1440–1443. [CrossRef]
14. Boukrouh, S.; Bensaha, R.; Bourgeois, S.; Finot, E.; Marco de Lucas, M.C. Reactive direct current magnetron sputtered TiO2 thin films with amorphous to crystalline structures. *Thin Solid Films* **2008**, *516*, 6353–6358. [CrossRef]
15. Chiu, S.-M.; Chen, Z.-S.; Yang, K.-Y.; Hsu, Y.-L.; Gan, D. Photocatalytic activity of doped TiO2 coatings prepared by sputtering deposition. *J. Mater. Process. Technol.* **2007**, *192–193*, 60–67. [CrossRef]
16. Wang, Y.-H.; Rahman, K.H.; Wu, C.-C.; Chen, K.-C. A review on the pathways of the improved structural characteristics and photocatalytic performance of titanium dioxide (TiO2) thin films fabricated by the magnetron-sputtering technique. *Catalysts* **2020**, *10*, 598. [CrossRef]
17. Zhang, W.; Li, Y.; Zhu, S.; Wang, F. Influence of argon flow rate on TiO2 photocatalyst film deposited by dc reactive magnetron sputtering. *Surf. Coat. Technol.* **2004**, *182*, 192–198. [CrossRef]
18. Huang, C.H.; Tsao, C.C.; Hsu, C.Y. Study on the photocatalytic activities of TiO2 films prepared by reactive RF sputtering. *Ceram. Int.* **2011**, *37*, 2781–2788. [CrossRef]
19. Chiou, A.H.; Kuo, C.G.; Huang, C.H.; Wu, W.F.; Chou, C.P.; Hsu, C.Y. Influence of oxygen flow rate on photocatalytic TiO2 films deposited by RF magnetron sputtering. *J. Mater. Sci. Mater. Electron.* **2011**, *23*, 589–594. [CrossRef]
20. Machda, F.; Ogawa, T.; Okumura, H.; Ishihara, K.N. Damp heat durability of Al-doped ZnO transparent electrodes with different crystal growth orientations. *ECS J. Solid State Sci. Technol.* **2019**, *8*, Q240–Q244. [CrossRef]
21. Wang, L.-Q.; Baer, D.R.; Engelhard, M.H. Creation of variable concentrations of defects on TiO2(110) using low-density electron beams. *Surf. Sci.* **1994**, *320*, 295–306. [CrossRef]
22. Yu, J.-G.; Yu, H.-G.; Cheng, B.; Zhao, X.-J.; Yu, J.C.; Ho, W.-K. The effect of calcination temperature on the surface microstructure and photocatalytic activity of TiO2 thin films prepared by liquid phase deposition. *J. Phys. Chem. B* **2003**, *107*, 13871–13879. [CrossRef]
23. Paul, A.; John, J.C.; Augustine, S.; Sebastian, T.; Joy, A.; Joy, J.; Maria Michael, T. Comparison of photocatalytic efficiency of TiO2 and In2S3 thin films under UV and visible light irradiance. *Adv. Mater. Lett.* **2020**, *11*, 1–7. [CrossRef]
24. Bharti, B.; Kumar, S.; Lee, H.N.; Kumar, R. Formation of oxygen vacancies and Ti3+ state in TiO2 thin film and enhanced optical properties by air plasma treatment. *Sci. Rep.* **2016**, *6*, 32355. [CrossRef]
25. Farfan-Arribas, E.; Madix, R.J. Role of defects in the adsorption of aliphatic alcohols on the TiO2(110) surface. *J. Phys. Chem. B* **2002**, *106*, 10680–10692. [CrossRef]

26. Löbl, P.; Huppertz, M.; Mergel, D. Nucleation and growth in TiO2 films prepared by sputtering and evaporation. *Thin Solid Films* **1994**, *251*, 72–79. [CrossRef]
27. Zeman, P.; Takabayashi, S. Effect of total and oxygen partial pressures on structure of photocatalytic TiO2 films sputtered on unheated substrate. *Surf. Coat. Technol.* **2002**, *153*, 93–99. [CrossRef]
28. Schiller, S.; Beister, G.; Sieber, W.; Schirmer, G.; Hacker, E. Influence of deposition parameters on the optical and structural properties of TiO2 films produced by reactive D.C. Plasmatron sputtering. *Thin Solid Films* **1981**, *83*, 239–245. [CrossRef]
29. Cheng, H.; Selloni, A. Surface and subsurface oxygen vacancies in anatase TiO2 and differences with rutile. *Phys. Rev. B* **2009**, *79*, 092101 .
30. Li, H.; Guo, Y.; Robertson, J. Calculation of TiO2 surface and subsurface oxygen vacancy by the screened exchange functional. *J. Phys. Chem. C* **2015**, *119*, 18160–18166. [CrossRef]
31. Ohno, T.; Tokieda, K.; Higashida, S.; Matsumura, M. Synergism between rutile and anatase TiO2 particles in photocatalytic oxidation of naphthalene. *Appl. Catal. A* **2003**, *244*, 383–391. [CrossRef]
32. Kho, Y.K.; Iwase, A.; Teoh, W.Y.; Mädler, L.; Kudo, A.; Amal, R. Photocatalytic H2 evolution over TiO2 nanoparticles. The synergistic effect of anatase and rutile. *J. Phys. Chem. C* **2010**, *114*, 2821–2829. [CrossRef]
33. Su, R.; Bechstein, R.; Sø, L.; Vang, R.T.; Sillassen, M.; Esbjörnsson, B.; Palmqvist, A.; Besenbacher, F. How the anatase-to-rutile ratio influences the photoreactivity of TiO2. *J. Phys. Chem. C* **2011**, *115*, 24287–24292. [CrossRef]
34. Nowotny, M.K.; Sheppard, L.R.; Bak, T.; Nowotny, J. Defect chemistry of titanium dioxide. Application of defect engineering in processing of TiO2-based photocatalysts. *J. Phys. Chem. C* **2008**, *112*, 5275–5300. [CrossRef]
35. Wang, J.; Liu, P.; Fu, X.; Li, Z.; Han, W.; Wang, X. Relationship between oxygen defects and the photocatalytic property of ZnO nanocrystals in Nafion membranes. *Langmuir* **2009**, *25*, 1218–1223. [CrossRef]
36. Pan, X.; Yang, M.Q.; Fu, X.; Zhang, N.; Xu, Y.J. Defective TiO2 with oxygen vacancies: Synthesis, properties and photocatalytic applications. *Nanoscale* **2013**, *5*, 3601–3614. [CrossRef]
37. Sclafani, A.; Herrmann, J.M. Comparison of the photoelectronic and photocatalytic activities of various anatase and rutile forms of titania in pure liquid organic phases and in aqueous solutions. *J. Phys. Chem.* **1996**, *100*, 13655–13661. [CrossRef]
38. Lee, S.-H.; Yamasue, E.; Ishihara, K.N.; Okumura, H. Photocatalysis and surface doping states of N-doped TiOx films prepared by reactive sputtering with dry air. *Appl. Catal. B* **2010**, *93*, 217–226. [CrossRef]

MDPI

Article

Effect of Photolysis on Zirconium Amino Phenoxides for the Hydrophosphination of Alkenes: Improving Catalysis

Bryan T. Novas, Jacob A. Morris, Matthew D. Liptak and Rory Waterman *

Department of Chemistry, University of Vermont, 82 University Place, Burlington, VT 05401, USA; Bryan.Novas@uvm.edu (B.T.N.); Jacob.Morris@uvm.edu (J.A.M.); Matthew.Liptak@uvm.edu (M.D.L.)
* Correspondence: Rory.Waterman@uvm.edu

Abstract: A comparative study of amino phenoxide zirconium catalysts in the hydrophosphination of alkenes with diphenylphosphine reveals enhanced activity upon irradiation during catalysis, with conversions up to 10-fold greater than reactions in ambient light. The origin of improved reactivity is hypothesized to result from substrate insertion upon an n→d charge transfer of a Zr–P bond in the excited state of putative phosphido (Zr–PR_2) intermediates. TD-DFT analysis reveals the lowest lying excited state in the proposed active catalysts are dominated by a P 3p→Zr 4d MLCT, presumably leading to enhanced catalysis. This hypothesis follows from triamidoamine-supported zirconium catalysts but demonstrates the generality of photocatalytic hydrophosphination with d^0 metals.

Keywords: hydrophosphination; photocatalysis; zirconium; phosphines

Citation: Novas, B.T.; Morris, J.A.; Liptak, M.D.; Waterman, R. Effect of Photolysis on Zirconium Amino Phenoxides for the Hydrophosphination of Alkenes: Improving Catalysis. *Photochem* **2022**, *2*, 77–87. https://doi.org/10.3390/photochem2010007

Academic Editor: Vincenzo Vaiano

Received: 29 November 2021
Accepted: 12 January 2022
Published: 18 January 2022

1. Introduction

Organophosphines have found extensive use in the areas of materials science, biology, agriculture, electronics, and especially, catalysis [1–7]. Despite their utility, responsible use of phosphorus is imperative as a result of dwindling supply [2,3,8]. Selective carbon–phosphorus bond formation has been an ongoing focus of research for these reasons [4,9–14]. A wide range of phosphine chemistry has been developed, with metal-catalyzed hydrophosphination being one of the most economic avenues for P–C bond formation. Though significant progress has occurred [1,7,15–24], challenges remain for this transformation, with catalyst and substrate scope being two key avenues for improvement [1,15,25–30].

Photolysis has been demonstrated to increase the activity of triamidoamine zirconium compounds for hydrophosphination catalysis while also unlocking reactivity with previously inert substrates [31–33]. This methodology, irradition during catlaysis or photocatalysis, has been extended to another group of 15 substrates, leading to improved hydroarsination catalysis with primary arsines [34]. Photolysis plays a key role in these reactions, where reactions are sluggish if not inactive under the strict exclusion of light and distinct from photoactivation where light is only needed to develop an active catalyst [27,31,32,34–36]. Analysis by time-dependent density functional theory (TD-DFT) suggests the enhanced reactivity under photolysis is due to the population of a charge-transfer state that exhibits significant σ* character and weakening of the Zr–P bond that allows more facile substrate insertion [31]. A question arose from the triamidoamine-supported zirconium studies: is this photocatalysis general? The particular geometry and frontier orbital arrangement of triamidoamine zirconium may result in exclusive photocatalytic activity. To test for general photocatalysis, other known hydrophosphination catalysts with different geometries and donor ligands must be screened.

Yao and coworkers reported a library of zirconium complexes bearing amino phenoxide ligands for the hydrophosphination of alkenes and heterocumulenes [37]. These compounds gave modest turnovers of the hydrophosphination of several styrene derivatives with diphenylphosphine under ambient conditions and low catalyst loadings [37], and a similar study was reported with primary phosphine substrates [38]. Their successful zirconium catalysts with pseudo-octahedral geometries and N_xO_y donor ligand sets were ideal to test hypothesis that photocatalytic enhancement is general. Yao and coworkers' most active hydrophosphination catalyst bearing an N_2O (N_2O=O-2,4-$^tBu_2C_6H_2$-6-$CH_2N(CH_2CH_2NMe_2)CH_2$-2-MeO-3,5-$^tBu_2C_6H_2$) donor set was chosen along with a less active analog bearing an N_2O_2 (N_2O_2=1,4-bis(O-2,4-$^tBu_2$-6-CH_2)piperazine) donor set (Figure 1). The study of these systems under photocatalytic conditions reulted in substantial enchancement of activity versus ambient light conditions.

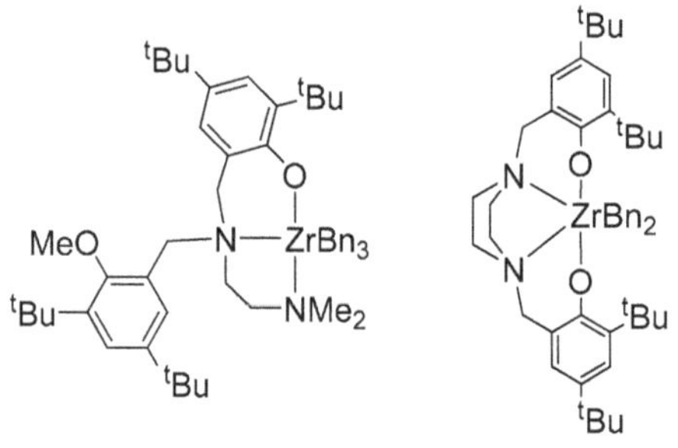

(1) (2)

Figure 1. Molecular structure of compounds **1** and **2**.

2. Results and Discussion

2.1. Photocatalytic Hydrophosphination

Styrene was treated with 1 equiv. of Ph_2PH and 5 mol % of $(N_2O)ZrBn_3$ (**1**) at ambient temperature under visible light irradiation to afford 83% conversion to the corresponding hydrophosphination product in 2 h (Table 1, Entry 1). Performing the same reaction under ambient light provided 8% conversion after 2 h (Table 1, Entry 2) and 87% conversion after a period of 24 h (Table 1, Entry 4). Reactions run in the dark showed severely reduced product formation, with a scarcely observable (~1%) product, namely phosphine, formed in 2 h (Table 1, Entry 3) and only 4% conversion after 24 h (Table 1, Entry 5). Catalysis was also expanded to *para*-substituted styrenes. Reaction of 4-*tert*-butyl styrene with Ph_2PH under identical conditions led to 70% conversion to the phosphine product (Table 1, Entry 6). Treatment of 4-bromo styrene with Ph_2PH under identical conditions gave 91% conversion in 2 h (Table 1, Entry 7). Similar reactions were successful with non-styrene substrates. Hydrophosphination with 2,3-dimethyl butadiene resulted in 65% conversion in 2 h (Table 1, Entry 8). Methyl acrylate, a commonly active hydrophosphination substrate, gave 90% conversion to the hydrophosphination product under identical conditions (Table 1, Entry 9). *Trans*-chalcone, a typical model substrate in asymmetric hydrophosphination [29,30], showed 68% conversion in a modest 24 h period (Table 1, Entry 10).

Table 1. Intermolecular hydrophosphination of alkenes and Ph_2PH catalyzed by **1**.

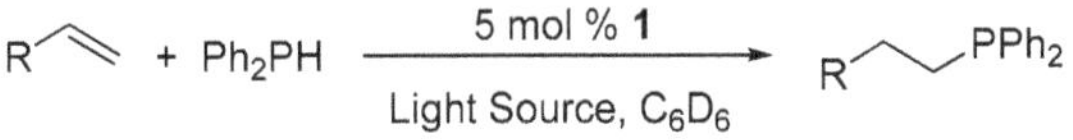

R + Ph_2PH → (5 mol % **1**; Light Source, C_6D_6) R PPh_2

Entry	Substrate	Product	Light Source	Time	Conversion
1	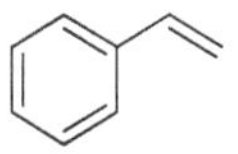	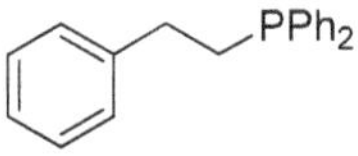 PPh_2	LED	2 h	83%
2	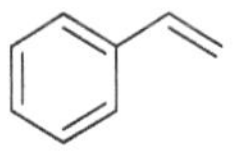	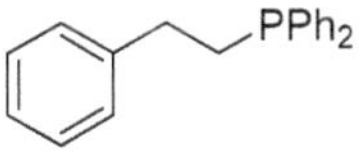 PPh_2	Ambient	2 h	8%
3	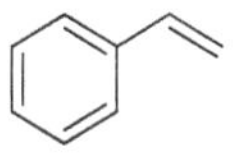	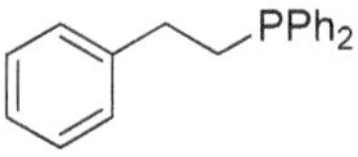 PPh_2	Dark	2 h	>1%
4	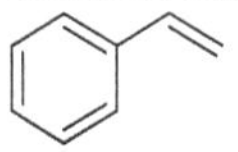	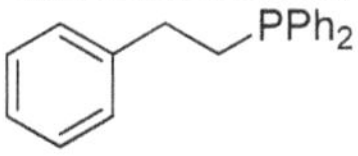 PPh_2	Ambient	24 h	87%
5	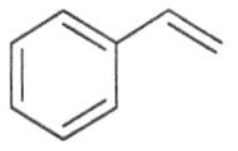	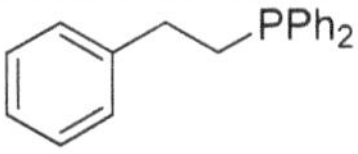 PPh_2	Dark	24 h	4%
6	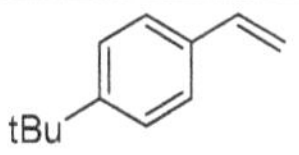 tBu	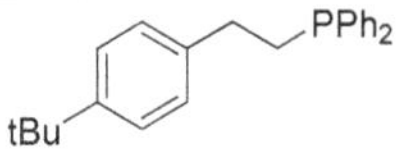 tBu, PPh_2	LED	2 h	70%
7	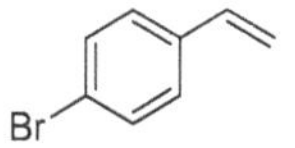 Br	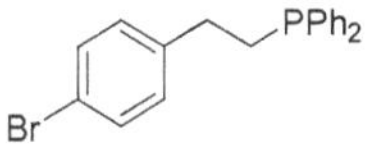 Br, PPh_2	LED	2 h	91%
8		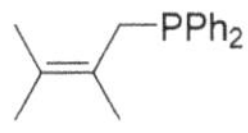 PPh_2	LED	2 h	65%
9	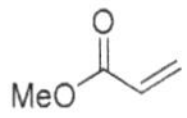 MeO, O	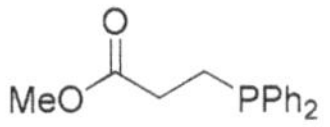 MeO, O, PPh_2	LED	2 h	90%
10	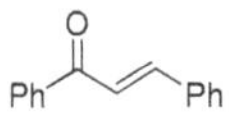 Ph, O, Ph	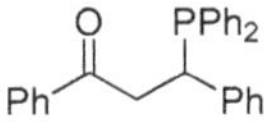 Ph, O, PPh_2, Ph	LED	24 h	68%

Light sources include ambient light from commercial fluorescent overhead lighting and direct irradiation by an LED in the form of a commercial bulb, as described in the Supporting Information.

Greater conversions were observed, even at lower catalyst loadings, for all styrene substrates through photolysis, complementing the progress made by Yao and co-workers in identifying this compound for hydrophosphination catalysis [37]. It is clear from these results that photolysis can serve to improve hydrophosphination catalysis for **1** using secondary phosphines.

Yao and co-workers demonstrated activity with primary phosphines as well [38]. In that report, neither **1** or **2** were used, but the reported catalysts resemble those investigated in the study and their prior work. Given the enhanced activity of **1** and **2** under photocatalytic conditions, expansion of the research to primary phosphines was explored. The reaction of styrene with $PhPH_2$ and 5 mol % of **1** resulted in quantitative consumption of styrene at 2 h of irradiation (Table 2, Entry 1). The same reaction under ambient light resulted in 21% conversion (Table 2, Entry 2). Extending the reaction period to 24 h resulted in 69% conversion (Table 2, Entry 4). Performing this reaction in the dark resulted in a severely diminished 2% conversion after 2 h (Table 2, Entry 3) and 4% conversion after 24 h (Table 2, Entry 4). Low conversion under ambient light suggests a reason why **1** was not reported in Yao's 2018 study [38], but it affirms the impact of photolysis on d^0 hydrophosphination catalysts.

Table 2. Intermolecular hydrophosphination of styrene and $PhPH_2$ catalyzed by **1**.

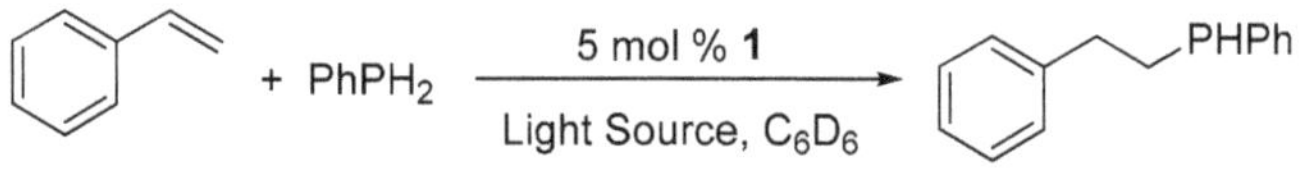

+ $PhPH_2$ —5 mol % **1**, Light Source, C_6D_6→ PHPh

Entry	Product	Light Source	Time	Conversion
1	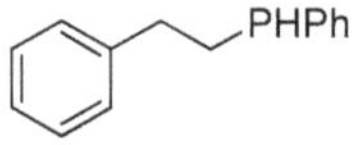 PHPh	LED	2 h	>99%
2	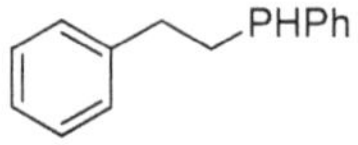 PHPh	Ambient	2 h	21%
3	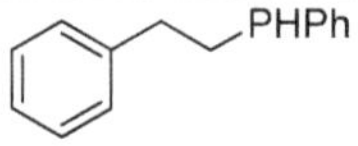 PHPh	Dark	2 h	2%
4	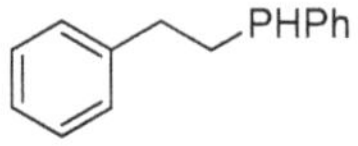 PHPh	Ambient	24 h	69%
5	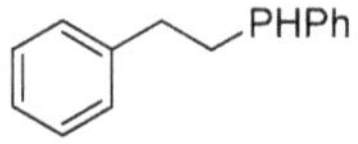 PHPh	Dark	24 h	4%

The change in geometry and lower relative reactivity of $(N_2O_2)ZrBn_2$ (**2**) as compared to **1** prompted exploration under photocatalytic conditions. The reaction of styrene with Ph_2PH and 5 mol % of **2** resulted in 91% product formation after 2 h (Table 3, Entry 1). Conversion under ambient light was behind, providing 12% conversion after 2 h (Table 3, Entry 2), and 92% conversion after an extended 24 h (Table 3, Entry 4). As expected, reactions run rigorously in the dark afforded barely detectable conversion after 2 h (Table 3, Entry 3), and ~1% conversion after 24 h (Table 3, Entry 5). Substituted styrenes showed a similar trend to **1**. However, slightly greater conversions were observed under photolysis as compared to ambient light. A conversion of 88% was observed for the reaction of 4-*tert*-butyl styrene (Table 3, Entry 6), and quantitative conversion was observed for the reaction of 4-bromo styrene after 2 h (Table 3, Entry 4). Under the same conditions with 2,3-dimethyl butadiene as substrate, 66% conversion was observed. (Table 3, Entry 6). Quantitative conversion was seen when using methyl acrylate as substrate (Table 3, Entry 7), and 83% conversion was observed with pro-chiral *trans*-chalcone over a period of 24 h (Table 3, Entry 8).

Table 3. Intermolecular hydrophosphination of alkenes and Ph_2PH catalyzed by **2**.

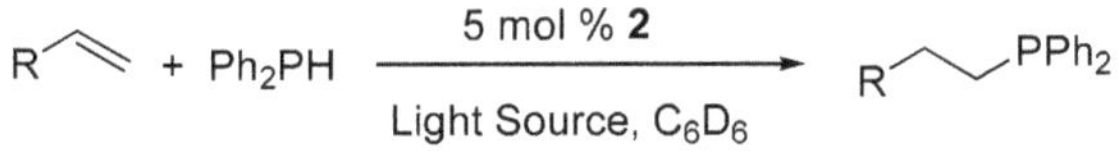

R + Ph_2PH —(5 mol % **2**; Light Source, C_6D_6)→ R–PPh_2

Entry	Substrate	Product	Light Source	Time	Conversion
1	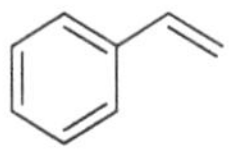	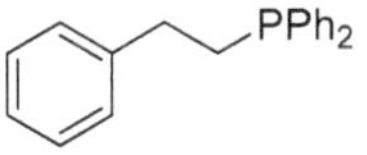 PPh_2	LED	2 h	91%
2	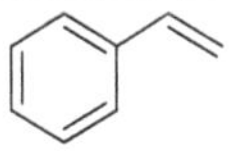	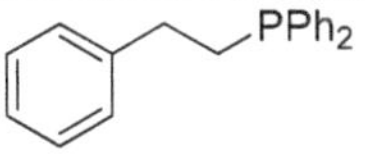 PPh_2	Ambient	2 h	12%
3	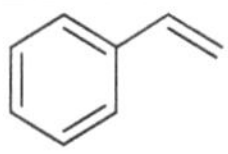	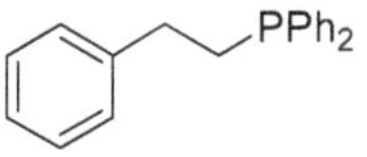 PPh_2	Dark	2 h	>1%
4	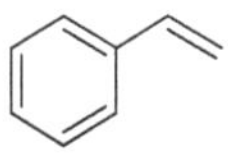	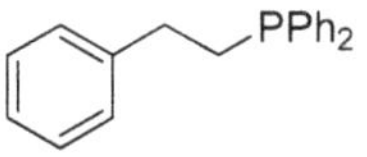 PPh_2	Ambient	24 h	92%
5	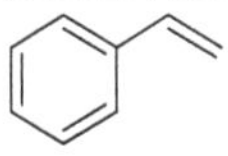	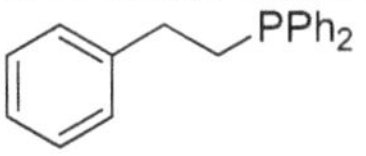 PPh_2	Dark	24 h	1%
6	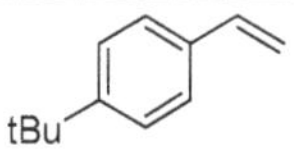 tBu	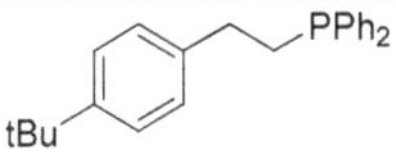 tBu, PPh_2	LED	2 h	88%
7	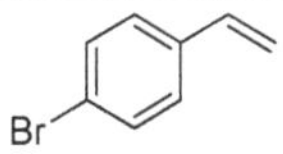 Br	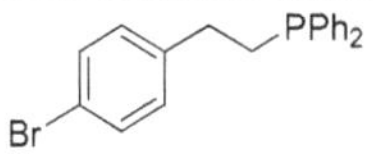 Br, PPh_2	LED	2 h	>99%
8		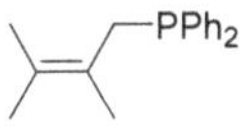 PPh_2	LED	2 h	66%
9	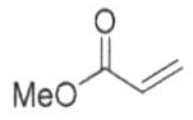 MeO, O	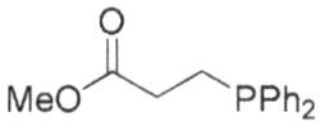 MeO, O, PPh_2	LED	2 h	>99%
10	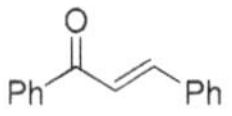 Ph, O, Ph	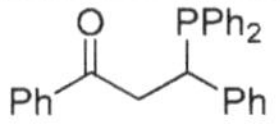 Ph, O, PPh_2, Ph	LED	24 h	83%

As with **1**, the reactivity of **2** in hydrophosphination with $PhPH_2$ was explored. The reaction of styrene with $PhPH_2$ and 5 mol % of **2** resulted in the quantitative conversion of styrene to the reaction's product (Table 4, Entry 1). Using these conditions under ambient light availed 17% conversion in 2 h (Table 4, Entry 2) and 76% conversion in 24 h (Table 4, Entry 4). Running this reaction in the absence of light reduced the conversion to ~2% after 24 h (Table 4, Entry 5).

Table 4. Intermolecular hydrophosphination of styrene and $PhPH_2$ catalyzed by **2**.

Styrene + $PhPH_2$ → (5 mol % **2**; Light Source, C_6D_6) → $PhCH_2CH_2PHPh$

Entry	Product	Light Source	Time	Conversion
1	$PhCH_2CH_2PHPh$	LED	2 h	>99%
2	$PhCH_2CH_2PHPh$	Ambient	2 h	17%
3	$PhCH_2CH_2PHPh$	Dark	2 h	>1%
4	$PhCH_2CH_2PHPh$	Ambient	24 h	76%
5	$PhCH_2CH_2PHPh$	Dark	24 h	2%

2.2. *Computational Analysis*

Spectroscopic and computational analysis indicated an n→d charge transfer in hydrophosphination catalysis using triamidoamine-supported zirconium, which led to improved reactivity as a result of promoted insertion [31]. It was previously hypothesized that pre-existing catalysts could be enhanced by photolysis, and this was confirmed by experimental results using **1** and **2**. To further elucidate whether enhanced catalysis is a result of accessing an excited state in potential intermediates where insertion is promoted, TD-DFT modeling was utilized.

All efforts to produce phosphido complexes of **1** and **2** failed, leading to the employment of the crystal structure of **2** to construct a structural model [39]. The geometry of the structural model was optimized using density functional theory (DFT) with the B3LYP functional and the def2-TZVP basis set [40–43]. The modeling employed the RIJCOSX approximation and tight SCF convergence criteria [44]. The conductor-like polarizable continuum model (CPCM) was used to define a solvent through its dielectric constant and refractive index. The root-mean-square deviation (RMSD) of the optimized geometry of **2** compared to the crystal structure was 0.813 Å. Visually, the DFT-optimized geometry has more exposed benzyl groups (Figure 2).

The electronic structure of **2** was probed via TD-DFT. The first ten electronic transitions were calculated with an expansion space of 60 vectors using the B3LYP functional and def2-TZVP basis set, again employing the RIJCOSX approximation and tight SCF convergence criteria. Solvent was simulated with the CPCM solvation model. The orca_mapspc was used to convolute the transitions through Gaussians with a full-width half-max (FWHM) of 1500 cm^{-1} [45]. This was compared with an experimental absorption spectrum of **2** in diethyl ether, revealing a low-energy, low-intensity shoulder and a higher-energy, higher-intensity peak around 30,000 cm^{-1} in both the experimental and predicted spectra (Figure 3). The predicted spectrum was slightly redshifted compared to the experimental spectrum; a common phenomenon that was also observed in the modeling of triamidoamine zirconium [31].

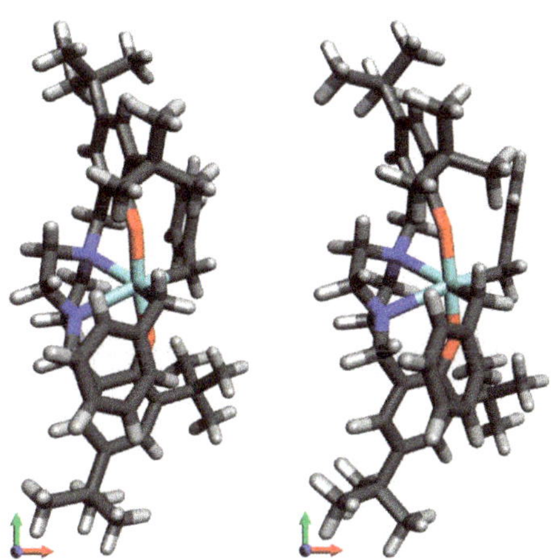

Figure 2. The crystal structure (**left**) and DFT-optimized geometry (**right**) of **2**.

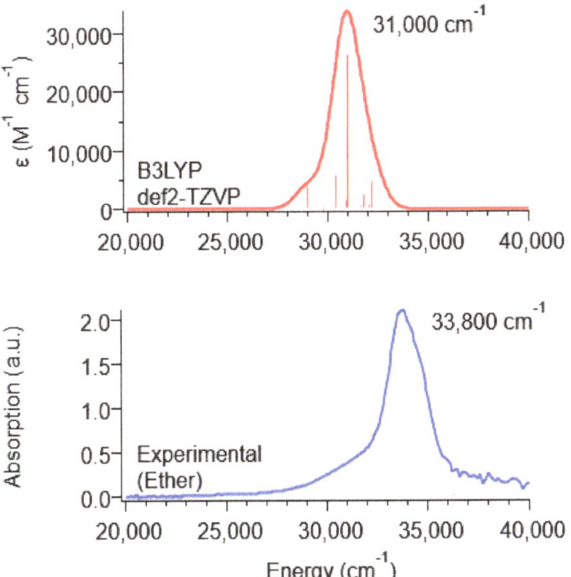

Figure 3. TD-DFT-predicted absorbance spectrum at the B3LYP/def2-TZVP level of theory in the gas phase for **2** (red spectrum) and experimental absorbance spectrum for **2** in diethyl ether (blue spectrum).

It is important to consider that from the absorbance spectra, B3LYP/def2-TZVP predicts a consistent electronic structure for **2** assuming the zirconium oxidation state and molecule charge does not change when forming the active catalyst ($(N_2O_2)ZrBn_x(PPh_2)_y$). If this is the case, the computational model will still be accurate. However, we cannot conclusively prove this without experimental spectra of the active catalyst.

Structural models of the active catalysts (hereafter **A**, **B**, and **AB**) were prepared from the B3LYP-optimized geometry of **2** using the program Avogadro [46]. Either one (**A**), the other (**B**), or both (**AB**) benzyl substituents were replaced with PPh_2 substituents. The geometry of **A**, **B**, and **AB** were optimized using DFT at the B3LYP/def2-TZVP level of theory, employing the RIJCOSX approximation, tight SCF convergence criteria, and simulating benzene solvent using CPCM (Figure 4).

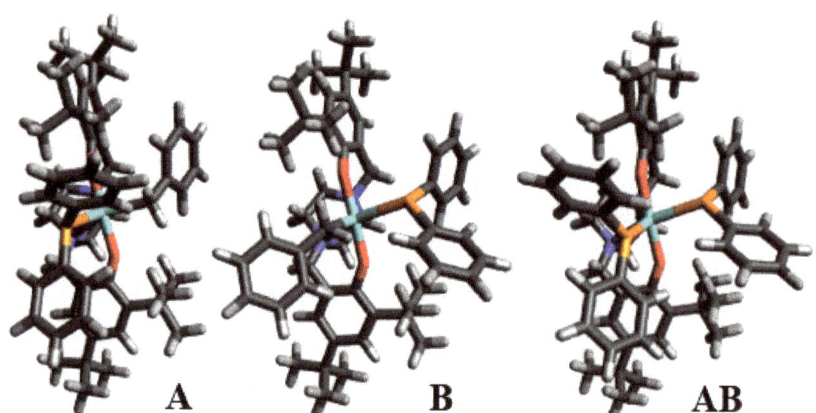

Figure 4. DFT optimized geometry of **A** (left), **B** (middle), and **AB** (right) at the B3LYP/def2-TZVP level of theory.

Geometric optimizations give a final Single Point Energy (SPE). The SPE is related to the number of atoms, and so only **A** and **B**, which have the same number of atoms, can be compared. There is no energetic preference for replacing Bn **A** or **B** with PPh_2 over the other.

TD-DFT calculations at the B3LYP/def2-TZVP level of theory, with the RIJCOSX approximation, tight SCF convergence criteria, and simulating benzene with CPCM were carried out to probe the electronic transitions of compounds **A**, **B**, and **AB**. Without experimental spectra, it cannot be conclusively stated that the computational electronic structure is consistent, but there is good reason to believe it would be. Regardless, the predicted absorbance spectra were mapped using orca_mapspc to convolute Gaussians with a FWHM of 1500 cm^{-1} (Figure in Supplementary Materials). The first 10 electronic transitions were calculated with an expansion space of 60 vectors.

It was hypothesized that excitation into high-lying excited states is followed by relaxation to the lowest-lying excited state, following Kasha's Rule, from which catalysis was proposed to occur [31,47]. The lowest three excited states for compound **A** were found to be dominated by transitions that exhibit donation from a P *3p* orbital to a Zr *4d* orbital, consistent with work using triamidoamine zirconium [31]. The excited states for compound **B** were largely the same, to the extent of having the same orbital numbers. In compound **AB**, the lowest four excited states were primarily P→Zr donation, because of the second PPh_2 moiety.

The charge and formal oxidation state of zirconium did not change when **2** became **A**, **B**, or **AB**, and thus, it can be assumed that the computational model will remain accurate. In all of the active catalyst models (**A**, **B**, or **AB**), the lowest lying excited state—where photochemistry is proposed to occur via Kasha's Rule—was dominated by a charge transfer from the P *3p* orbital to the Zr *4d* orbital. These were P n→Zr d transitions. Based on prior results [31], we can assume that this charge transfer is correlated with the elongation of the Zr–P bond in the lowest-lying excited state, thereby weakening the bond to facilitate insertion chemistry. This hypothesis is corroborated by our experimental results in photocatalytic hydrophosphination using **1** and **2**.

3. Conclusions

Irradiation serves to enhance intermolecular hydrophosphination catalysis with **1** and **2** for both secondary and primary phosphines. An accurate computational model of the electronic structure of **2** was determined. The lowest lying excited states in compounds **A**, **B**, and **AB** were found to be dominated by P n→Zr d transitions, likely promoting insertion chemistry. In sum, these findings establish that photocatalytic hydrophosphination is not restricted to triamidoamine-supported zirconium, five-coordinate zirconium, or nitrogen donors. It is general to other zirconium catalysts equipped with distinct geometries and donor ligand sets, and these results suggest that this enhancement may be broadly applicable to d^0 metals though a similar mechanism.

4. Synthetic, Spectroscopic, and Catalytic Methods

All air-sensitive manipulations were performed according to a previously published literature procedure [31,32]. Diphenylphosphine was synthesized according to a modified literature procedure [48]. In addition, **1** and **2** were synthesized according to modified literature procedures [37,39]. All other chemicals were obtained from commercial suppliers and dried by conventional means.

NMR spectra were recorded with a Bruker AXR 500 MHz spectrophotometer in benzene-d_6 solution and reported with reference to residual solvent signal (δ = 7.16 ppm) in ^{1}H NMR spectra. Absorption spectra were recorded with an Agilent Technologies Cary 100 Bio UV-Visible Spectrophotometer (Santa Clara, CA, USA) as ether solutions. $(N_2O_2)ZrBn_2$ was excited between 700 and 200 nm and the excitation slits were set to 2 nm.

Hydrophosphination of alkenes was carried out in a PTFE-sealed J-Young style NMR tube charged with 0.1 mmol alkene, 0.1 mmol phosphine (1.0 M benzene-d_6 solvent stock solution), and 5 mol % of catalyst (0.04 M benzene-d_6 stock solution). The solutions were reacted at ambient temperature for the noted periods under irradiation. The consumption of substrate to product was monitored by ^{1}H NMR and ^{31}P{^{1}H} NMR spectroscopy. Reactions run in new NMR tubes showed identical conversions as those run in reused, washed NMR tubes.

Supplementary Materials: The following supporting information can be downloaded at https://www.mdpi.com/article/10.3390/photochem2010007/s1.

Author Contributions: Conceptualization, B.T.N. and R.W.; synthesis, B.T.N.; catalysis, B.T.N.; computation, J.A.M. and M.D.L.; draft preparation, B.T.N., J.A.M. and R.W.; supervision, M.D.L. and R.W.; funding acquisition, R.W. All authors have read and agreed to the published version of the manuscript.

Funding: This research was funded by the National Science Foundation through CHE-2101766.

Data Availability Statement: Computational details, Cartesian coordinates, and NMR data is provided in the Supplementary Information. Original data is available at https://www.uvm.edu/~waterman/.

Acknowledgments: The authors would like to thank Christine Bange for her seminal contributions in light-driven, zirconium-catalyzed hydrophosphination.

Conflicts of Interest: The authors declare no conflict of interest. The funders had no role in the design of the study; in the collection, analyses, or interpretation of data; in the writing of the manuscript, or in the decision to publish the results.

References

1. Bange, C.A.; Waterman, R. Challenges in Catalytic Hydrophosphination. *Chem. Eur. J.* **2016**, *22*, 12598–12605. [CrossRef]
2. Slootweg, J.C. Sustainable Phosphorus Chemistry: A Silylphosphide Synthon for the Generation of Value-Added Phosphorus Chemicals. *Angew. Chem. Int. Ed. Engl.* **2018**, *57*, 6386–6388. [CrossRef]
3. Øgaard, A.; Brod, E. Efficient Phosphorus Cycling in Food Production: Predicting the Phosphorus Fertilization Effect of Sludge from Chemical Wastewater Treatment. *J. Agric. Food Chem.* **2016**, *64*, 4821–4829. [CrossRef] [PubMed]
4. Troev, K.D. Chapter 2—Reactivity of P–H Group of Phosphines. In *Reactivity of P-H Group of Phosphorus Based Compounds*; Troev, K.D., Ed.; Academic Press: Cambridge, MA, USA, 2018; pp. 19–144.
5. Greenberg, S.; Stephan, D.W. Phosphines Bearing Alkyne Substituents: Synthesis and Hydrophosphination Polymerization. *Inorg. Chem.* **2009**, *48*, 8623–8631. [CrossRef]

6. Kovacik, I.; Wicht, D.K.; Grewal, N.S.; Glueck, D.S.; Incarvito, C.D.; Guzei, I.A.; Rheingold, A.L. Pt(Me-Duphos)-Catalyzed Asymmetric Hydrophosphination of Activated Olefins: Enantioselective Synthesis of Chiral Phosphines. *Organometallics* **2000**, *19*, 950–953. [CrossRef]
7. Koshti, V.; Gaikwad, S.; Chikkali, S.H. Contemporary Avenues in Catalytic PH Bond Addition Reaction: A Case Study of Hydrophosphination. *Coord. Chem. Rev.* **2014**, *265*, 52–73. [CrossRef]
8. Bange, C.A. Exploration of Zirconium-Catalyzed Intermolecular Hydrophosphination with Primary Phosphines: Photocatalytic Single and Double Hydrophosphination. Ph.D. Thesis, University of Vermont, Burlington, VT, USA, 2018; pp. 1–339.
9. Gladysz, J.A.; Bedford, R.B.; Fujita, M.; Gabbaï, F.P.; Goldberg, K.I.; Holland, P.L.; Kiplinger, J.L.; Krische, M.J.; Louie, J.; Lu, C.C.; et al. Organometallics Roundtable 2013–2014. *Organometallics* **2014**, *33*, 1505–1527. [CrossRef]
10. King, A.K.; Gallagher, K.J.; Mahon, M.F.; Webster, R.L. Markovnikov versus anti-Markovnikov Hydrophosphination: Divergent Reactivity Using an Iron(II) β-Diketiminate Pre-Catalyst. *Chem. Eur. J.* **2017**, *23*, 9039–9043. [CrossRef]
11. Kamitani, M.; Itazaki, M.; Tamiya, C.; Nakazawa, H. Regioselective Double Hydrophosphination of Terminal Arylacetylenes Catalyzed by an Iron Complex. *J. Am. Chem. Soc.* **2012**, *134*, 11932–11935. [CrossRef]
12. Bange, C.A.; Waterman, R. Zirconium-Catalyzed Intermolecular Double Hydrophosphination of Alkynes with a Primary Phosphine. *ACS Catal.* **2016**, *6*, 6413–6416. [CrossRef]
13. Mimeau, D.; Gaumont, A.-C. Regio- and Stereoselective Hydrophosphination Reactions of Alkynes with Phosphine−Boranes: Access to Stereodefined Vinylphosphine Derivatives. *J. Org. Chem.* **2003**, *68*, 7016–7022. [CrossRef]
14. Basalov, I.V.; Dorcet, V.; Fukin, G.K.; Carpentier, J.-F.; Sarazin, Y.; Trifonov, A.A. Highly Active, Chemo- and Regioselective YbII and SmII Catalysts for the Hydrophosphination of Styrene with Phenylphosphine. *Chem. Eur. J.* **2015**, *21*, 6033–6036. [CrossRef] [PubMed]
15. Rosenberg, L. Mechanisms of Metal-Catalyzed Hydrophosphination of Alkenes and Alkynes. *ACS Catal.* **2013**, *3*, 2845–2855. [CrossRef]
16. Wang, C.; Huang, K.; Ye, J.; Duan, W.-L. Asymmetric Synthesis of P-Stereogenic Secondary Phosphine-Boranes by an Unsymmetric Bisphosphine Pincer-Nickel Complex. *J. Am. Chem. Soc.* **2021**, *143*, 5685–5690. [CrossRef] [PubMed]
17. Lapshin, I.V.; Basalov, I.V.; Lyssenko, K.A.; Cherkasov, A.V.; Trifonov, A.A. CaII, YbII and SmII Bis(Amido) Complexes Coordinated by NHC Ligands: Efficient Catalysts for Highly Regio- and Chemoselective Consecutive Hydrophosphinations with PH3. *Chem. Eur. J.* **2019**, *25*, 459–463. [CrossRef] [PubMed]
18. Sadeer, A.; Kojima, T.; Ng, J.S.; Gan, K.; Chew, R.J.; Li, Y.; Pullarkat, S.A. Catalytic Access to Ferrocenyl Phosphines Bearing both Planar and Central Chirality—A Kinetic Resolution Approach via Catalytic Asymmetric P(III)–C Bond Formation. *Tetrahedron* **2020**, *76*, 131259. [CrossRef]
19. Isley, N.A.; Linstadt, R.T.H.; Slack, E.D.; Lipshutz, B.H. Copper-Catalyzed Hydrophosphinations of Styrenes in Water at Room Temperature. *Dalton Trans.* **2014**, *43*, 13196–13200. [CrossRef]
20. Li, J.; Lamsfus, C.A.; Song, C.; Liu, J.; Fan, G.; Maron, L.; Cui, C. Samarium-Catalyzed Diastereoselective Double Addition of Phenylphosphine to Imines and Mechanistic Studies by DFT Calculations. *ChemCatChem* **2017**, *9*, 1368–1372. [CrossRef]
21. Moglie, Y.; González-Soria, M.J.; Martín-García, I.; Radivoy, G.; Alonso, F. Catalyst- and Solvent-Free Hydrophosphination and Multicomponent Hydrothiophosphination of Alkenes and Alkynes. *Green Chem.* **2016**, *18*, 4896–4907. [CrossRef]
22. Teo, R.H.X.; Chen, H.J.; Li, Y.; Pullarkat, S.A.; Leung, P.-H. Asymmetric Catalytic 1,2-Dihydrophosphination of Secondary 1,2-Diphosphines—Direct Access to Free P*- and P*,C*-Diphosphines. *Adv. Synth. Catal.* **2020**, *362*, 2373–2378. [CrossRef]
23. Garner, M.E.; Parker, B.F.; Hohloch, S.; Bergman, R.G.; Arnold, J. Thorium Metallacycle Facilitates Catalytic Alkyne Hydrophosphination. *J. Am. Chem. Soc.* **2017**, *139*, 12935–12938. [CrossRef] [PubMed]
24. Waterman, R. Triamidoamine-Supported Zirconium Compounds in Main Group Bond-Formation Catalysis. *Acc. Chem. Res.* **2019**, *52*, 2361–2369. [CrossRef] [PubMed]
25. Trifonov, A.A.; Basalov, I.V.; Kissel, A.A. Use of Organolanthanides in the Catalytic Intermolecular Hydrophosphination and Hydroamination of Multiple C–C Bonds. *Dalton Trans.* **2016**, *45*, 19172–19193. [CrossRef]
26. Webster, R.L. β-Diketiminate Complexes of the First Row Transition Metals: Applications in Catalysis. *Dalton Trans.* **2017**, *46*, 4483–4498. [CrossRef] [PubMed]
27. Dannenberg, S.G.; Waterman, R. A Bench-Stable Copper Photocatalyst for the Rapid Hydrophosphination of Activated and Unactivated Alkenes. *Chem. Comm.* **2020**, *56*, 14219–14222. [CrossRef]
28. Sarazin, Y.; Carpentier, J.-F. Calcium, Strontium and Barium Homogeneous Catalysts for Fine Chemicals Synthesis. *Chem. Rec.* **2016**, *16*, 2482–2505. [CrossRef]
29. Seah, J.W.K.; Teo, R.H.X.; Leung, P.-H. Organometallic Chemistry and Application of Palladacycles in Asymmetric Hydrophosphination Reactions. *Dalton Trans.* **2021**, *50*, 16909–16915. [CrossRef]
30. Pullarkat, S.A. Recent Progress in Palladium-Catalyzed Asymmetric Hydrophosphination. *Synthesis* **2016**, *48*, 493–503. [CrossRef]
31. Bange, C.A.; Conger, M.A.; Novas, B.T.; Young, E.R.; Liptak, M.D.; Waterman, R. Light-Driven, Zirconium-Catalyzed Hydrophosphination with Primary Phosphines. *ACS Catal.* **2018**, *8*, 6230–6238. [CrossRef]
32. Novas, B.T.; Bange, C.A.; Waterman, R. Photocatalytic Hydrophosphination of Alkenes and Alkynes Using Diphenylphosphine and Triamidoamine-Supported Zirconium. *Eur. J. Inorg. Chem.* **2019**, *2019*, 1640–1643. [CrossRef]
33. Cibuzar, M.P.; Novas, B.T.; Waterman, R. Zirconium Complexes. In *Comprehensive Coordination Chemistry III*; Constable, E.C., Parkin, G., Que, L., Jr., Eds.; Elsevier: Oxford, UK, 2021; pp. 162–196.

34. Bange, C.A.; Waterman, R. Zirconium-Catalyzed Hydroarsination with Primary Arsines. *Polyhedron* **2018**, *156*, 31–34. [CrossRef]
35. Cibuzar, M.P.; Dannenberg, S.G.; Waterman, R. A Commercially Available Ruthenium Compound for Catalytic Hydrophosphination. *Isr. J. Chem.* **2020**, *60*, 446–451. [CrossRef]
36. Ackley, B.J.; Pagano, J.K.; Waterman, R. Visible-Light and Thermal Driven Double Hydrophosphination of Terminal Alkynes Using a Commercially Available Iron Compound. *Chem. Comm.* **2018**, *54*, 2774–2776. [CrossRef]
37. Zhang, Y.; Qu, L.; Wang, Y.; Yuan, D.; Yao, Y.; Shen, Q. Neutral and Cationic Zirconium Complexes Bearing Multidentate Aminophenolato Ligands for Hydrophosphination Reactions of Alkenes and Heterocumulenes. *Inorg. Chem.* **2018**, *57*, 139–149. [CrossRef]
38. Zhang, Y.; Wang, X.; Wang, Y.; Yuan, D.; Yao, Y. Hydrophosphination of Alkenes and Alkynes with Primary Phosphines Catalyzed by Zirconium Complexes Bearing Aminophenolato Ligands. *Dalton Trans.* **2018**, *47*, 9090–9095. [CrossRef] [PubMed]
39. Zhang, Y.; Sun, Q.; Wang, Y.; Yuan, D.; Yao, Y.; Shen, Q. Intramolecular Hydroamination Reactions Catalyzed by Zirconium Complexes Bearing Bridged Bis(phenolato) Ligands. *RSC Adv.* **2016**, *6*, 10541–10548. [CrossRef]
40. Becke, A.D. Density-Functional Exchange-Energy Approximation with Correct Asymptotic Behavior. *Phys. Rev. A* **1988**, *38*, 3098–3100. [CrossRef] [PubMed]
41. Becke, A.D. Density-Functional Thermochemistry. III. The Role of Exact Exchange. *J. Chem. Phys.* **1993**, *98*, 5648–5652. [CrossRef]
42. Lee, C.; Yang, W.; Parr, R.G. Development of the Colle-Salvetti Correlation-Energy Formula into a Functional of the Electron Density. *Phys. Rev. B* **1988**, *37*, 785–789. [CrossRef]
43. Weigend, F.; Ahlrichs, R. Balanced Basis Sets of Split Valence, Triple Zeta Valence and Quadruple Zeta Valence Quality for H to Rn: Design and Assessment of Accuracy. *Phys. Chem. Chem. Phys.* **2005**, *7*, 3297–3305. [CrossRef]
44. Izsák, R.; Neese, F. An Overlap Fitted Chain of Spheres Exchange Method. *J. Chem. Phys.* **2011**, *135*, 144105. [CrossRef] [PubMed]
45. Neese, F. The ORCA Program System. *WIREs Comp. Mol. Sci.* **2012**, *2*, 73–78. [CrossRef]
46. Hanwell, M.D.; Curtis, D.E.; Lonie, D.C.; Vandermeersch, T.; Zurek, E.; Hutchison, G.R. Avogadro: An Advanced Semantic Chemical Editor, Visualization, and Analysis Platform. *J. Cheminform.* **2012**, *4*, 17. [CrossRef]
47. Kasha, M. Characterization of Electronic Transitions in Complex Molecules. *Discuss. Faraday Soc.* **1950**, *9*, 14–19. [CrossRef]
48. Waterman, R. Selective Dehydrocoupling of Phosphines by Triamidoamine Zirconium Catalyst. *Organometallics* **2007**, *26*, 2492–2494. [CrossRef]

photochem

Article

Remazol Black Decontamination Study Using a Novel One-Pot Synthesized S and Co Co-Doped TiO_2 Photocatalyst

Riska Dwiyanna , Roto Roto and Endang Tri Wahyuni *

Department of Chemistry, Faculty of Mathematics and Natural Sciences, Universitas Gadjah Mada-Sekip Utara, Bulaksumur, Yogyakarta 55281, Indonesia; riska.dwiyanna@mail.ugm.ac.id (R.D.); roto05@ugm.ac.id (R.R.)

* Correspondence: endang_triw@ugm.ac.id; Tel.: +62-81328892114

Abstract: This study investigated the decolorization of Remazol Black (RBB) using a TiO_2 photocatalyst modified by S and Co co-doped TiO_2 (S-Co-TiO_2) from a single precursor. X-ray diffraction, Fourier transform infrared spectroscopy, scanning electron microscopy, and UV–Vis specular reflectance spectroscopy were used to characterize the photocatalysts. The results revealed that the band-gap energy of the doped and co-doped TiO_2 decreased, with the S-Co-TiO_2 8% showing the greatest one, and was found to be 2.78 eV while undoped TiO_2 was 3.20 eV. The presence of S and Co was also identified through SEM-EDX. An activity study on RBB removal revealed that the S-Co-TiO_2 photocatalyst showed the best result compared to undoped TiO_2, S-TiO_2, and Co-TiO_2. The S-Co-TiO_2 8% photocatalyst reduced RBB concentration (20 mg L^{-1}) up to 96% after 90 min of visible light irradiation, whereas S-TiO_2, Co-TiO_2, and undoped TiO_2 reduced it to 89%, 56%, and 39%, respectively. A pH optimization study showed that the optimum pH of RBB decolorization by S-Co-TiO_2 was 3.0, the optimum mass was 0.6 g L^{-1}, and reuse studies show that S-Co-TiO_2 8% has the potential to be used repeatedly to remove colored pollutants. The results obtained indicate that the modification of S, Co co-doped titania synthesized using a single precursor has been successfully carried out and showed excellent characteristics and activity compared to undoped or doped TiO_2.

Keywords: cobalt; decolorization; photocatalyst; Remazol Black; sulfur; TiO_2

Citation: Dwiyanna, R.; Roto, R.; Wahyuni, E.T. Remazol Black Decontamination Study Using a Novel One-Pot Synthesized S and Co Co-Doped TiO_2 Photocatalyst. *Photochem* **2021**, *1*, 488–504. https://doi.org/10.3390/photochem1030032

Received: 4 November 2021
Accepted: 23 November 2021
Published: 26 November 2021

Publisher's Note: MDPI stays neutral with regard to jurisdictional claims in published maps and institutional affiliations.

1. Introduction

Dyestuff waste is a significant source of contamination in the aquatic environment. Among various synthetic dyes, Remazol Black (RBB) is the most widely used because of its low energy consumption during the dyeing process, water fastness, color brightness, and good fixation characteristics on fabric fibers [1,2]. However, due to its complex chemical structure, this dye is stable and difficult to biodegrade, so the concentration in the environment does not tend to decrease [3–6]. The release of this dye into the environment is extremely hazardous because it can produce toxic, mutagenic, and harmful by-products of oxidation, hydrolysis, and other chemical reactions that occur in the wastewater mixture, which are toxic, mutagenic, and harmful to microorganisms, aquatic life, and humans [5]. Therefore, an efficient method is needed to remove the dye concentration from wastewater before being discharged into the environment.

One of the effective methods for dealing with various organic pollutants such as dyestuff waste is advanced oxidation processes (AOPs) using heterogeneous photocatalysts [7,8]. Compared to other semiconductors, TiO_2 is the most widely used due to its advantages such as high photocatalytic activity, stability, non-toxicity, and low cost, and it has been widely used for environmental pollution control, especially water pollution by dyestuff waste [9,10]. However, due to its large band gap (anatase 3.2 eV), TiO_2 can only be activated by UV radiation, which is only 3–4% of the available total solar radiation. To improve the efficiency of photocatalyst activity in the visible light region, TiO_2 should be further modified [11–13].

Modification with dopants has been shown to reduce the band-gap energy and increase the responsivity of TiO_2 to the visible light region. The use of single dopants has been widely studied. However, there are several drawbacks to the effects of single doping, including that high doping levels are almost unattainable due to the mismatch of ionic charge and/or atomic radius of the dopants with TiO_2, thermal instability, and the rate of electron recombination being increased compared to undoped conditions [14], which can inhibit the photocatalytic activity of TiO_2. Co-doping can be an alternative to solve this problem. The co-doping technique is an effective method for lowering the band gap energy of TiO_2, allowing TiO_2 to be responsive in the visible light range [15–18]. Meanwhile, the combination of metallic and non-metallic dopants proved to be effective in reducing the band gap energy of TiO_2 significantly, reducing electron and hole recombination that might result from a single doping effect, thereby increasing photocatalytic activity in the visible spectrum for efficient utilization of sunlight [19].

Compared to other non-metallic dopants such as N and C, sulfur is considered advantageous due to its ability to narrow the band gap, high thermal stability, and enhanced photocatalytic activity [10,20–22]. Meanwhile, incorporation of 3D transition metals into TiO_2 is also an effective approach to reducing the band gap energy. The unfilled d-electron structure in transition metals can accommodate more electrons, allowing transition dopants to act as a photogenerated electron–hole pair trap, reducing the occurrence of electron–hole pair recombination on the photocatalyst surface [23]. Among other transition metals such as V, Cr, Mn, Fe, and Co, cobalt is considered a good candidate for TiO_2 doping due to the similarity of the ionic radius of Co^{2+} to Ti^{4+}, and it has been shown to increase the photocatalytic activity of TiO_2 in the visible region [24–26]. Co-doping modification of TiO_2 with S and Co has been reported. However, to the best of our knowledge, no studies have reported the use of a single precursor as a source of S and Co dopant, as well as activity assays for photocatalytic decontamination of Remazol Black under visible light exposure. Furthermore, we developed light-emitting diodes (LED) as a light source due to their long lifetime and high energy efficiency compared to typical light sources such as Xe lamps and Hg-Xe lamps [27].

2. Materials and Methods

2.1. Materials

Titanium(IV) isopropoxide (TTIP, 97%) was purchased from Hangzhou Jiu Peng Material Co., Ltd. (Zhejiang, China). Ethanol (C_2H_5OH, 99.5%), cobalt(II) sulfate ($CoSO_4$), thiourea (CH_4N_2S), cobalt(II) chloride hexahydrate ($CoCl_2{\cdot}6H_2O$), hydrochloric acid (HCl, 36%), sodium hydroxide (NaOH), and Remazol Black (RBB, $C_{26}H_{21}O_{19}N_5S_6Na_4$, MW = 991.82 g mol^{-1}) were obtained from Merck, and deionized water was used in this work. The chemical structure of RBB is shown in Figure S1.

2.2. Synthesis of Photocatalysts

The sol-gel method was used to synthesize all of the photocatalysts. Under magnetic stirring, titanium(IV) isopropoxide (97%) was stoichiometrically dissolved in 20 mL ethanol. In separate bins, 0.24 g of thiourea was dissolved in a mixture of distilled water and ethanol (1:1 volume ratio) to obtain a dopant concentration of 10% S (w/w). The prepared TTIP solution was added dropwise to this solution under magnetic stirring, and the pH of the solution was adjusted to 3 with the addition of 1 M HCl. Stirring was continued for 2 h, and then the mixture was allowed to stand for 24 h for the gel ripening process. Then, the gel that formed was dried in an oven at 80 °C for 4 h, and the solid obtained was calcined at 450 °C for 3 h. The catalyst obtained was labeled S-TiO_2.

To synthesize Co-doped TiO_2, a similar method to the previous one was used. $CoCl_2{\cdot}6H_2O$ 4% (w/w) as a source of Co dopant was dissolved into a mixture of distilled water and ethanol to obtain a solution of Co. The prepared titanium solution was added dropwise into this dopant solution, followed by a similar procedure for the S-TiO_2 synthesis, and then the catalyst was marked as Co-TiO_2.

To synthesize S and Co co-doped TiO_2, a stoichiometric amount of titanium(IV) isopropoxide was dissolved in ethanol absolute under magnetic stirring (solution A). At the same time, 47.91 mg of $CoSO_4$ was dissolved in a mixture of distilled water and ethanol (volume ratio 1:1) to obtain the desired ratio of 4% (w/w) Ti: $CoSO_4$ dopant concentration (solution B). Solution A was added dropwise to solution B under magnetic stirring, then the solution pH was adjusted to 3.0 with the addition of 1 M HCl, and stirring continued for 6 h, followed by aging for 24 h. Afterward, the gels were dried at 80 °C for 4 h to evaporate water and organic materials. Then, dry gels were calcined at 450 °C for 3 h to control the crystal phase of the catalyst. The final catalyst was thoroughly milled and labeled as S-Co-TiO_2 2%. A similar procedure was followed to prepare S-Co-TiO_2 with concentration ratios of 6%, 8%, and pure TiO_2 without adding dopants.

2.3. Photocatalyst Characterization

The composition and crystalline phase of the photocatalyst was identified by X-ray diffractometer (XRD Shimadzu 6000, Cu Kα radiation λ = 0.15406 nm as the source of X-rays, operated at 40 kV, 30 mA, the angular range of 2θ = 5–90° and nickel as the filter). The crystallite size of the prepared photocatalyst was estimated by Scherrer equation (Equation (1)):

$$D = \frac{k\lambda}{\beta\ cos\theta} \tag{1}$$

where k is a shape factor (0.94), λ is the wavelength of Cu Kα source used, β is the full width at half maximum (FWHM), and θ is the angle of diffraction.

A Fourier-Transform Infrared (FTIR) spectrophotometer (Shimadzu Prestige 21) was used to verify the functional groups and chemical bonds in the photocatalyst in the wavenumber 400–4000 cm^{-1}. Scanning Electron Microscopy (SEM) equipped with Energy Dispersive X-ray Analysis (EDX) was used to determine nanoparticle morphology and nanoparticle composition, and to identify the photocatalyst light absorption profiles, UV–Vis Specular Reflectance Spectroscopy (UV–Vis SRS) was used.

2.4. Photocatalytic Activity

The photocatalytic activity of the synthesized S-Co-TiO_2 was studied for Remazol Black (RBB) decolorization using a batch system in a closed reactor equipped with 4 UV lamps (@20 W, intensity 200 lm/m^2) and 4 visible lamps (TL-D, intensity @20 W, 2000 lm/m^2), which can be adjusted for use (Figure S2). For comparison, photocatalytic activity tests were performed under the same conditions on other prepared photocatalysts TiO_2, S-TiO_2, and Co-TiO_2. The variables studied included the type of photocatalyst, light source, initial pH, photocatalyst dose, dye concentration, and photocatalyst reuse test. In the study of light source parameters, visible light or UV light can be adjusted as needed, so the visible and UV tests were carried out under different conditions. At the beginning of the photocatalytic reaction, the prepared photocatalyst was dispersed in RBB solution and then magnetically stirred for 30 min in the dark without irradiation to achieve adsorption–desorption equilibrium conditions. The following process is a photocatalytic reaction initiated by contacting the solution with visible/UV light under continuous stirring. The photocatalyst was separated from the solution by centrifugation at 5000 rpm for 10 min at certain time intervals. Furthermore, the residual concentration after the photocatalytic process was investigated using a UV–Vis spectrophotometer at a wavelength of 598 nm (λ_{max} = 598 nm). The efficiency of RBB removal by photocatalyst was determined using Equation (3):

$$\text{The removal efficiency of RBB } (\%) = \frac{Ci - Cf}{Ci}\ 100 \tag{2}$$

where Ci was the initial RBB concentration (mg L^{-1}) and Cf was the final RBB concentration (mg L^{-1}). Each experiment was repeated four times.

2.5. pH Point of Zero Charge (pH_{PZC}) of TiO_2-S-Co

pH_{PZC} of S-Co-TiO_2 was determined using the procedures from previous work [28] with a few modifications. A 0.1 g measure of catalyst was added to 25 mL 0.1 M $NaNO_3$. The pH of the solution was adjusted by adding HCl or NaOH solution to obtain pH values in the range of 2–11. The final pH values of the solution were determined after constant shaking for 2 h at 293 K and left for 24 h.

3. Results and Discussion

3.1. Photocatalyst Characterization

3.1.1. XRD Analysis

The crystal phase of the photocatalyst was analyzed using XRD, and the XRD pattern is shown in Figure 1. The diffraction pattern of pure TiO_2, S-TiO_2, Co-TiO_2, and co-doped S-Co-TiO_2 showed the major peaks of the crystal planes for (101), (200), (105), (211), (204), and (215) corresponded to the anatase phase of TiO_2 (JCPDS reference No. 21-1272), and no other phases such as rutile or brookite appeared in the samples.

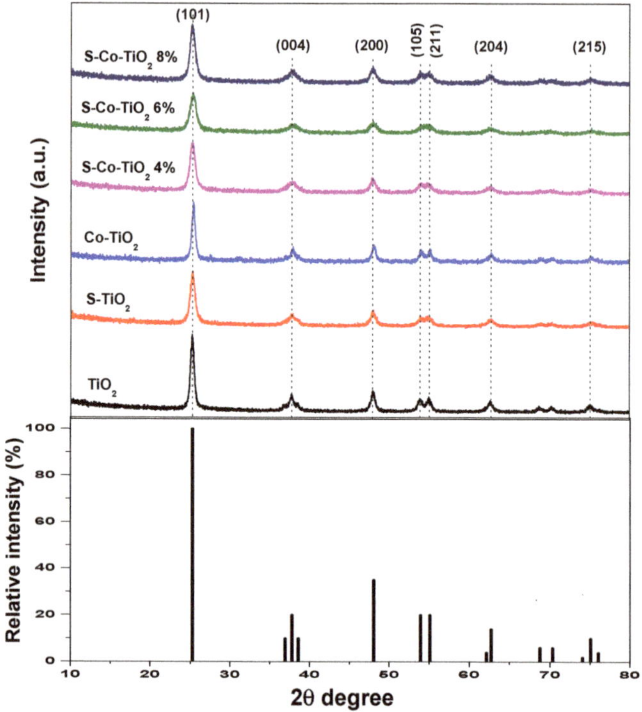

Figure 1. XRD pattern of undoped, doped, and co-doped TiO_2 photocatalyst.

There were no other crystallite peaks for S and Co ions in any samples, indicating that the S and Co dopants were evenly distributed on the titania surface [15]. Furthermore, the diffraction peaks of doped and co-doped samples shifted narrowly to a larger diffraction angle, followed by a decrease in peak intensity compared to undoped TiO_2. The difference was also observed in the average crystallite size (D) (Table 1), which is estimated using Scherrer's equation (Equation (1)). It was shown that the presence of dopant reduced the average crystallite size, with the most significant decrease attributed to the co-doping effect. As previously reported, the shift in diffraction peaks accompanied by a decrease in intensity and the average crystallite size after co-doping can be attributed to the presence of new bonds formed after the addition of sulfur and cobalt dopant to titania, in the form of Ti–O–S or Ti–O–Co bonds or both, which suppresses titania crystal growth and causes distortion of the crystal structure [25,29,30].

Table 1. The average crystallite sizes and the band gap energy of undoped TiO_2 and S-TiO_2, Co-TiO_2, and S-Co-TiO_2.

Photocatalyst	*D* (nm)	Band Gap (eV)
TiO_2	5.154	3.20
S-TiO_2	3.504	3.16
Co-TiO_2	5.170	3.15
S-Co-TiO_2 4%	2.405	2.82
S-Co-TiO_2 6%	2.079	2.83
S-Co-TiO_2 8%	2.445	2.78

3.1.2. FTIR Analysis

The functional groups formed on the photocatalyst before and after doped and co-doping were identified using the FTIR spectrum (Figure 2). Based on Figure 2, the spectrum of undoped TiO_2 displayed a broad absorption band at 3400–3600 and 1635 cm^{-1}, which were identified as stretching and bending vibrations of the hydroxyl groups on the TiO_2 surface [20]. The adsorption band at 2924 cm^{-1} was defined as the Ti–OH_2 group vibration, and the strong absorption band at 500–800 cm^{-1} was defined for Ti–O strain vibrations [31]. In the S-TiO_2 and Co-TiO_2 doped samples, the Ti–O absorption band widens and shifts slightly to a lower wavenumber due to the substitution of Ti^{4+} ions by Co^{2+} ions [32] and Ti^{4+} by S^{6+}/S^{4+} ions [33,34]. A more significant shift occurred in the S-Co-TiO_2 sample, which was associated with the substitution of the Ti^{4+} ion by the two dopants S^{6+} and Co^{2+} ions. This replacement caused more defects, particularly oxygen vacancies, due to the charge neutrality after the replacement of Ti^{4+} by Co^{2+} or/and $S^{6+}S^{4+}$ [32,35]. Then, in the S-TiO_2 and S-Co-TiO_2 samples, a new peak appeared at 1047–1080 cm^{-1}, which corresponds to the bending vibration of Ti–O–S. This bond is formed from the substitution of Ti^4+ by S^{6+}/S^{4+}, which also causes the Ti–O bond to weaken [36]. Furthermore, the peak at 1120–1125 cm^{-1} confirmed the S–O bending vibration in the form of $SO_4{}^{2-}$, indicating that S was presented in cationic species (S^{6+}/S^{4+}) since the substitution of Ti^{4+} by S^{6+}/S^{4+} is chemically more favorable than the substitution of O^{2-} by S^{2-} (anionic sulfur), because of the ionic radius of S^{2-} (1.7 Å) is quite larger than O^{2-} (1.22 Å), so this substitution is difficult to achieve compared to the substitution by the cationic sulfur [20,34]. Meanwhile, the shift in the Ti–O absorption peak was greater in the co-doped sample than the doped sample, indicating that more Ti ions were replaced by S and Co dopants [15,18].

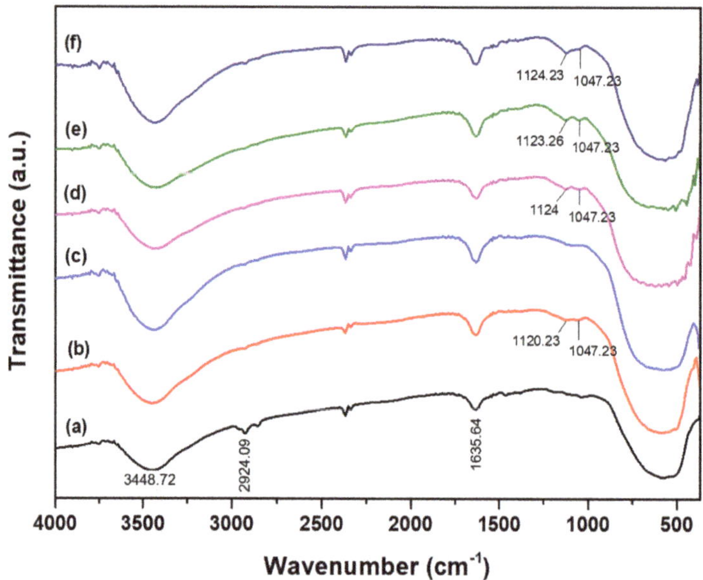

Figure 2. FTIR spectra of (**a**) pure TiO_2, (**b**) S-TiO_2, (**c**) Co-TiO_2, (**d**) S-Co-TiO_2 4%, (**e**) S-Co-TiO_2 6%, and (**f**) S-Co-TiO_2 8%.

3.1.3. SEM Analysis

SEM analysis was used to identify the surface morphology of the photocatalysts, and the SEM micrographs of undoped, doped, and co-doped TiO_2 are presented in Figure 3. The surface of pure TiO_2 showed a uniform spherical morphology and size homogeneity. After being modified with Co dopant, the photocatalyst showed a surface with large and inhomogeneous agglomerates, while after co-doped with S and Co, the particles formed homogeneous spherical agglomerates with smaller sizes.

EDX was employed to classify the elemental composition of the photocatalyst. The EDX pattern confirmed the presence of S, Co, or both dopants in the S-TiO_2, Co-TiO_2, or co-doped TiO_2 samples in addition to the main elements Ti and O (Figure 3). A small amount of sulfur could not be identified in the S-Co-TiO_2 4% sample, which may be due to the dopants' content being too small. Other than that, the EDX pattern confirmed that all the targeted elements are in the TiO_2 co-doped sample (Table 2).

Table 2. The elemental composition of undoped TiO_2, S-TiO_2, Co-TiO_2, and S-Co-TiO_2.

Element	% Atom					
	TiO_2	**S-TiO_2**	**Co-TiO_2**	**S-Co-TiO_2 4%**	**S-Co-TiO_2 6%**	**S-Co-TiO_2 8%**
Ti	27.28	22.53	29.07	21.26	20.93	22.94
O	72.72	77.37	69.35	78.37	78.60	76.25
S	-	0.10	-	-	0.09	0.17
Co	-	-	1.58	0.38	0.39	0.63

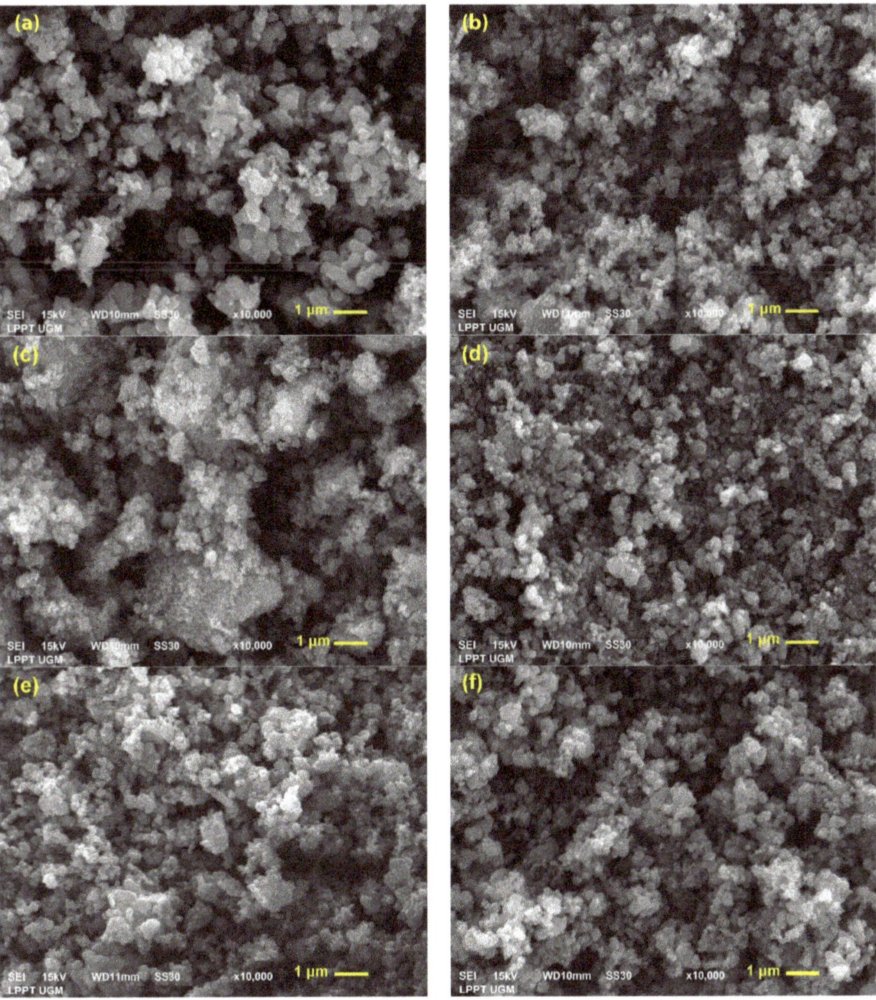

Figure 3. SEM images of (**a**) pure TiO_2, (**b**) S-TiO_2, (**c**) Co-TiO_2, (**d**) S-Co-TiO_2 4%, (**e**) S-Co-TiO_2 6%, and (**f**) S-Co-TiO_2 8%.

3.1.4. UV–Vis SRS Analysis

The UV–Vis SRS spectrum revealed the optical properties and band gap energy of the photocatalyst. Figure 4 exhibits the UV–Vis SRS spectrum and Tauc's plot of undoped, doped, and co-doped TiO_2. As shown in Figure 4a, TiO_2 shows an ultraviolet absorption semiconductor profile (200–400 nm) due to the large band gap energy of anatase TiO_2 (3.2 eV). A small red-shift to the visible region was observed in the S-TiO_2 sample due to the formation of a new sub-band gap in the valence band (VB) from the formation of the Ti–O–S bond after S doped to titania [37]. The new sub-band gap is formed by mixing the S_{3p} orbital with the O_{2p} orbital, which narrows the band gap of TiO_2 [25]. The distinctive red-shift was also observed in the Co-TiO_2 sample, indicating that Co-TiO_2 has an absorption in the visible region. The optical absorption mechanism of undoped TiO_2 is the transition energy from the valence band to the conduction band. The sp-d exchange interaction between the band electrons and the localized d-electrons of the Co^{2+} ion replacing the Ti^{4+} cation is responsible for the red-shift in the Co-TiO_2 sample. This interaction between sp and pd causes a downward shift in the conduction band (CB) and an upward shift in VB, narrowing the band gap system [38].

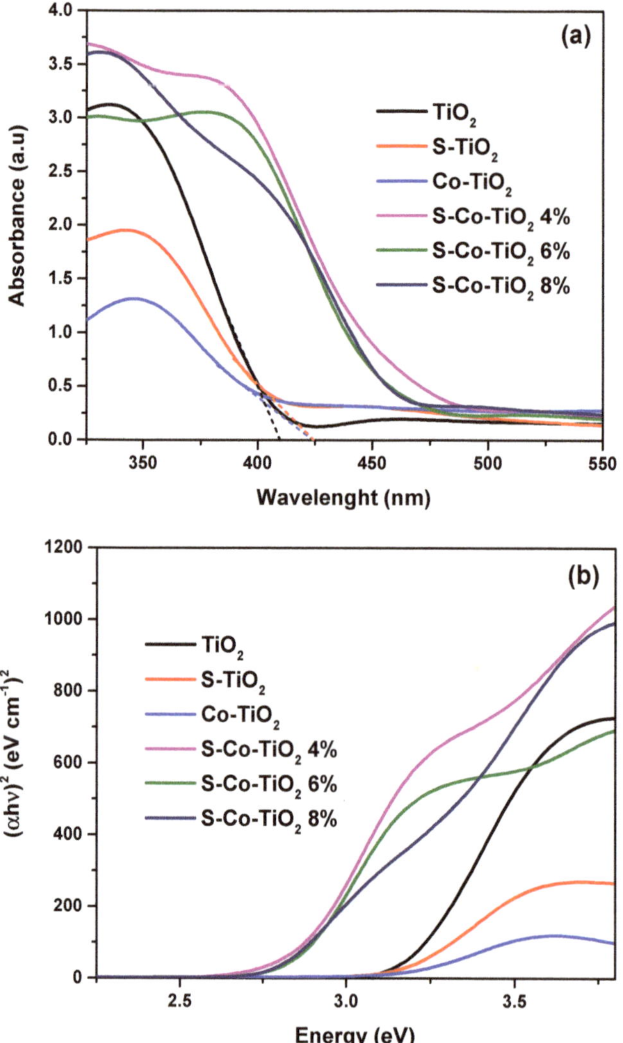

Figure 4. (**a**) UV–Vis SRS spectrum and (**b**) Tauc's plot of prepared photocatalysts.

Meanwhile, the red shift in the co-doping sample (S-Co-TiO_2) was the most significant compared to the previous two doping samples (S-TiO_2 or Co-TiO_2). The S-Co-TiO_2 sample displayed a high adsorption profile in the ultraviolet region, even higher than undoped and doped TiO_2. Then, the absorption bands are significantly widened to the visible light region, implying that the photocatalyst activity in visible light has increased due to the impurity of S and Co co-dopants in TiO_2. The SRUV–Vis absorption data were implemented into a Tauc's plot between $(\alpha h\upsilon)^2$ versus $h\upsilon$ to determine the band gap energy before and after modification, and the results are presented in Figure 4b. The band gap energy of undoped TiO_2 was identified to be 3.2 eV. When impurity S is present in TiO_2, the band gap is reduced to 3.16 eV, while doping by Co lowered the band gap to 3.15 eV. The decrease in band gap energy was more significant when TiO_2 was co-doped by S and Co. The addition of 4% and 6% $CoSO_4$ as a single precursor of co-dopants caused a decrease in the band gap to 2.82 and 2.83, respectively. The most significant decrease occurred at the S-Co-TiO_2 8% sample, resulting in a band-gap width of 2.78 eV (Table 1). The more significant band-gap narrowing in the co-doped sample may be due to the formation of two impurity states close to the VB and CB. One is formed by mixing the S_{3p} orbitals with the O_{2p} orbitals of the S dopant, and the other is formed by exchange interactions between the sp on the electrons band and d-electrons of the Co^{2+} ion that replaces the Ti^{4+} ion [25,38,39].

3.2. Photocatalytic Performance on RBB Removal

3.2.1. Effect of Different Photocatalyst

The prepared photocatalysts TiO_2, S-TiO_2, Co-TiO_2, and S-Co-TiO_2, were used in the photocatalytic removal of RBB under visible or UV-light irradiation separately to evaluate the photocatalytic activity before and after modification. In this experiment, 20 mg L^{-1} of RBB solution was added with the prepared photocatalyst and then irradiated with visible/UV light for a certain time. The results of RBB removal using various photocatalysts are shown in Figure 5.

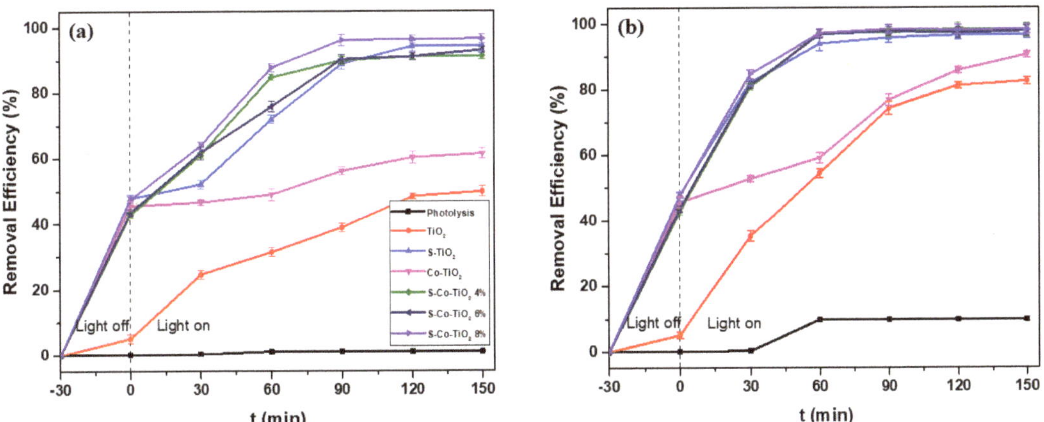

Figure 5. The effect of various photocatalysts on the removal efficiency of RBB using (**a**) visible, and (**b**) UV–light irradiation, RBB concentration: 20 mg L^{-1}, photocatalyst dosage: 1 g L^{-1}, and pH: 5.5.

In order to determine the photolysis properties of RBB, the removal concentration without the addition of the photocatalyst was also investigated. The results showed that RBB did not undergo photolysis after 150 min of UV or visible light exposure.

Figure 5a shows the photocatalytic activity of various prepared photocatalysts under visible light. A reasonably good result is shown in the S-TiO_2 sample. The undoped TiO_2 was only able to degrade 49% of RBB after 150 min of visible light exposure. Meanwhile, after dopant addition, the removal efficiency of RBB increased to 94% and 61% for S-TiO_2 and Co-TiO_2 after 120 min of exposure, respectively. However, the most effective removal of RBB was obtained from the TiO_2 co-doped by S and Co 8% (S-Co-TiO_2 8%) photocatalyst. Although S-TiO_2 reduced 94% of dye concentration, time efficiency can be achieved in co-doped samples (S-Co-TiO_2 8%), which can remove 96% of RBB in 90 min of visible light irradiation. Thus, the S-Co-TiO_2 8% demonstrated the best photocatalytic performance compared to other prepared photocatalysts.

Under UV irradiation, the S-Co-TiO_2 8% sample also demonstrated the highest photocatalytic activity compared to other photocatalysts. During 90 min of UV irradiation, 98% of RBB could be removed, while S-TiO_2, Co-TiO_2, and pure TiO_2 removed 95%, 76%, and 74% of RBB concentration, respectively. These findings confirm that S-and-Co-co-doped

TiO_2 (S-Co-TiO_2) exhibits the most significant increase in photocatalytic activity in visible or UV light.

Pure TiO_2 displayed significant photocatalytic activity in UV light but not in visible light since the energy from visible light is insufficient to activate TiO_2. The UV light energy corresponds to the expansive band-gap energy of TiO_2 (3.2 eV). However, the presence of S and Co as impurity dopants in the titania structure can enhance the photocatalytic activity in visible light due to the formation of new sub-levels energy in the titania structure, resulting in a narrowing of the TiO_2 band gap energy [25]. The new energy level formed by the S atom was attributed to the substitution of Ti^{4+} by S^{6+} from mixing the S_{3p} orbital with the O_{2p} orbital close to the valence band (VB), which narrowed the band gap of TiO_2 and increased the absorption of visible light [13,25,29]. On the other hand, Co dopant contributes to lowering the band gap energy from the substitution of Ti^{4+} by Co^{2+}. Co dopant directs the sp-d exchange interaction between the band electrons and localized d-electrons of the Co^{2+} ion that replace the Ti^{4+} ion. This interaction causes a downward shift of the conduction band (CB), resulting in a narrowing of the band gap system [25,30,38]. The synergetic effect of S and Co dopants forming two new energy levels on the S-Co-TiO_2 sample resulted in the most optima in decreasing the band gap energy and encouraging the visible-light absorption compared to S-TiO_2 or Co-TiO_2 samples.

Although both were doped, S-TiO_2 had a much higher photocatalytic activity than Co-TiO_2. According to the literature, the higher activity of S-TiO_2 can be attributed to the presence of hydroxyl groups on the surface of the sulfur oxide photocatalyst, which can create an acidic environment on the surface. Then, the substitution of Ti^{4+} by S^{6+} ions in the TiO_2 lattice causes a charge imbalance in the structure of TiO_2. The extra positive charge formed will attract anions from the solution, such as hydroxide ions, to the surface of the photocatalyst, thereby neutralizing the charge imbalance [40]. Furthermore, this extra positive charge can adsorb the $h_{vb}{}^{+}$ formed by the photocatalyst induced by light to produce hydroxyl radicals, which have strong oxidizing power to decompose the organic compounds such as RBB dyes. The difference in electronegativity between the S and O atoms in the Ti-O-S bond causes electron transfer from the less electronegative S atom to the more electronegative O atom. This causes the sulfur atom to be deprived of electrons and will retain e^{-} generated by the photocatalyst. As a result, the electron and hole recombination rate decreases, and $h_{vb}{}^{+}$ will be abundantly available to generate more hydroxyl radicals on the surface of the photocatalyst [13]. The abundant availability of hydroxyl radicals will increase the decontamination efficiency of organic compounds such as RBB dyes.

The co-doped samples demonstrated the most superior photocatalytic activity in visible or UV-light due to a variety of factors, including a better narrowing in band gap, a decrease in particle size, and an increase in surface-area-reduced electron and hole recombination [15,41]. This result is in accordance with the band gap energy (Eg) obtained from the UV–vis SRS data, such that S-Co-TiO_2 8% produces the most significant decrease in Eg, which is 2.78 eV and corresponds to the energy of visible light irradiation.

The decolorization process by the photocatalyst starts with the surface adsorption process and continues with the photocatalytic reaction. In order to evaluate the surface adsorption properties of the S-Co-TiO_2 8% photocatalyst as well as to evaluate whether the adsorption process continued along with the photocatalytic process, the adsorption of the S-Co-TiO_2 8% catalyst was carried out in the dark for 150 min without irradiation as shown in Figure 6. The undoped TiO_2 adsorption test was also used for comparison purposes. The results indicate that adsorption takes place in the first 15 min and the concentration of the dye tends not to decrease, which implies that the next process is photocatalytic decolorization after a contact period of 15 min.

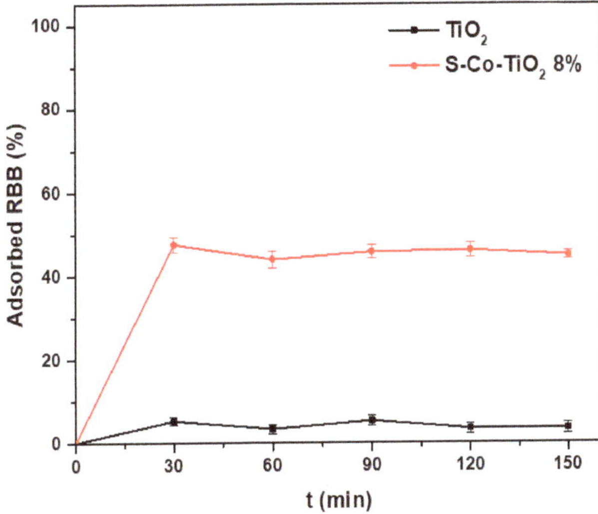

Figure 6. Adsorption of RBB by TiO_2 and S-Co-TiO_2 8% in the dark condition at RBB concentration: 20 mg L^{-1}, photocatalyst dosage: 1 g L^{-1}, and initial pH: 5.5.

3.2.2. Effect of pH

The effect of pH is a critical parameter in the efficiency of the decolorization process because it can affect the photocatalyst surface, dye characteristics, and the rate of decolorization. As shown in Figure 7, the effect of the initial pH on RBB removal by the S-CoTiO_2 8% catalyst under 60 min of visible light exposure was evaluated in the range of 2.0–10.0.

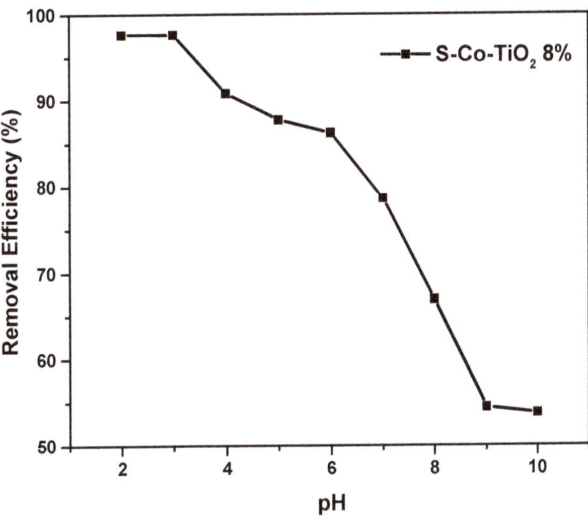

Figure 7. Influence of pH on photocatalytic decolorization of RBB under 60 min of visible light irradiation by the S-Co-TiO_2 8% photocatalyst, RBB concentration: 20 mg L^{-1}, photocatalyst dosage: 1 g L^{-1}.

As shown in Figure 7, when the initial pH was reduced from 5.0 to 3.0, the decolorization efficiency of RBB increased from 88% to 97% at 60 min of irradiation. However, there was no increase in decolorization when the pH was reduced to 2.0. However, increasing the solution pH to 8.0, 9.0, and 10.0 reduced the decolorization efficiency to 67%, 54%, and 53%, respectively. This result can be explained by the fact that TiO_2 has a positive or negative charge depending on the pH of the solution:

$$pH < pH_{ZPC}: Ti\text{-}OH + H^+ \rightarrow TiOH_2^+ \quad (3)$$

$$pH > pH_{ZPC}: Ti\text{-}OH + OH^- \rightarrow TiO^- + H_2O \quad (4)$$

The optimum pH for RBB decolorization was found to be 3.0, and the pH_{PZC} of S-Co-TiO_2 8% was found to be 6.41 (Figure 8). When the solution pH is below the pH_{PZC}, the TiO_2 surface will be positively charged. At acidic medium (pH = 3.0), RBB has sulfonate (SO_3^-) and sulfite (SO_3^{2-}) groups, which can be adequately adsorbed by the positive charge on the surface of TiO_2 through electrostatic force at the beginning of the decolorization process [3,42]. As a result, the optimal pH of RBB adsorption on the catalyst surface is 3.0. When the pH of the solution is lowered from 7.0 to 10.0, the removal efficiency decreases as well. It can be explained because at alkaline pH conditions, the surface of TiO_2 is negatively charged, leading to charge repulsion with the RBB, which is also negative.

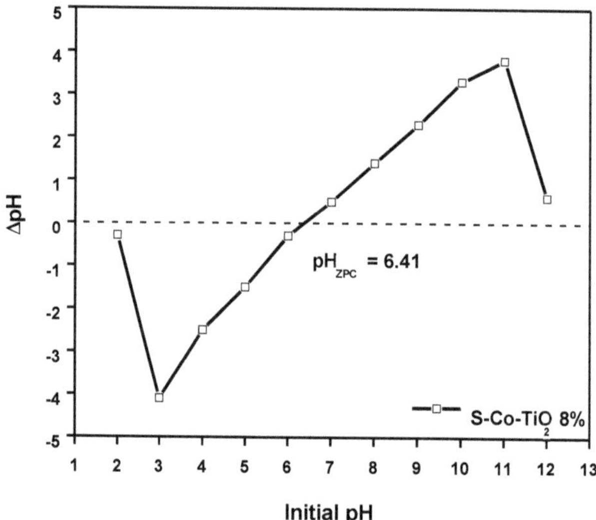

Figure 8. pH_{PZC} of the S-Co-TiO_2 8% photocatalyst.

3.2.3. Effect of Catalyst Dosage

The effect of photocatalyst amount on the RBB removal under visible light irradiation was investigated by varying the mass of the S-Co-TiO_2 8% from 0.2 to 1.0 g L^{-1} (C_0 = 20 mg L^{-1}, pH = 3.0). The obtained results are shown in Figure 9.

As shown in Figure 9, increasing the photocatalyst mass from 0.2 to 1.0 mg L^{-1} resulted in additional RBB removal, which was associated with an increased adsorption rate on the catalyst surface and increased hydroxyl radical formation. The photocatalyst mass of 1.0 g L^{-1} provided the highest decolorization efficiency. However, the amount of dye adsorbed on the surface of the photocatalyst was very high at the beginning of the reaction, whereas the photocatalytic reaction became insignificant. It can be seen that after 60 min of irradiation, the photocatalytic reaction did not continue since an excessive amount of photocatalyst had increased the number of active sites on the surface of the photocatalyst, allowing adsorption to dominate at the start of the process while the photocatalytic reaction becomes insignificant.

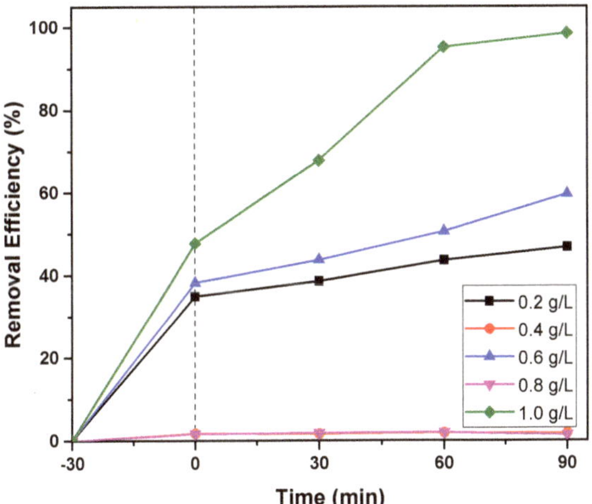

Figure 9. The effect of the S-Co-TiO_2 8% photocatalyst dosage on RBB removal at RBB concentration: 20 mg L^{-1} and initial pH: 3.0.

The photocatalyst mass of 0.6 mg L^{-1} was chosen for further investigation due to its high photocatalytic efficiency after 30 min of adsorption–desorption. When the photocatalyst dosage was increased from 0.6 to 1.0 mg L^{-1}, decolorization was almost insignificantly different.

3.2.4. Effect of Dye Concentration

The color removal process is affected by the initial concentration of pollutants. Therefore, the effect of RBB concentration on the efficiency of decolorization by the S-Co-TiO_2 8% photocatalyst was carried out by varying the RBB concentration from 20 to 50 mg L^{-1}, and the results obtained are presented in Figure 10. The decolorization efficiency decreased from 98% to 57% as the RBB concentration increased from 20 mg L^{-1} to 50 mg L^{-1}. This is due to the fact that as the dye concentration increases, the number of molecules adsorbed on the photocatalyst surface increases, obstructing direct contact with the holes and inhibiting the formation of hydroxyl radicals. Furthermore, at higher concentrations, the dye molecules adsorbed more photons, thereby reducing the light intensity and decolorization efficiency [28,43].

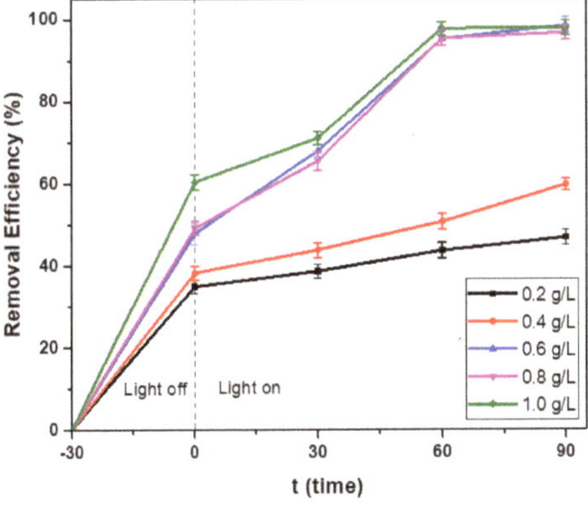

Figure 10. The effect of RBB concentration on S-Co-TiO_2 8% dose 0.6 g L^{-1}, initial pH: 3.0, and 90 min of visible light irradiation.

3.2.5. The Recyclable Ability of S-Co-TiO_2 8% Photocatalyst

The stability and reusability of photocatalysts are critical parameters in the study of photocatalysts for use in sustainable wastewater treatment. The S-Co-TiO_2 8% photocatalyst was tested for stability and reuse four times. In each test, S-Co-TiO_2 8% was centrifuged and then washed and dried before being used in the next cycle under the same experimental conditions for 90 min of visible light irradiation. The reuse test results for S-Co-TiO_2 8% photocatalyst are shown in Figure 11. The results showed that the S-Co-TiO_2 8% sample had high decolorization efficiency after four reuse cycles, but there was a 44% decrease in decolorization efficiency after four uses. This could be due to the loss of photocatalyst during the iteration process. These findings suggest that the co-doped sample S-Co-TiO_2 8% photocatalyst has promising potential and can be used repeatedly to remove dye pollutants.

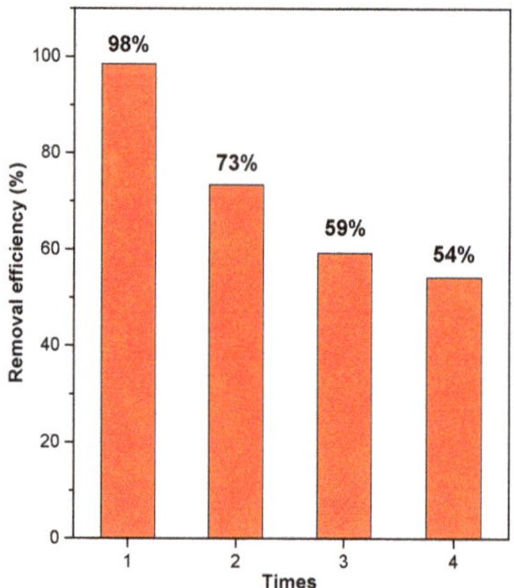

Figure 11. Recyclability of S-Co-TiO_2 8% for RBB removal under visible light illumination, RBB con-centration: 20 mg L^{-1}, initial pH: 3.0, and photocatalyst dosage: 0.6 g L^{-1}.

3.3. Mechanism of RBB Photocatalytic Decolorization

Photocatalytic removal reactions can occur when TiO_2 absorbs photons with energies equal to or greater than their band gap energy. S and Co co-dopants create impurity states close to VB and CB, resulting in a narrowing of the TiO_2 band gap, thereby increasing photocatalytic visible light activity. The schematic of the S-Co-TiO_2 photocatalyst mechanism is presented in Figure S6. The characterization results using XRD, FTIR, SEM-EDX, and SRS UV–vis indicate that the modification of the addition of S and Co co-dopants to TiO_2 has been successfully carried out.

When S-Co-TiO_2 is exposed to visible light radiation, electrons in the VB are excited towards the CB (e_{CB}^-), producing holes (h_{VB}^+) in the VB. These excited electron–hole pairs are commonly referred to as excitons [44]. The mechanism that occurs is as follows:

$$\text{S-Co-TiO}_2 + h\nu \rightarrow h_{VB}^+ + e_{CB}^- \quad (5)$$

The holes can react with adsorbed hydroxyl anion on the TiO_2 surface to form hydroxyl radicals, which have high oxidizing power and can decompose organic pollutants, and excited electrons (e_{CB}^-) can reduce O_2 to form oxygen radicals, which can be converted into hydroxyl radicals in water as follows:

$$OH^- + h_{VB}^+ \rightarrow \bullet OH \quad (6)$$

$$\bullet OH + RBB \rightarrow \text{degradation pollutant } (Cl^-, NO_3^-, SO_4^{2-}, CO_2 \text{ and } H_2O) \quad (7)$$

4. Conclusions

The decolorization of Remazol Black (RBB) was investigated using an S-Co-TiO_2 photocatalyst. This photocatalyst was prepared through one-pot synthesis using a single precursor as a source of S and Co dopants. Characterization using XRD, FT-IR, SEM-EDX, and UV–Vis SRS showed that S-and-Co co-doped TiO_2 was successfully synthesized and degraded RBB under visible light better than TiO_2 or S-TiO_2 and Co-TiO_2. Eight percent (w/w) of $CoSO_4$ as a single precursor of S and Co provides the best reduction in band gap energy to 2.78 eV and is supported by the best photocatalytic activity when compared to other photocatalysts. After 90 min of temporary visible light, the S-Co-TiO_2 8% can degrade 96% of RBB, whereas S-TiO_2, Co-TiO_2, and undoped TiO_2 can decrease to 89%, 56%, and 39%, respectively, and the optimum decolorization results are obtained at pH 3.0. Reuse studies show that S-Co-TiO_2 8% has the potential to be used repeatedly for the removal of dye pollutants in the environment.

Supplementary Materials: The following are available online at https://www.mdpi.com/article/10.3390/photochem1030032/s1, Figure S1: Chemical structure of Remazol Black (RBB); Figure S2: Schematic diagram of photoreactor for photocatalytic experiment; Figure S3: EDX pattern of (a) TiO_2, (b) S-TiO_2, (c) Co-TiO_2, (d) S-Co-TiO_2 4%, (e) S-Co-TiO_2 6%, and (f) S-Co-TiO_2 8%; Figure S4: Absorption spectrum of Remazol Black analyzed by spectrophotometry, Remazol Black concentration: 20 mg L^{-1}; Figure S5: Spectrum of Remazol Black after decolorization by S-Co-TiO_2 8% photocatalyst for 150 min visible light irradiation, photocatalyst dose: 1.0 g L^{-1}, initial pH: 5.5, Remazol Black concentration: 20 mg L^{-1}, Figure S6: Schematic diagram of photocatalytic decolorization of RBB by S-Co-TiO_2 photocatalyst.

Author Contributions: Conceptualization, E.T.W.; methodology, E.T.W. and R.D.; software, R.D.; validation, E.T.W., R.R. and R.D.; formal analysis, R.D., E.T.W. and R.R.; investigation, R.D., E.T.W. and R.R.; resources, R.D. and E.T.W.; data curation, R.D. and E.T.W.; writing—original draft preparation, R.D.; writing—review and editing, E.T.W., R.R. and R.D.; visualization, R.D., E.T.W. and R.R.; supervision, R.D.; project administration, E.T.W. and R.D.; funding acquisition, E.T.W. and R.D. All authors have read and agreed to the published version of the manuscript.

Funding: This research was financially supported by Kemenristekdikti (Ministry of Research, Technology and Higher Education) of Indonesia through the Pendidikan Magister menuju Doktor untuk Sarjana Unggul (PMDSU) Scholarship Program (contract No. 2335/UN1/DITLIT/DIT-LIT/PT/2021).

Institutional Review Board Statement: Not applicable.

Informed Consent Statement: Not applicable.

Data Availability Statement: All the data presented in this study are available in the article.

Acknowledgments: The first author is grateful to the Ministry of Research, Technology, and Higher Education of Indonesia (Kemenristekdikti) for providing a comprehensive scholarship to enable this research to be carried out, as well as to the Department of Chemistry, Universitas Gadjah Mada, which has provided this research facilities.

Conflicts of Interest: The authors declare no conflict of interest, and the funders had no role in the design of the study; in the collection, analyses, or interpretation of data; in the writing of the manuscript, or in the decision to publish the results.

References

1. Marques, S.M.; Tavares, C.J.; Oliveira, L.F.; Oliveira-Campos, A.M.F. Photocatalytic degradation of C.I. Reactive Blue 19 with nitrogen-doped TiO_2 catalysts thin films under UV/visible light. *J. Mol. Struct.* **2010**, *983*, 147–152. [CrossRef]
2. Ahmad, M.A.; Rahman, N.K. Equilibrium, kinetics and thermodynamic of Remazol Brilliant Orange 3R dye adsorption on coffee husk-based activated carbon. *Chem. Eng. J.* **2011**, *170*, 154–161. [CrossRef]
3. Muruganandham, M.; Sobana, N.; Swaminathan, M. Solar assisted photocatalytic and photochemical degradation of Reactive Black 5. *J. Hazard. Mater.* **2006**, *137*, 1371–1376. [CrossRef]
4. Sahel, K.; Perol, N.; Dappozze, F.; Bouhent, M.; Derriche, Z.; Guillard, C. Photocatalytic degradation of a mixture of two anionic dyes: Procion Red MX-5B and Remazol Black 5 (RB5). *J. Photochem. Photobiol. A Chem.* **2010**, *212*, 107–112. [CrossRef]

5. Chen, C.Y.; Cheng, M.C.; Chen, A.H. Photocatalytic decolorization of Remazol Black 5 and Remazol Brilliant Orange 3R by mesoporous TiO_2. *J. Environ. Manag.* **2012**, *102*, 125–133. [CrossRef]
6. de Oliveira Pereira, L.; Marques Sales, I.; Pereira Zampiere, L.; Silveira Vieira, S.; do Rosário Guimarães, I.; Magalhães, F. Preparation of magnetic photocatalysts from TiO_2, activated carbon and iron nitrate for environmental remediation. *J. Photochem. Photobiol. A Chem.* **2019**, *382*, 111907. [CrossRef]
7. Behnajady, M.A.; Modirshahla, N.; Shokri, M.; Rad, B. Enhancement of photocatalytic activity of TiO_2 nanoparticles by Silver doping: Photodeposition versus liquid impregnation methods. *Glob. Nest J.* **2008**, *10*, 1–7. [CrossRef]
8. Wu, M.C.; Wu, P.Y.; Lin, T.H.; Lin, T.F. Photocatalytic performance of Cu-doped TiO_2 nanofibers treated by the hydrothermal synthesis and air-thermal treatment. *Appl. Surf. Sci.* **2018**, *430*, 390–398. [CrossRef]
9. Murcia, J.J.; Hidalgo, M.C.; Navío, J.A.; Araña, J.; Doña-Rodríguez, J.M. Study of the phenol photocatalytic degradation over TiO_2 modified by sulfation, fluorination, and platinum nanoparticles photodeposition. *Appl. Catal. B Environ.* **2015**, *179*, 305–312. [CrossRef]
10. Lin, Y.H.; Hsueh, H.T.; Chang, C.W.; Chu, H. The visible light-driven photodegradation of dimethyl sulfide on S-doped TiO_2: Characterization, kinetics, and reaction pathways. *Appl. Catal. B Environ.* **2016**, *199*, 1–10. [CrossRef]
11. Ramacharyulu, P.V.R.K.; Praveen Kumar, J.; Prasad, G.K.; Sreedhar, B. Sulphur doped nano TiO_2: Synthesis, characterization and photocatalytic degradation of a toxic chemical in presence of sunlight. *Mater. Chem. Phys.* **2014**, *148*, 692–698. [CrossRef]
12. McManamon, C.; O'Connell, J.; Delaney, P.; Rasappa, S.; Holmes, J.D.; Morris, M.A. A facile route to synthesis of S-doped TiO_2 nanoparticles for photocatalytic activity. *J. Mol. Catal. A Chem.* **2015**, *406*, 51–57. [CrossRef]
13. Bakar, S.A.; Ribeiro, C. A comparative run for visible-light-driven photocatalytic activity of anionic and cationic S-doped TiO_2 photocatalysts: A case study of possible sulfur doping through chemical protocol. *J. Mol. Catal. A Chem.* **2016**, *421*, 1–15. [CrossRef]
14. Chen, C.; Ma, W.; Zhao, J. Photocatalytic Degradation of Organic Pollutants by Co-Doped TiO_2 Under Visible Light Irradiation. *Curr. Org. Chem.* **2010**, *14*, 630–644. [CrossRef]
15. Nasir, M.; Xi, Z.; Xing, M.; Zhang, J.; Chen, F.; Tian, B.; Bagwasi, S. Study of synergistic effect of Ce- and S-codoping on the enhancement of visible-light photocatalytic activity of TiO_2. *J. Phys. Chem. C* **2013**, *117*, 9520–9528. [CrossRef]
16. Sun, H.; Zhou, G.; Liu, S.; Ang, H.M.; Tadé, M.O.; Wang, S. Visible light responsive titania photocatalysts codoped by nitrogen and metal (Fe, Ni, Ag, or Pt) for remediation of aqueous pollutants. *Chem. Eng. J.* **2013**, *231*, 18–25. [CrossRef]
17. Wang, F.; Ma, Z.; Ban, P.; Xu, X. C, N and S codoped rutile TiO_2 nanorods for enhanced visible-light photocatalytic activity. *Mater. Lett.* **2017**, *195*, 143–146. [CrossRef]
18. Singaram, B.; Varadharajan, K.; Jeyaram, J.; Rajendran, R.; Jayavel, V. Preparation of cerium and sulfur codoped TiO_2 nanoparticles based photocatalytic activity with enhanced visible light. *J. Photochem. Photobiol. A Chem.* **2017**, *349*, 91–99. [CrossRef]
19. Sinhmar, A.; Setia, H.; Kumar, V.; Sobti, A.; Toor, A.P. Enhanced photocatalytic activity of nickel and nitrogen codoped TiO_2 under sunlight. *Environ. Technol. Innov.* **2020**, *18*, 100658. [CrossRef]
20. Niu, Y.; Xing, M.; Tian, B.; Zhang, J. Improving the visible light photocatalytic activity of nano-sized titanium dioxide via the synergistic effects between sulfur doping and sulfation. *Appl. Catal. B Environ.* **2012**, *115–116*, 253–260. [CrossRef]
21. Yan, X.; Yuan, K.; Lu, N.; Xu, H.; Zhang, S.; Takeuchi, N.; Kobayashi, H.; Li, R. The interplay of sulfur doping and surface hydroxyl in band gap engineering: Mesoporous sulfur-doped TiO_2 coupled with magnetite as a recyclable, efficient, visible light active photocatalyst for water purification. *Appl. Catal. B Environ.* **2017**, *218*, 20–31. [CrossRef]
22. Kumar, M.K.; Godavarthi, S.; Karthik, T.V.K.; Mahendhiran, M.; Hernandez-Eligio, A.; Hernandez-Como, N.; Agarwal, V.; Martinez Gomez, L. Green synthesis of S-doped rod shaped anatase TiO_2 microstructures. *Mater. Lett.* **2016**, *183*, 211–214. [CrossRef]
23. Zhang, D.; Chen, J.; Xiang, Q.; Li, Y.; Liu, M.; Liao, Y. Transition-Metal-Ion (Fe, Co, Cr, Mn, Etc.) Doping of TiO_2 Nanotubes: A General Approach. *Inorg. Chem.* **2019**, *58*, 12511–12515. [CrossRef] [PubMed]
24. Nguyen, T.M.H.; Bark, C.W. Synthesis of Cobalt-Doped TiO_2 Based on Metal-Organic Frameworks as an Effective Electron Transport Material in Perovskite Solar Cells. *ACS Omega* **2020**, *5*, 2280–2286. [CrossRef] [PubMed]
25. Siddiqa, A.; Masih, D.; Anjum, D.; Siddiq, M. Cobalt and sulfur co-doped nano-size TiO_2 for photodegradation of various dyes and phenol. *J. Environ. Sci.* **2015**, *37*, 100–109. [CrossRef]
26. El Mragui, A.; Logvina, Y.; da Silva, L.P.; Zegaoui, O.; da Silva, J.C.G.E. Synthesis of fe-and co-doped TiO_2 with improved photocatalytic activity under visible irradiation toward carbamazepine degradation. *Materials* **2019**, *12*, 3874. [CrossRef]
27. Eskandari, P.; Farhadian, M.; Solaimany Nazar, A.R.; Jeon, B.H. Adsorption and Photodegradation Efficiency of TiO_2/Fe_2O_3/PAC and TiO_2/Fe_2O_3/Zeolite Nanophotocatalysts for the Removal of Cyanide. *Ind. Eng. Chem. Res.* **2019**, *58*, 2099–2112. [CrossRef]
28. Isari, A.A.; Payan, A.; Fattahi, M.; Jorfi, S.; Kakavandi, B. Photocatalytic degradation of rhodamine B and real textile wastewater using Fe-doped TiO_2 anchored on reduced graphene oxide (Fe-TiO_2 /rGO): Characterization and feasibility, mechanism and pathway studies. *Appl. Surf. Sci.* **2018**, *462*, 549–564. [CrossRef]
29. Yi, C.; Liao, Q.; Deng, W.; Huang, Y.; Mao, J.; Zhang, B.; Wu, G. The preparation of amorphous TiO_2 doped with cationic S and its application to the degradation of DCFs under visible light irradiation. *Sci. Total Environ.* **2019**, *684*, 527–536. [CrossRef]
30. Samet, L.; Ben Nasseur, J.; Chtourou, R.; March, K.; Stephan, O. Heat treatment effect on the physical properties of cobalt doped TiO_2 sol-gel materials. *Mater. Charact.* **2013**, *85*, 1–12. [CrossRef]

31. Chen, X.; Sun, H.; Zhang, J.; Guo, Y.; Kuo, D.H. Cationic S-doped TiO_2/SiO_2 visible-light photocatalyst synthesized by co-hydrolysis method and its application for organic degradation. *J. Mol. Liq.* **2019**, *273*, 50–57. [CrossRef]
32. Das, K.; Sharma, S.N.; Kumar, M.; De, S.K. Morphology dependent luminescence properties of Co doped TiO_2 nanostructures. *J. Phys. Chem. C* **2009**, *113*, 14783–14792. [CrossRef]
33. Han, C.; Andersen, J.; Likodimos, V.; Falaras, P.; Linkugel, J.; Dionysiou, D.D. The effect of solvent in the sol-gel synthesis of visible light-activated, sulfur-doped TiO_2 nanostructured porous films for water treatment. *Catal. Today* **2014**, *224*, 132–139. [CrossRef]
34. Devi, L.G.; Kavitha, R. Enhanced photocatalytic activity of sulfur doped TiO_2 for the decomposition of phenol: A new insight into the bulk and surface modification. *Mater. Chem. Phys.* **2014**, *143*, 1300–1308. [CrossRef]
35. Mugundan, S.; Rajamannan, B.; Viruthagiri, G.; Shanmugam, N.; Gobi, R.; Praveen, P. Synthesis and characterization of undoped and cobalt-doped TiO_2 nanoparticles via sol–gel technique. *Appl. Nanosci.* **2015**, *5*, 449–456. [CrossRef]
36. Chen, Z.; Ma, J.; Yang, K.; Feng, S.; Tan, W.; Tao, Y.; Mao, H.; Kong, Y. Preparation of S-doped TiO_2-three dimensional graphene aerogels as a highly efficient photocatalyst. *Synth. Met.* **2017**, *231*, 51–57. [CrossRef]
37. Abu Bakar, S.; Ribeiro, C. An insight toward the photocatalytic activity of S doped 1-D TiO_2 nanorods prepared via novel route: As promising platform for environmental leap. *J. Mol. Catal. A Chem.* **2016**, *412*, 78–92. [CrossRef]
38. Naseem, S.; Khan, W.; Khan, S.; Husain, S.; Ahmad, A. Consequences of (Cr/Co) co-doping on the microstructure, optical and magnetic properties of microwave assisted sol-gel derived TiO_2 nanoparticles. *J. Lumin.* **2019**, *205*, 406–416. [CrossRef]
39. Olowoyo, J.O.; Kumar, M.; Jain, S.L.; Shen, S.; Zhou, Z.; Mao, S.S.; Vorontsov, A.V.; Kumar, U. Reinforced photocatalytic reduction of CO_2 to fuel by efficient S-TiO_2: Significance of sulfur doping. *Int. J. Hydrogen Energy* **2018**, *43*, 17682–17695. [CrossRef]
40. Yu, J.C.; Ho, W.; Yu, J.; Yip, H.; Po, K.W.; Zhao, J. Efficient visible-light-induced photocatalytic disinfection on sulfur-doped nanocrystalline titania. *Environ. Sci. Technol.* **2005**, *39*, 1175–1179. [CrossRef]
41. Chen, X.Q.; Su, Y.L.; Zhang, X.W.; Lei, L.C. Fabrication of visible-light responsive S-F-codoped TiO_2 nanotubes. *Chin. Sci. Bull.* **2008**, *53*, 1983–1987. [CrossRef]
42. Ghoreishian, S.M.; Badii, K.; Norouzi, M.; Rashidi, A.; Montazer, M.; Sadeghi, M.; Vafaee, M. Decolorization and mineralization of an azo reactive dye using loaded nano-photocatalysts on spacer fabric: Kinetic study and operational factors. *J. Taiwan Inst. Chem. Eng.* **2014**, *45*, 2436–2446. [CrossRef]
43. Ali, M.H.H.; Al-Afify, A.D.; Goher, M.E. Preparation and characterization of graphene—TiO_2 nanocomposite for enhanced photodegradation of Rhodamine-B dye. *Egypt. J. Aquat. Res.* **2018**, *44*, 263–270. [CrossRef]
44. Islam, M.T.; Dominguez, A.; Turley, R.S.; Kim, H.; Sultana, K.A.; Shuvo, M.A.I.; Alvarado-Tenorio, B.; Montes, M.O.; Lin, Y.; Gardea-Torresdey, J.; et al. Development of photocatalytic paint based on TiO_2 and photopolymer resin for the degradation of organic pollutants in water. *Sci. Total Environ.* **2020**, *704*, 135406. [CrossRef] [PubMed]

Editorial

Visible-Light-Active Photocatalysts for Environmental Remediation and Organic Synthesis

Vincenzo Vaiano 

Department of Industrial Engineering, University of Salerno, Via Giovanni Paolo II, 132, 84084 Fisciano, Italy; vvaiano@unisa.it

In recent years, the formulation of innovative photocatalysts activated by visible or solar light has been attracting increasing attention because of their notable potential for environmental remediation and use in organic synthesis reactions. Generally, the strategies for the development of visible-light-active photocatalysts are mainly focused on enhancing degradation efficiency (in the case of environmental remediation) or increasing selectivity toward the desired product (in the case of organic synthesis). These goals can be achieved by doping the semiconductor lattice with metal and/or non-metal elements in order to reduce band gap energy, thereby providing the semiconductor with the ability to absorb light at a wavelength higher than the UV range. Other interesting options are the formulation of different types of heterojunctions (to increase visible absorption properties and to reduce the recombination rate of charge carriers) and the development of innovative catalytic materials with semiconducting properties. This Special Issue is focused on visible-light-active photocatalysts for environmental remediation and organic synthesis, featuring the state of the art as well as advances in this field.

Currently, this issue has collected six papers addressing the preparation and characterization of novel photocatalytic materials and their use in the visible (or solar) light-driven photocatalytic removal of pollutants from liquid phases and in the inactivation of bacteria [1–6]. The photocatalytic efficiencies of Ag_3PO_4 nanoparticles [2], $Bi_{12}NiO_{19}$ sillenite crystals [3], g-C_3N_4/nanodiamond heterostructures [4], Ag/Cu_2O [5], and activated carbon/TiO_2 [6] are shown. An improvement in photocatalytic activity toward the simultaneous removal of phenol and Cr(VI) under visible light is also evidenced by MoS_2 decorated on a g-C_3N_4 heterostructure catalyst [1].

Moreover, a review article highlighting recent progress in the development of visible-light-active heterogeneous photocatalysts based on the design of sustainable synthetic methodologies and the use of biomass and waste as sources of chemicals embedded in the final photoactive materials [7] is also included in this Special Issue.

I sincerely hope that additional papers and/or review articles will be submitted to this interesting Special Issue in order to achieve a collection covering all of the aspects related to the preparation and chemical–physical characterization of different types of photocatalyst to be studied both for environmental purposes and in the selective synthesis of organic compounds (e.g., phenol from benzene, aniline from nitrobenzene, and methanol from methane).

The Guest Editor would like to express their sincere appreciation to all of the authors and reviewers for their contribution to this Special Issue of *Photochem*. Special thanks should also be given to the MDPI assistant team for their constant and kind support.

Funding: This research received no external funding.

Conflicts of Interest: The author declares no conflict of interest.

Citation: Vaiano, V. Visible-Light-Active Photocatalysts for Environmental Remediation and Organic Synthesis. *Photochem* **2021**, *1*, 460–461. https://doi.org/10.3390/photochem1030029

Received: 4 November 2021
Accepted: 8 November 2021
Published: 11 November 2021

References

1. Rapti, I.; Bairamis, F.; Konstantinou, I. g-C3N4/MoS2 Heterojunction for Photocatalytic Removal of Phenol and Cr (VI). *Photochem* **2021**, *1*, 358–370. [CrossRef]
2. Panthi, G.; Park, M. Electrospun Carbon Nanofibers Decorated with Ag3PO4 Nanoparticles: Visible-Light-Driven Photocatalyst for the Photodegradation of Methylene Blue. *Photochem* **2021**, *1*, 345–357. [CrossRef]
3. Brahimi, B.; Kenfoud, H.; Benrighi, Y.; Baaloudj, O. Structural and Optical Properties of Bi12NiO19 Sillenite Crystals: Application for the Removal of Basic Blue 41 from Wastewater. *Photochem* **2021**, *1*, 319–329. [CrossRef]
4. Kublik, N.; Gomes, L.E.; Plaça, L.F.; Lima, T.H.; Abelha, T.F.; Ferencz, J.A.; Caires, A.R.; Wender, H. Metal-Free g-C3N4/Nanodiamond Heterostructures for Enhanced Photocatalytic Pollutant Removal and Bacteria Photoinactivation. *Photochem* **2021**, *1*, 302–318. [CrossRef]
5. Chen, W.-S.; Chen, J.-Y. Photocatalytic Decomposition of Nitrobenzene in Aqueous Solution by Ag/Cu2O Assisted with Persulfate under Visible Light Irradiation. *Photochem* **2021**, *1*, 220–236. [CrossRef]
6. Mondol, B.; Sarker, A.; Shareque, A.; Dey, S.C.; Islam, M.T.; Das, A.K.; Shamsuddin, S.M.; Molla, M.; Islam, A.; Sarker, M. Preparation of activated carbon/TiO2 nanohybrids for photodegradation of reactive red-35 dye using sunlight. *Photochem* **2021**, *1*, 54–66. [CrossRef]
7. Zuliani, A.; Cova, C.M. Green Synthesis of Heterogeneous Visible-Light-Active Photocatalysts: Recent Advances. *Photochem* **2021**, *1*, 147–166. [CrossRef]

Article

g-C_3N_4/MoS_2 Heterojunction for Photocatalytic Removal of Phenol and Cr(VI)

Ilaeira Rapti [1], Feidias Bairamis [1] and Ioannis Konstantinou [1,2,*]

1 Department of Chemistry, University of Ioannina, 45100 Ioannina, Greece; il_rapti@windowslive.com (I.R.); bairamisfeidias@gmail.com (F.B.)
2 Institute of Environment and Sustainable Development, University Research Center of Ioannina (URCI), 45110 Ioannina, Greece
* Correspondence: iokonst@uoi.gr

Abstract: In this study, molybdenum disulfide (MoS_2) decorated on graphitic carbon nitride (g-C_3N_4) heterostructure catalysts at various weight ratios (0.5%, 1%, 3%, 10%, *w*/*w*) were successfully prepared via a two-step hydrothermal synthesis preparation method. The properties of the synthesized materials were studied by X-ray diffraction (XRD), attenuated total reflectance–Fourier transform infrared spectroscopy (ATR-FT-IR), UV–Vis diffuse reflection spectroscopy (DRS), scanning electron microscopy (SEM) and N_2 porosimetry. MoS_2 was successfully loaded on the g-C_3N_4 forming heterojunction composite materials. N_2 porosimetry results showed mesoporous materials, with surface areas up to 93.7 m^2g^{-1}, while determined band gaps ranging between 1.31 and 2.66 eV showed absorption over a wide band of solar light. The photocatalytic performance was evaluated towards phenol oxidation and of Cr(VI) reduction in single and binary systems under simulated sunlight irradiation. The optimum mass loading ratio of MoS_2 in g-C_3N_4 was 1%, showing higher photocatalytic activity under simulated solar light in comparison with bare g-C_3N_4 and MoS_2 for both oxidation and reduction processes. Based on scavenging experiments a type-II photocatalytic mechanism is proposed. Finally, the catalysts presented satisfactory stability (7.8% loss) within three catalytic cycles. Such composite materials can receive further applications as well as energy conversion.

Keywords: C_3N_4; MoS_2; composite; photocatalytic oxidation; Cr(VI) reduction

Citation: Rapti, I.; Bairamis, F.; Konstantinou, I. g-C_3N_4/MoS_2 Heterojunction for Photocatalytic Removal of Phenol and Cr(VI). *Photochem* **2021**, *1*, 358–370. https://doi.org/10.3390/photochem1030023

Academic Editor: Vincenzo Vaiano

Received: 22 September 2021
Accepted: 12 October 2021
Published: 15 October 2021

1. Introduction

Energy crisis and environmental pollution are currently among the most serious problems for humans [1]. Emerging and priority environmental pollutants comprise several chemical categories such as pharmaceuticals, pesticides and plasticizers. Phenol and in general phenolics may be present in several wastewater streams from different production processes as well as degradation products from the previously mentioned pollutants. In addition, toxic metal ions such as Cr(VI) may be present in combined wastewater streams; thus, treatment methods should be capable of removing both organic and inorganic pollutants. However, conventional treatment methods present either low removal of certain organic pollutants or are incapable of simultaneous removal of organics and toxic metal ions [2]. Therefore, there has been an increased interest in the development of more efficient and environmentally friendly methods for the removal of pollutants from wastewater, and for this reason the advanced oxidation processes (AOPs) are widely used. Their effectiveness for organics removal is based on the production of hydroxyl radicals, which are powerful oxidizing agents. Heterogeneous photocatalysis, which presents advantages over other removal techniques as it can be applied for both oxidation and reduction of pollutants in ambient conditions in the presence of solar radiation and for organic pollutants, can lead to complete mineralization [2–6].

Graphitic carbon nitride (g-C_3N_4) has drawn widespread attention and become a hotspot in various scientific disciplines as a metal-free semiconductor due to its advantages. It has an appealing electronic structure with a 2.7 eV band gap, it can be fabricated with low-cost precursors (e.g., urea, thiourea, melamine) [3,7,8] and it is environmentally friendly. However, the rapidly photogenerated electron–hole recombination diminishes the photocatalytic efficiency. The coupling of g-C_3N_4 with other semiconductors with an appropriate band gap was considered to increase photocatalytic efficiency.

Molybdenum sulfide (MoS_2) has attracted attention as an emerging photocatalytic cocatalyst material due to its distinctive features, such as high stability and hardness, reliability, non-toxicity and low cost. It possesses a narrow band gap (1.2–1.9 eV), which offers a wide spectral absorption range and provides high speed channels for the interfacial charge transfer in heterostructures [3]. Previous studies have explored the application of MoS_2 in Cr(VI) reduction. Photoreduction and chemical reduction (redox between MoS_2 and Cr(VI)) were determined through a collaborative contribution to reducing Cr(VI) by MoS_2 under light conditions [9–11]. The coupling of two semiconductors (MoS_2, g-C_3N_4) could significantly reduce the charge carrier recombination and highly improve photocatalytic activity compared to those bare materials. MoS_2/g-C_3N_4 has been synthesized and tested for energy conversion, as a supercapacitor and for organic dye degradation [3,12–15]; however, its application to the simultaneous oxidation/reduction of environmental pollutants has not been studied so far.

Based on the above, the present study deals with the preparation of a series of MoS_2/g-C_3N_4 heterostructure composite catalysts and their application for the removal of phenol and Cr(VI) for aqueous solutions. There are few works exploring the synthesis of MoS_2/g-C_3N_4 heterojunction structure via various methods such as hydrothermal treatment, impregnation-sulfidation and photodeposition [16–19]. Here, a two-step hydrothermal synthesis combined with a sonication–liquid exfoliation preparation method is proposed.

2. Materials and Methods

2.1. Materials and Chemicals

Urea (99.5%, M_W: 60.06 $gmol^{-1}$), thiourea (>99%, M_W: 76.12 $gmol^{-1}$) and 1,5-diphenyl carbazide (98% (M_W: 242.28) were obtained from Acros Organics (Geel, Belgium). Ammonium molybdate tetrahydrate (99%, M_W: 1235.86 $gmol^{-1}$) was supplied by Janssen Chimica. Potassium dichromate (KCr_2O_7, M_W: 294.18 $gmol^{-1}$) was obtained from Sigma-Aldrich (St. Louis, MO, USA). Phenol (M_W: 94.11 $gmol^{-1}$) was purchased by Merck kGaA (Darmstadt, Germany). Methanol (MeOH), sulfuric acid >95% (H_2SO_4), methanol HPLC grade (MeOH, M_W: 32.04 $gmol^{-1}$) and water HPLC grade (M_W: 18.02 $gmol^{-1}$) were supplied by Fischer Scientific (Loughborough, Leics, UK). Distilled water was used throughout the experimental procedures of the study.

2.2. Preparation of g-C_3N_4, MoS_2 and Composite MoS_2/g-C_3N_4 Photocatalysts

2.2.1. Synthesis of Bare Carbon Nitride (g-C_3N_4) and Molybdenum Disulfide (MoS_2)

Graphitic carbon nitride (g-C_3N_4) was synthesized by thermal treatment of urea as a precursor compound. In a typical synthesis of g-C_3N_4, 30 g of urea was put in a crucible. After being dried at 90 °C it was heated to 500 °C at a heating rate of 10 °C/min and kept at this temperature for 4 h. The resulting yellow powder was collected and ground into a powder.

Molybdenum disulfide was prepared by hydrothermal synthesis. Hexaammonium heptamolybdate tetrahydrate and thiourea were dissolved in 10 mL of deionized water by stirring for 1 h at room temperature. The resulting solution was then placed in a Teflon-lined autoclave reactor and heated in an oven at 200 °C for 12 h.

2.2.2. Synthesis of $MoS_2/g\text{-}C_3N_4$ Heterostructure

$MoS_2/g\text{-}C_3N_4$ heterostructures with 0.5%, 1%, 3%, 10% (wt%) ratios were synthesized. An appropriate amount of bulk $g\text{-}C_3N_4$ was added in isopropanol into a beaker, and the mixture was sonicated for 2 h at room temperature. In a separate beaker, MoS_2 nanoparticles was added in a water/isopropanol (2:1) solution, and the mixture was sonicated for 2 h at room temperature. The solutions were mixed together and sonicated for another 2 h at room temperature. The product was cleaned with water and ethanol, collected by centrifuging and dried at 60 °C [18]. Then it was calcinated in air at 200 °C for 4 h.

2.3. *Material Characterization*

The crystal forms of the photocatalysts were analyzed from their X-ray diffraction (XRD) patterns using a Bruker Advance D8 X-ray powder diffractometer (Billerica, MA, USA) working with Cu-K_a (λ = 1.5406 Å) radiation. Diffractograms were scanned from 2θ 10° to 90°. The patterns were assigned to crystal phases with the use of the International Center for Diffraction Data (ICCD).

Attenuated total reflectance–Fourier transform infrared spectroscopy (ATR-FT-IR) was conducted to characterize the functional groups of different catalysts. A Shimadzu IR Spirit-T spectrophotometer (Kyoto, Japan) with a QATR-S single-reflection ATR measurement attachment equipped with diamond prism was used. The photocatalysts were scanned in the range 4000–400 cm^{-1}.

The UV–Vis diffuse reflectance spectra (DRS) of the catalyst powders were recorded using a Shimadzu 2600 spectrophotometer bearing an IRS-2600 integrating sphere (Kyoto, Japan) in the wavelength of 200–800 nm at room temperature using $BaSO_4$ (Nacalai Tesque, extra pure reagent, Kyoto, Japan) as a reference sample.

The morphology of the catalysts was examined by scanning electron microscopy (SEM) using a JEOL JSM 5600 (Tokyo, Japan) working at 20 kV.

Nitrogen adsorption–desorption isotherms were collected by porosimetry using a Quantachrome Autosorb-1 instrument (Bounton Beach, FL, USA) at 77K. The samples ($\approx$80 mg) were degassed at 150 °C for 3 h. The Brunauer–Emmet–Teller (BET) method was used to calculate the specific area (SSA) of each material at relative pressure between 0.05 and 0.3. The adsorbed amount of nitrogen at relative pressure P/P_0 = 0.95 was used in order to calculate the total pore volume (V_{TOT}). The Barrett, Joyner and Halenda (BJH) method was used to determine the pore size distribution (PSD) of the photocatalysts.

2.4. *Photocatalytic Experiments*

Photocatalytic experiments were carried out using Atlas Suntest XLS+ (Atlas, Germany) solar simulator apparatus equipped with a xenon lamp (2.2 kW) and special filters in place to prevent the transmission of wavelengths below 290 nm. During the experiments the irradiation intensity was maintained at 500 W m^{-2}.

Aqueous solutions (100 mL) and the catalyst (10 mg) were transferred into a double-walled Pyrex glass reactor, thermostated by water flowing in the double-skin of the reactor. In binary systems the pH was adjusted to ca. 2 with sulfuric acid in order to achieve high adsorption efficiency of Cr(VI) onto the catalysts and avoid $Cr(OH)_3$ deposition on the catalyst surface, based on previous studies [11,20]. Before exposure to light, the suspension was stirred in the dark for 30 min to ensure the establishment of adsorption–desorption equilibrium onto the catalyst surface. The obtained aliquot (5 mL) was withdrawn at different time intervals and was filtered through 0.45 μm filters in order to remove the catalyst's particles before further analysis. Quantification of phenol for the kinetic studies in aquatic solutions was carried out by high performance liquid chromatography (HPLC) (Schimadzu, LC 10AD, Diode Array Detector SPD-M10A, Degasser DGU-14A). The column oven (Schimadzu, CTO-10A) was set at 40 °C. The mobile phase used was a mixture of water HPLC grade (50%) and methanol HPLC grade (50%). The concentration of Cr(VI) was determined by the diphenylcarbazide colorimetric method measuring the absorbance at the wavelength of 540 nm using a UV–Vis spectrophotometer (Jasco-V630, Tokyo, Japan).

3. Results and Discussion

3.1. Characterization of the Prepared Photocatalysts

3.1.1. Structural Characterization

The XRD patterns for g-C_3N_4, MoS_2 and MoS_2/g-C_3N_4 composites are shown in Figure 1. For g-C_3N_4 two characteristic intense peaks were observed at 13.1° and 27.4°. A broad peak was observed corresponding to the (100) plane, arising from in-plane stacking of the tri-s-triazine motif, and to the (002) interlayer stacking of the conjugated aromatic rings. For MoS_2 three characteristic intense peaks at 13.4°, 33.5° and 57.8° were observed, which can be indexed to the (002), (100) and (110) planes, respectively, corresponding to the hexagonal phase of MoS_2 (JCPDS 87-1526) [21–24].

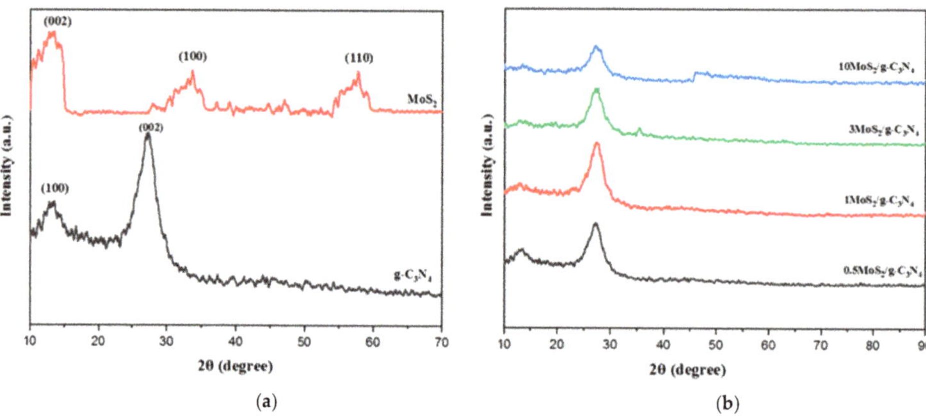

Figure 1. XRD patterns of (**a**) g-C_3N_4, MoS_2, (**b**) composite catalysts.

For the prepared composite catalysts no diffraction peaks corresponding to MoS_2 can be recognized, which may be because of the small quantity of MoS_2 contents and high dispersion in g-C_3N_4 photocatalysts.

The ATR-FT-IR spectra of g-C_3N_4, MoS_2 and g-C_3N_4/MoS_2 composite are shown in Figure 2. The prepared catalysts presented the main characteristic peaks of both g-C_3N_4 and MoS_2 suggesting the formation of composite heterostructures. The peaks of g-C_3N_4 can be observed at 1573, 1465, 1403, 1319 and 1241 cm^{-1} confirming the stretching vibration of C–N (–C)–C or C–NH–C heterocycles. The broad peaks between 3300 and 3000 cm^{-1} are related to the stretching vibration of N-H, and the peak at 1637 cm^{-1} corresponds to the C=N bending vibration. The peak at 810 cm^{-1} corresponds to heptazine ring vibration. The peak at 1400 cm^{-1} confirms the presence of amino groups. The peaks of MoS_2 can be observed at 689 cm^{-1}, confirming the stretching vibration of Mo–O. The peak at 1112 cm^{-1} corresponds to the stretching vibration of S=O [25].

3.1.2. Morphology Surface Analysis

The morphological evaluation of all catalysts was carried out by scanning electron microscopy (Figure 3). Figure 3a shows the morphology of g-C_3N_4 consisting of packed sheets. Figure 3b presents aggregation of flower-like spherical particles of MoS_2 [26]. Figure 3c (1MoS_2/g-C_3N_4) shows a morphology that seems to be close to that of g-C_3N_4; this may be because of the small quantity of MoS_2 contents on the surface of g-C_3N_4. The 10MoS_2/g-C_3N_4 (Figure 3d) reveals MoS_2 spheres attached on the surface of the g-C_3N_4 sheets.

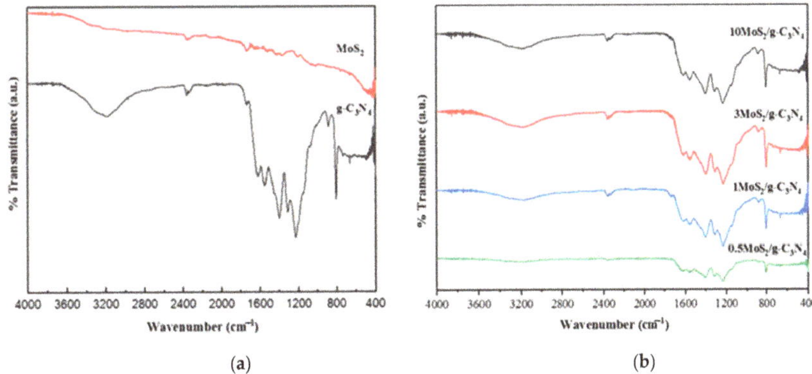

Figure 2. ATR-FT-IR spectra of (**a**) g-C_3N_4, MoS_2, (**b**) composite catalysts.

Figure 3. SEM images of (**a**) g-C_3N_4, (**b**) MoS_2 and (**c**) 1MoS_2/g-C_3N_4 and (**d**) 10MoS_2/g-C_3N_4.

The nitrogen adsorption–desorption isotherms for the prepared photocatalysts are presented in Figure 4.

All composites samples exhibited typical type IV(a) adsorption and desorption isotherms with H3 hysteresis loop according to the IUPAC classification, suggesting mesoporous materials [27]. BET specific surface area ranged between 2.0 m^2g^{-1} for MoS_2 and 93.7 m^2g^{-1} for g-C_3N_4. MoS_2 presented a non-porous structure with some porosity owed to the aggregation of spherical particles. The calculated specific surface area, average pore diameter and total pore volume of the catalysts are shown in Table 1.

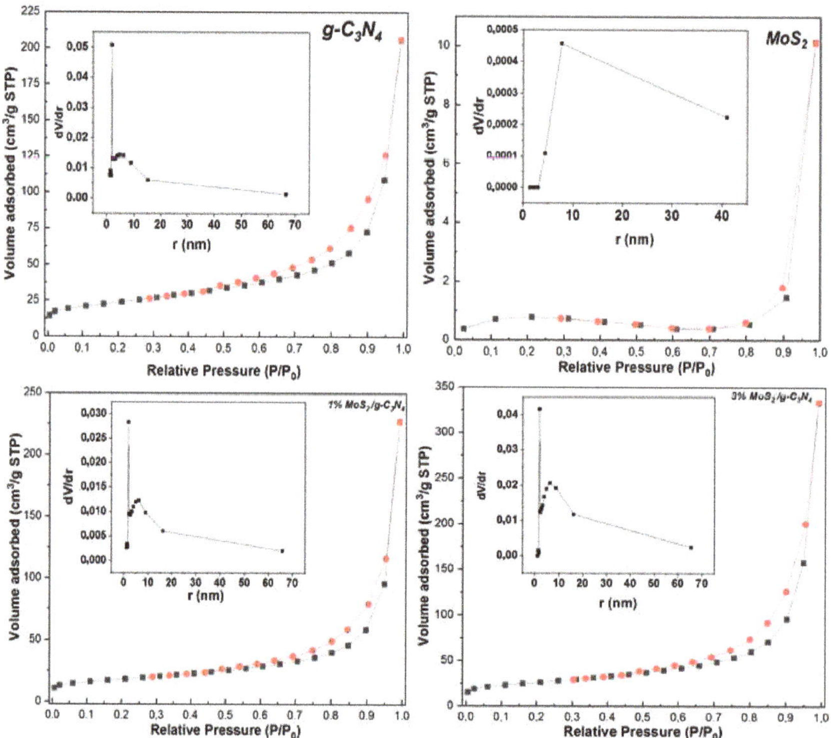

Figure 4. Nitrogen adsorption–desorption isotherms and pore size distributions for g-C_3N_4, MoS_2, 1MoS_2/g-C_3N_4, 3MoS_2/g-C_3N_4.

Table 1. Specific surface area, average pore diameter and total pore volume of prepared catalysts.

Catalyst	Specific Surface Area S_{BET} (m^2g^{-1})	Average Pore Diameter (nm)	V_{TOT} ($cm^3\ g^{-1}$)
MoS_2	2.0	31	0.016
g-C_3N_4	81.3	8.3	0.169
0.5% MoS_2/g-C_3N_4	93.7	10.2	0.238
1% MoS_2/g-C_3N_4	62.2	9.5	0.148
3% MoS_2/g-C_3N_4	90.2	10.9	0.245
10% MoS_2/g-C_3N_4	92.1	10.5	0.239

3.1.3. UV–Vis Spectroscopy and Band Gap Determination

The diffuse reflectance spectroscopy (DRS) results and the energy band gap (E_g), calculated with the Kübelka–Munk function, of all photocatalysts are presented in Figure 5. The absorption edge of MoS_2 was determined at 946 nm, while the absorption edge of g-C_3N_4 was at 480 nm indicating visible light response. The absorption edges of the prepared composite catalysts varied between 466–946 nm and the energy band gap 1.31–2.66 eV. The band gap energy and corresponding absorption edges are shown in Table 2.

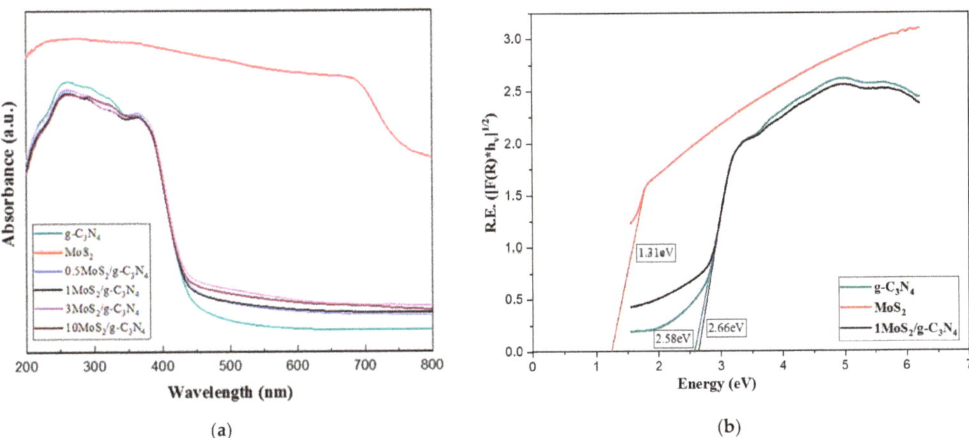

Figure 5. (**a**) UV–Vis absorption spectra of prepared catalysts, (**b**) Kübelka–Munk plots for g-C_3N_4, MoS_2 and 1MoS_2/g-C_3N_4 for indirect band gap semiconductors.

Table 2. Energy band gap and absorption edge of prepared catalysts.

Catalyst	Energy Band Gap (eV)	Absorption Edge λ (nm)
g-C_3N_4	2.58	480
MoS_2	1.31	946
0.5% MoS_2/g-C_3N_4	2.56	484
1% MoS_2/g-C_3N_4	2.66	466
3% MoS_2/g-C_3N_4	2.64	469
10% MoS_2/g-C_3N_4	2.64	469

MoS_2/g-C_3N_4 heterojunction samples presented enhanced light absorption in the visible region as compared to g-C_3N_4. The absorption intensity of the prepared samples strengthened with increasing MoS_2 contents, which agrees with the color of the prepared samples that vary from light yellow to grey.

The band edge position of conduction band (CB) and valence band (VB) of prepared catalysts was determined applying Equations (1) and (2).

$$E_{VB} = X - E_0 + 0.5E_g \tag{1}$$

$$E_{CB} = E_{VB} - E_g \tag{2}$$

where E_{VB} is the VB edge potential, E_{CB} is the CB edge potential, E_g is the band gap energy of the catalysts, X is the electronegativity of g-C_3N_4 (4.67 eV) and MoS_2 (5.33 eV) [28] and E_0 is the energy of free electrons on the hydrogen scale (4.5 eV). The E_{VB} and E_{CB} of g-C_3N_4 was calculated at 1.46 and −1.12 eV, respectively, while for MoS_2 the corresponding values were 1.48 and 0.17 eV.

3.2. Photocatalytic Study

The photocatalytic performance of composite catalysts was studied against the oxidation of phenol (single system: C_{ph} = 10 mg L^{-1}, C_{cat} = 100 mg L^{-1}) as well as the simultaneous photocatalytic reduction of Cr(VI) and oxidation of phenol (binary system: phenol/Cr(VI) = 1/10, C_{cat} = 100 mg L^{-1}, pH = 2 [29,30]) under simulated sunlight irradiation. The results of all photocatalytic experiments are shown in Figure 6 and Tables 3 and 4.

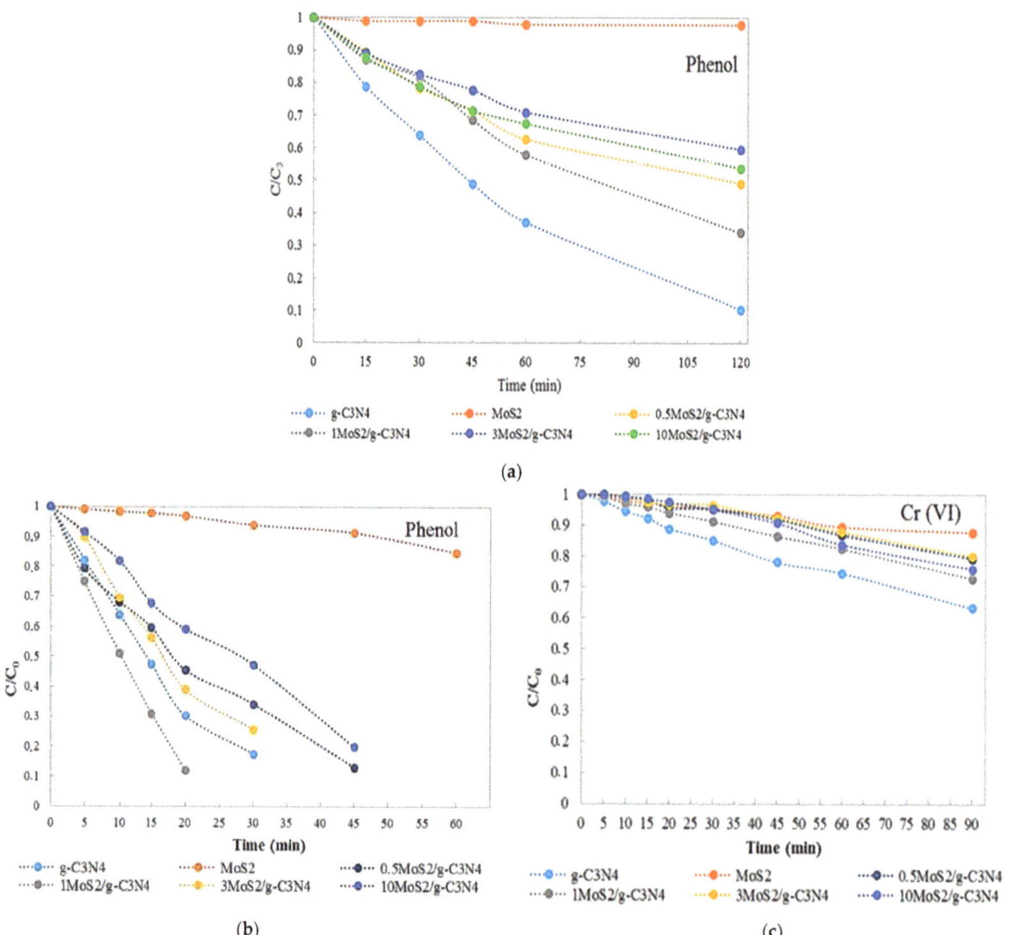

Figure 6. Photocatalytic degradation kinetics of (**a**) phenol in single system, (**b**) phenol and (**c**) Cr(VI) in binary system under solar light irradiation.

Table 3. Kinetic parameters (first-order kinetic constants (k), half-life ($t_{1/2}$) and correlation coefficients (R^2) of phenol in single system.

Catalyst	k (min^{-1})	$t_{1/2}$ (min)	R^2
g-C_3N_4	0.022	31.5	0.9401
MoS_2	0.0002	3465.0	0.5016
0.5MoS_2/g-C_3N_4	0.006	116.0	0.9528
1MoS_2/g-C_3N_4	0.009	77.0	0.9940
3MoS_2/g-C_3N_4	0.005	139.0	0.9250
10MoS_2/g-C_3N_4	0.006	116.0	0.9158

The first-order apparent reaction rates for phenol photocatalytic degradation in single system follows the sequence $k_{g\text{-}C_3N_4} > k_{1MoS_2/g\text{-}C_3N_4} > k_{0.5MoS2/g\text{-}C3N4} > k_{10MoS_2/g\text{-}C_3N_4} > k_{3MoS_2/g\text{-}C_3N_4} > k_{MoS_2}$. In the binary system, reaction rates of phenol oxidation were as follows: $k_{1MoS_2/g\text{-}C_3N_4} > k_{g\text{-}C_3N_4} > k_{3MoS_2/g\text{-}C_3N_4} > k_{0.5MoS_2/g\text{-}C_3N_4} > k_{10MoS_2/g\text{-}C_3N_4} > k_{MoS_2}$, and as to the reduction of Cr(VI) it was as follows: $k_{g\text{-}C_3N_4} > k_{1MoS_2/g\text{-}C_3N_4} > k_{10MoS_2/g\text{-}C_3N_4} > k_{0.5MoS_2/g\text{-}C_3N_4} > k_{3MoS_2/g\text{-}C_3N_4} > k_{MoS_2}$.

In comparison, the prepared catalysts exhibit faster kinetics of phenol and Cr(VI) removal in binary than in single systems.

Table 4. Kinetic parameters (first-order kinetic constants (k), half-life ($t_{1/2}$) and correlation coefficients (R^2)) of phenol and Cr(VI) in binary system.

Binary	Phenol			Cr(VI)		
Catalyst	k (min^{-1})	$t_{1/2}$ (min)	R^2	k (min^{-1})	$t_{1/2}$ (min)	R^2
$g\text{-}C_3N_4$	0.068	10.2	0.9978	0.003	231.0	0.9477
MoS_2	0.002	346.5	0.9373	0.002	346.5	0.9694
$0.5MoS_2/g\text{-}C_3N_4$	0.042	16.5	0.9738	0.002	346.5	0.9552
$1MoS_2/g\text{-}C_3N_4$	0.091	7.6	0.9246	0.003	231.0	0.9740
$3MoS_2/g\text{-}C_3N_4$	0.042	16.5	0.9513	0.002	346.5	0.9478
$10MoS_2/g\text{-}C_3N_4$	0.031	22.4	0.9374	0.003	231.0	0.9325

3.3. Photocatalytic Mechanism for the Composite Photocatalysts

Different photoinduced reactive species, such as hydroxyl radicals (OH·), superoxide radicals (O_2^-) and holes (h^+), may be involved in the photocatalytic process after electron–hole pairs are generated. Thus, photocatalytic experiments in the presence of isopropanol (IPA), superoxide dismutase (SOD) and triethanolamine (TEOA) were used to scavenge •OH, O_2^- and h^+, respectively, during the photodegradation of phenol in order to investigate the photocatalytic mechanism [3]. As shown in Table 5, the degradation rate constant (k) for phenol was 0.009 min^{-1} after exposure for 120 min under simulated sunlight without scavenger addition, and this changed after adding, IPA, SOD and TEOA, to 0.003, 0.004 and 0.002 min^{-1}, respectively. It is clearly observed that the degradation of phenol was inhibited with the addition of scavengers, though TEOA was the most obvious. The degradation kinetics of phenol in the presence of scavengers are shown in Figure 7 and Table 5.

Table 5. Kinetic parameters (first-order kinetic constants (k), half-life ($t_{1/2}$) and correlation coefficients (R^2)) of phenol in scavenging experiments.

Scavenger	k (min^{-1})	$t_{1/2}$ (min)	R^2
IPA	0.003	231	0.9742
SOD	0.004	173.2	0.9590
TEOA	0.002	346.5	0.9553
no scavenger	0.009	77.0	0.9940

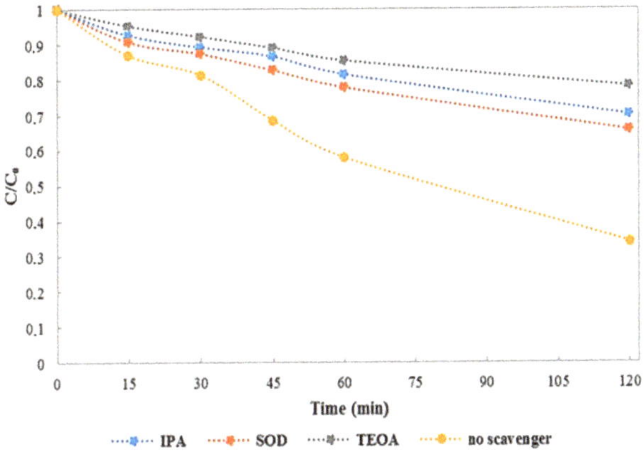

Figure 7. Photocatalytic degradation kinetics of phenol in single system in the presence of scavengers and $1MoS_2/g\text{-}C_3N_4$ photocatalyst under solar light irradiation.

According to the above analysis, the proposed photocatalytic mechanism (Type II), involving the carrier transfer and photodegradation of phenol was illustrated as shown in Figure 8. The photogenerated electron–hole pairs were induced in the conduction band (CB) and valence band (VB) of g-C_3N_4 and MoS_2 under irradiation. The photogenerated electrons transferred from the CB of g-C_3N_4 to that of MoS_2, while holes moved from the VB of MoS_2 to that of g-C_3N_4, leading the photogenerated electron–hole pair to be separated and transferred effectively. The holes (h^+) in the VB of g-C_3N_4 participate in the degradation process. Simultaneously, electrons (e^-) in the CB together with H^+ can reduce O_2 to H_2O_2, which forms hydroxyl radicals. The e^- on the CB of g-C_3N_4 and MoS_2 participates in the reduction of Cr(VI) to Cr(III). The proposed mechanism provides additional explanation for the better removal performance in the binary system. Regarding the reduction process, the electrons on MoS_2 have weaker reduction ability; consequently they can reduce strong oxidation agents such as Cr(VI), enhancing the target reduction reaction without other reactions, such as O_2 reduction. As far as the oxidation pathway is concerned, the enhancement is due to the better charge separation. The results of the present study indicate that the MoS_2 could be a promising candidate as a non-noble metal co-catalyst for g-C_3N_4 photocatalysts. The composite catalysts can receive further applications either for degradation of pollutants or in energy conversion.

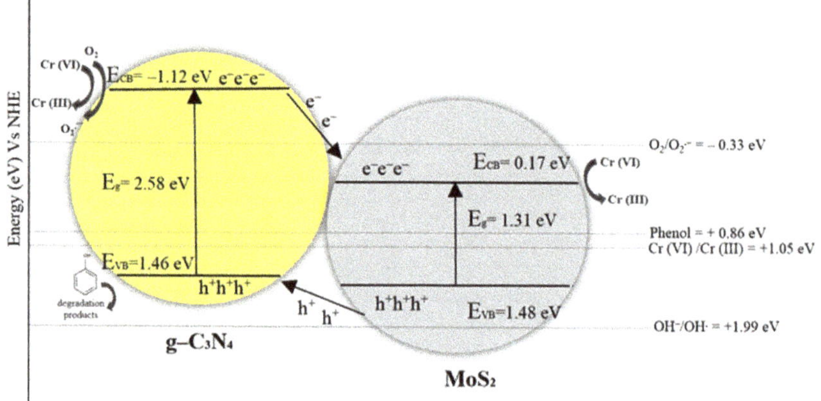

Figure 8. Schematic illustration of carrier transfer in the MoS_2/g-C_3N_4 heterojunction.

3.4. Recyclability of the Composite Photocatalysts

The stability of the prepared photocatalysts was investigated by examining the photocatalytic activity of the most efficient photocatalyst (1MoS_2/g-C_3N_4) against oxidation of phenol in binary system for three continuous cycles (Figure 9). Reaction constants of phenol oxidation were 0.090, 0.089 and 0.083 min^{-1} for the first, second and third cycle, respectively (Table 6). A total loss of photocatalytic activity about 7.8% after the third photocatalytic cycle showed the sufficient stability of the catalyst considering also the possible losses of catalyst particles during the recovery process after each cycle. In addition, the catalyst separated at the end of the third catalytic cycle was characterized by XRD, SEM and ATR-FT-IR (Figure 10a–c), revealing a nearly identical pattern to the starting material.

Table 6. Kinetic parameters (first-order kinetic constants (k), half-life ($t_{1/2}$) and correlation coefficients (R^2)) of phenol degradation by 1MoS_2/g-C_3N_4 catalyst after three consecutive cycles.

Cycle	k (min^{-1})	$t_{1/2}$ (min)	R^2
1	0.09	7.7	0.9510
2	0.089	7.8	0.9453
3	0.083	8.3	0.9496

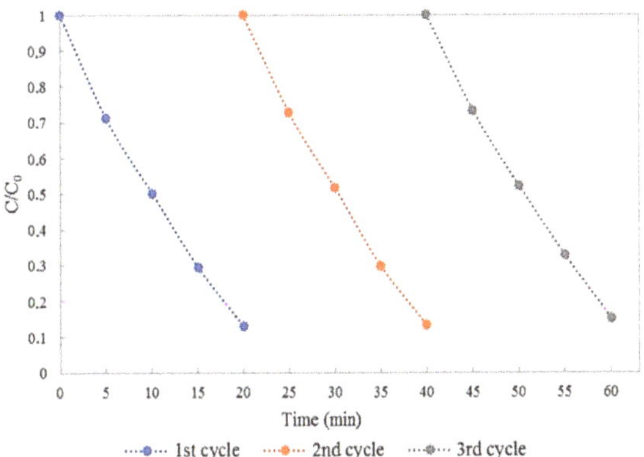

Figure 9. Reusability performance of $1MoS_2/g\text{-}C_3N_4$ photocatalyst for three continuous cycles.

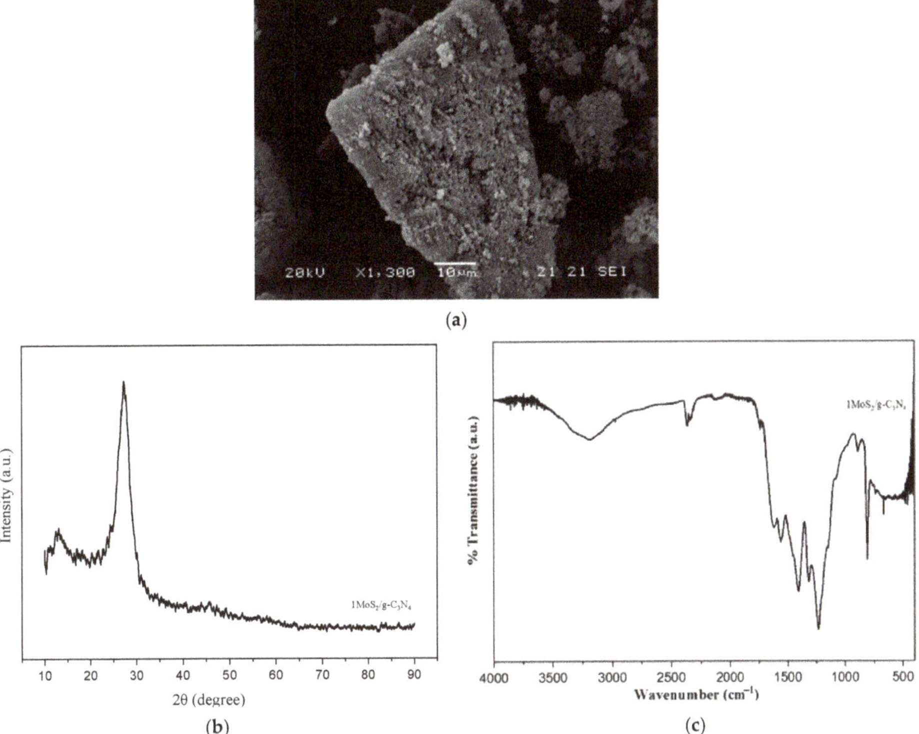

Figure 10. (**a**) SEM image, (**b**) XRD pattern, (**c**) ATR-FT-IR spectra of $1MoS_2/g\text{-}C_3N_4$ photocatalyst after the third photocatalyst cycle.

4. Conclusions

In this work, mesoporous $MoS_2/g\text{-}C_3N_4$ heterostructures were successfully synthesized via a two-step hydrothermal synthesis at various weight ratios (up to 10% *w*/*w* of MoS_2) and assayed for the simultaneous oxidation of phenol and Cr(VI) reduction. Composite materials show higher photocatalytic activity under solar light in comparison with bare $g\text{-}C_3N_4$ and MoS_2 for both oxidation and reduction processes, while the material with 1% MoS_2 loading was found as the catalyst with the best efficiency. A considerable enhancement of the photocatalytic efficiency was observed for the binary system compared to the single component systems, indicating a promoting synergistic effect. The three cycling experiments indicated that the catalysts presented a good stability. Finally, type-II band alignment was proposed as the photocatalytic mechanism for the composed catalysts.

Author Contributions: Conceptualization, I.R. and I.K.; methodology, I.R. and I.K.; formal analysis, I.R. and F.B.; investigation, I.R. and I.K.; resources, I.K.; data curation, I.R.; writing—original draft preparation, I.R. and I.K.; writing—review and editing I.K., I.R. and F.B.; visualization, I.R.; supervision, I.K.; project administration, I.R.; funding acquisition, I.K. All authors have read and agreed to the published version of the manuscript.

Funding: This research received no external funding.

Institutional Review Board Statement: Not applicable.

Informed Consent Statement: Not applicable.

Data Availability Statement: Not applicable.

Acknowledgments: The authors acknowledge the access in XRD and SEM units of the University of Ioannina. The authors would like to thank D. Petrakis for the collaboration in the Laboratory of Industrial Chemistry, Dept. of Chemistry, University of Ioannina.

Conflicts of Interest: The authors declare no conflict of interest.

References

1. Jiménez-Rangel, K.Y.; Lartundo-Rojas, L.; García-García, A.; Cipagauta-Díaz, S.; Mantilla, A.; Samaniego-Benítez, J.E. Hydrothermal Synthesis of a Two-Dimensional g-C_3N_4/MoS_2/MnOOH Composite Material and Its Potential Application as Photocatalyst. *J. Chem. Technol. Biotechnol.* **2019**, *94*, 3447–3456. [CrossRef]
2. Antonopoulou, M.; Giannakas, A.; Konstantinou, I. Simultaneous Photocatalytic Reduction of Cr(VI) and Oxidation of Benzoic Acid in Aqueous N-F-Codoped TiOsuspensions: Optimization and Modeling Using the Response Surface Methodology. *Int. J. Photoenergy* **2012**, *1*, 358–370. [CrossRef]
3. Zhang, X.; Zhang, R.; Niu, S.; Zheng, J.; Guo, C. Enhanced Photo-Catalytic Performance by Effective Electron-Hole Separation for MoS_2 Inlaying in g-C_3N_4 Hetero-Junction. *Appl. Surf. Sci.* **2019**, *475*, 355–362. [CrossRef]
4. Cuerda-Correa, E.M.; Alexandre-Franco, M.F.; Fernández-González, C. Advanced Oxidation Processes for the Removal of Antibiotics from Water. An Overview. *Water* **2020**, *12*, 102. [CrossRef]
5. Barrera-Díaz, C.; Cañizares, P.; Fernández, F.J.; Natividad, R.; Rodrigo, M.A.; Rosedal, E.; Toluca, E.; de México, M. Electrochemical Advanced Oxidation Processes: An Overview of the Current Applications to Actual Industrial Effluents. *J. Mex. Chem. Soc.* **2014**, *58*, 3. [CrossRef]
6. Malato, S.; Fernández-Ibáñez, P.; Maldonado, M.I.; Blanco, J.; Gernjak, W. Decontamination and Disinfection of Water by Solar Photocatalysis: Recent Overview and Trends. *Catal. Today* **2009**, *147*, 1–59. [CrossRef]
7. Zhang, G.; Zhang, J.; Zhang, M.; Wang, X. Polycondensation of Thiourea into Carbon Nitride Semiconductors as Visible Light Photocatalysts. *J. Mater. Chem.* **2012**, *22*, 8083–8091. [CrossRef]
8. Bairamis, F.; Konstantinou, I. WO3 Fibers/g-C_3N_4 Z-Scheme Heterostructure Photocatalysts for Simultaneous Oxidation/Reduction of Phenol/Cr(VI) in Aquatic Media. *Catalysts* **2021**, *11*, 792. [CrossRef]
9. Sun, K.; Jia, F.; Yang, B.; Lin, C.; Li, X.; Song, S. Synergistic Effect in the Reduction of Cr(VI) with Ag-MoS_2 as Photocatalyst. *Appl. Mater. Today* **2020**, *18*, 100453. [CrossRef]
10. Sun, H.; Wu, T.; Zhang, Y.; Ng, D.H.L.; Wang, G. Structure-Enhanced Removal of Cr(VI) in Aqueous Solutions Using MoS_2 Ultrathin Nanosheets. *New. J. Chem.* **2018**, *42*, 9006–9015. [CrossRef]
11. Wang, K.; Chen, P.; Nie, W.; Xu, Y.; Zhou, Y. Improved Photocatalytic Reduction of Cr(VI) by Molybdenum Disulfide Modified with Conjugated Polyvinyl Alcohol. *Chem. Eng. J.* **2019**, *359*, 1205–1214. [CrossRef]
12. Fageria, P.; Sudharshan, K.Y.; Nazir, R.; Basu, M.; Pande, S. Decoration of MoS_2 on g-C_3N_4 Surface for Efficient Hydrogen Evolution Reaction. *Electrochim. Acta* **2017**, *258*, 1273–1283. [CrossRef]
13. Qi, Y.; Liang, Q.; Lv, R.; Shen, W.; Kang, F.; Huang, Z.H. Synthesis and Photocatalytic Activity of Mesoporous g-C_3N_4/MoS_2 Hybrid Catalysts. *R. Soc. Open Sci.* **2018**, *5*, 180187. [CrossRef] [PubMed]
14. Samy, O.; el Moutaouakil, A. A Review on MoS_2 Energy Applications: Recent Developments and Challenges. *Energies* **2021**, *14*, 4568. [CrossRef]
15. Mamba, G.; Mishra, A.K. Graphitic Carbon Nitride (g-C_3N_4) Nanocomposites: A New and Exciting Generation of Visible Light Driven Photocatalysts for Environmental Pollution Remediation. *Appl. Catal. B Environ.* **2016**, *198*, 347–377. [CrossRef]
16. Zheng, D.; Zhang, G.; Hou, Y.; Wang, X. Layering MoS_2 on Soft Hollow g-C_3N_4 Nanostructures for Photocatalytic Hydrogen Evolution. *Appl. Catal. A Gen.* **2016**, *521*, 2–8. [CrossRef]

17. Huang, Q.; Liu, X.; Chen, Z.; Gong, S.; Huang, H. Surface Affinity Modulation of MoS_2 by Hydrothermal Synthesis and Its Intermediary Function in Interfacial Chemistry. *Chem. Phys. Lett.* **2019**, *730*, 608–611. [CrossRef]
18. Koutsouroubi, E.D.; Vamvasakis, I.; Papadas, I.T.; Drivas, C.; Choulis, S.A.; Kennou, S.; Armatas, G.S. Interface Engineering of MoS_2-Modified Graphitic Carbon Nitride Nano-Photocatalysts for an Efficient Hydrogen Evolution Reaction. *ChemPlusChem* **2020**, *85*, 1379–1388. [CrossRef] [PubMed]
19. Sivasankaran, R.P.; Rockstroh, N.; Kreyenschulte, C.R.; Bartling, S.; Lund, H.; Acharjya, A.; Junge, H.; Thomas, A.; Brückner, A. Influence of MoS_2 on Activity and Stability of Carbon Nitride in Photocatalytic Hydrogen Production. *Catalysts* **2019**, *9*, 695. [CrossRef]
20. Wei, H.; Zhang, Q.; Zhang, Y.; Yang, Z.; Zhu, A.; Dionysiou, D.D. Enhancement of the Cr(VI) Adsorption and Photocatalytic Reduction Activity of g-C_3N_4 by Hydrothermal Treatment in HNO_3 Aqueous Solution. *Appl. Catal. A Gen.* **2016**, *521*, 9–18. [CrossRef]
21. Tian, Y.; Ge, L.; Wang, K.; Chai, Y. Synthesis of Novel MoS_2/g-C_3N_4 Heterojunction Photocatalysts with Enhanced Hydrogen Evolution Activity. *Mater. Charact.* **2014**, *87*, 70–73. [CrossRef]
22. Visic, B.; Dominko, R.; Gunde, M.K.; Hauptman, N.; Skapin, S.D.; Remskar, M. Optical Properties of Exfoliated MoS_2 Coaxial Nanotubes—Analogues of Graphene. *Nanoscale Res. Lett.* **2011**, *6*, 593. [CrossRef] [PubMed]
23. Hu, K.H.; Hu, X.G.; Wang, J.; Xu, Y.F.; Han, C.L. Tribological Properties of MoS_2 with Different Morphologies in High-Density Polyethylene. *Tribol. Lett.* **2012**, *47*, 79–90. [CrossRef]
24. Konstas, P.S.; Konstantinou, I.; Petrakis, D.; Albanis, T. Synthesis, Characterization of g-C_3N_4/$SrTiO_3$ Heterojunctions and Photocatalytic Activity for Organic Pollutants Degradation. *Catalysts* **2018**, *8*, 554. [CrossRef]
25. Amini, M.; Ahmad Ramazani, S.A.; Faghihi, M.; Fattahpour, S. Preparation of Nanostructured and Nanosheets of MoS_2 Oxide Using Oxidation Method. *Ultrason. Sonochem.* **2017**, *39*, 188–196. [CrossRef]
26. Zhang, X.; Suo, H.; Zhang, R.; Niu, S.; Zhao, X.; Zheng, J.; Guo, C. Photocatalytic Activity of 3D Flower-like MoS_2 Hemispheres. *Mater. Res. Bull.* **2018**, *100*, 249–253. [CrossRef]
27. Thommes, M.; Kaneko, K.; Neimark, A.V.; Olivier, J.P.; Rodriguez-Reinoso, F.; Rouquerol, J.; Sing, K.S.W. Physisorption of Gases, with Special Reference to the Evaluation of Surface Area and Pore Size Distribution (IUPAC Technical Report). *Pure Appl. Chem.* **2015**, *87*, 1051–1069. [CrossRef]
28. Li, J.; Liu, E.; Ma, Y.; Hu, X.; Wan, J.; Sun, L.; Fan, J. Synthesis of MoS_2/g-C_3N_4 Nanosheets as 2D Heterojunction Photocatalysts with Enhanced Visible Light Activity. *Appl. Surf. Sci.* **2016**, *364*, 694–702. [CrossRef]
29. Barrera-Díaz, C.E.; Lugo-Lugo, V.; Bilyeu, B. A Review of Chemical, Electrochemical and Biological Methods for Aqueous Cr(VI) Reduction. *J. Hazard. Mater.* **2012**, *224*, 1–12. [CrossRef]
30. Du, X.; Yi, X.; Wang, P.; Deng, J.; Wang, C.C. Enhanced Photocatalytic Cr(VI) Reduction and Diclofenac Sodium Degradation under Simulated Sunlight Irradiation over MIL-100(Fe)/g-C_3N_4 Heterojunctions. *Cuihua Xuebao/Chin. J. Catal.* **2019**, *40*, 70–79. [CrossRef]

MDPI

Article

Electrospun Carbon Nanofibers Decorated with Ag_3PO_4 Nanoparticles: Visible-Light-Driven Photocatalyst for the Photodegradation of Methylene Blue

Gopal Panthi [1] and Mira Park [1,2,*]

[1] Carbon Composite Energy Nanomaterials Research Center, Woosuk University, Wanju, Chonbuk 55338, Korea; gopalpanthi2003@gmail.com

[2] Woosuk Institute of Smart Convergence Life Care (WSCLC), Woosuk University, Wanju, Chonbuk 55338, Korea

* Correspondence: wonderfulmira@woosuk.ac.kr; Tel.: +82-63-290-1536

Abstract: For the first time, heterostructures of electrospun carbon nanofibers decorated with Ag_3PO_4 nanoparticles (Ag_3PO_4/CNFs) were successfully fabricated by the combination of simple and versatile electrospinning technique followed by carbonization and incorporation of Ag_3PO_4 nanoparticles via colloidal and precipitation synthesis approaches. The as-fabricated heterostructures were characterized by FESEM with EDS, XRD, TEM with HRTEM, FTIR and UV-vis diffuse reflectance spectroscopy. Experimental results revealed that the heterostructure obtained by colloidal synthesis approach (Ag_3PO_4/CNFs-1) was decorated with small-sized (~20 nm) and uniformly distributed Ag_3PO_4 nanoparticles on the surface of CNFs without any evident agglomeration, while in the heterostructure obtained by the precipitation synthesis approach (Ag_3PO_4/CNFs-2), CNFs were decorated with agglomerated and bigger-sized Ag_3PO_4 nanoparticles. The visible-light-driven photocatalytic investigation signified that the Ag_3PO_4/CNFs-1 heterostructure can exhibit higher performance towards the photodegradation of MB dye solution compared to the Ag_3PO_4/CNFs-2 heterostructure, which could be attributed to the synergistic effect between the uniformity and small size of Ag_3PO_4 nanoparticles and CNFs that can serve as a conductivity network to prevent the recombination of charge carriers. Moreover, the mechanism of the photocatalytic activity as-prepared heterostructure is proposed.

Keywords: electrospinning; carbon composite nanofibers; water pollution; Ag_3PO_4; photocatalyst; visible light

Citation: Panthi, G.; Park, M. Electrospun Carbon Nanofibers Decorated with Ag_3PO_4 Nanoparticles: Visible-Light-Driven Photocatalyst for the Photodegradation of Methylene Blue. *Photochem* **2021**, *1*, 345–357. https://doi.org/10.3390/photochem1030022

Academic Editor: Vincenzo Vaiano

Received: 1 September 2021
Accepted: 8 October 2021
Published: 10 October 2021

1. Introduction

Photocatalysis, a "green technique", has been accepted as a potential remedy for the removal of organic pollutants from wastewater [1]. The discovery of the electrochemical photocatalysis of water at TiO_2 electrode by Honda and Fujishima [2] has inspired researchers working in the field of photocatalysis to design and synthesize semiconductor-based photocatalysts. Hence, a number of semiconductor-based photocatalysts such as metal oxides, sulfides and composites have been studied so far [3–6]. In particular, intense research works are focused on the fabrication of visible-light-driven photocatalysts to avoid the cost of UV light and create a comfortable environment for all living beings [7–10]. On that note, Ye and coworkers [11] carried out a breakthrough work to study the photocatalytic activity of silver orthophosphate (Ag_3PO_4) towards water splitting and decomposition of organic contaminants in aqueous solution under visible light irradiation. Afterwards, Ag_3PO_4 as a photocatalyst has attracted much more attention for the removal of various types of organic contaminants from wastewater [12–14]. Ag_3PO_4 is a promising semiconductor material being used as an efficient photocatalyst to achieve significantly higher photocatalytic activity utilizing visible light, but its application is limited due to photocorrosion under repetitive use when used without sacrificial reagent [11].

Indeed, the process of photocorrosion is the insurmountable problem that mostly occurs in silver-based catalysts [15,16]. In this connection, various attempts have been made for designing and developing a Ag_3PO_4-based photocatalyst with a sufficient charge separation ability and high photocatalytic stability in the visible light region either by doping SO_4^- ions [17] and coupling with other semiconductor/s [18–20] or by fabricating composites with graphene [21–23], carbon nanotubes [24,25], and non-metallic adsorbent [26]. However, the effect of the incorporation of Ag_3PO_4 nanoparticles on the surface of carbon nanofibers (CNFs) to enhance its photocatalytic performance has been rarely reported [27]. Due to its one-dimensional structure, high specific surface area, good electrical conductivity and high chemical stability, the use of CNFs in the form of scaffold/supporters as well as ideal electron pathways has attracted great interest in recent years [28,29]. Additionally, some reports have demonstrated that CNFs are capable of capturing electrons and serve as the conductivity network by transforming photogenerated electrons along their longitude conductive length [30,31]. Based on these unique properties of CNFs, fabrication of Ag_3PO_4/CNFs heterostructures could be an ideal strategy for hindering the recombination of photogenerated electrons and holes thereby improving the photocatalytic performance of Ag_3PO_4. The introduction of surface functional groups on CNFs that can act as anchoring sites for nanoparticles to obtain uniform dispersion is another important task. Therefore, various approaches have been proposed to introduce surface functional groups on CNFs, such as chemical oxidation [32], air oxidation [33], plasma treatment [34], and electrochemical oxidation [35]. Among these, the chemical oxidation of CNFs by acid treatment is being used widely to introduce oxygen containing functional groups under different conditions [36].

Herein, we report a facile approach to fabricate novel organic–inorganic heterostructures composed of functionalized CNFs and Ag_3PO_4 nanoparticles by the combination of carbonization of electrospun precursor PAN nanofibers and decoration of Ag_3PO_4 nanoparticles on the surface of functionalized CNFs following colloidal and precipitation synthesis approaches. To the best of our knowledge, there has been no report on the fabrication and visible light photocatalytic properties of Ag_3PO_4/CNFs heterostructures with sufficient electron–hole separation ability. The experimental results showed that the Ag_3PO_4/CNFs heterostructure obtained by the colloidal synthesis approach exhibited higher photocatalytic activity in the visible light region compared to the heterostructure obtained by the precipitation synthesis approach. Additionally, due to the high length-to-diameter ratio of CNFs, the as-fabricated Ag_3PO_4/CNFs heterostructures could be separated by the sedimentation process. Furthermore, we present a study on the effect of solution concentrations on the morphology of nanostructures (nanoagglomerates or nanoparticles) as well as their size.

2. Materials and Methods

2.1. Materials

Polyacrylonitrile (PAN, MW-150000), silver nitrate ($AgNO_3$), sodium phosphate dibasic dihydrate ($Na_2HPO_4.2H_2O$), and methylene blue (MB) were purchased from Sigma-Aldrich, St. Louis, MO, USA. *N,N*-dimethylformamide (DMF) and nitric acid (HNO_3, 65%) were purchased from Daejung Chemicals, Seoul, Korea. All the chemicals were used as received.

2.2. Fabrication of Carbon Nanofibers

A 10 wt% of PAN solution was prepared by dissolving required amount of PAN powder in DMF at room temperature under constant magnetic stirring for 12 h. The precursor solution was then loaded into a plastic syringe equipped with metallic needle connected to a high-voltage power supply (high voltage power supply, HV 30/ESN-HV30N, Nano NC, Seoul, Korea) capable of producing DC voltages from 0 to 30 kV and electrospun at 18 kV with a constant flow rate of 1 mL/h through a syringe pump. The nanofibers were collected on aluminum foil wrapped over a rotating drum collector at

15 cm apart from the needle tip. All experiments were conducted at room temperature and atmosphere pressure. The as-fabricated precursor PAN-nanofibers were vacuum dried at 70 °C for 12 h in order to remove the residual solvent. For carbonization, the vacuum-dried PAN-nanofibers were placed in a tube furnace and stabilized in air at 250 °C for 1 h with a heating rate of 1 °C/min, then carbonized in nitrogen up to 800 °C at a rate of 5 °C/min, and finally cooled to room temperature to the obtained electrospun carbon nanofibers (CNFs).

2.3. Fabrication of Ag_3PO_4/CNFs Heterostructures

At first, the oxidation of CNFs was carried out by treating with HNO_3 (65%) at 45 °C for 12 h followed by washing with distilled water until the pH became 7 then dried at 80 °C for 9 h. Afterwards, Ag_3PO_4/CNFs heterostructures were fabricated following two synthetic approaches, i.e., colloidal and precipitation synthesis approach simply by adjusting the concentrations of reactants ($Na_2HPO_4.2H_2O$ and $AgNO_3$) as described by Khan et al. [37]. In the colloidal synthesis approach, 0.1 g of acid-treated CNFs were added to 100 mL of (0.02 M) $AgNO_3$ solution and stirred for 30 min. Then, 100 mL of (0.02 M) Na_2HPO_4 solution was added to the CNFs/$AgNO_3$ mixture solution in a dropwise manner under constant stirring. The product was collected and washed with distilled water and ethanol followed by vacuum drying at 70 °C for 12 h. Likewise, in the precipitation synthetic approach, the concentration of reactants was adjusted to 0.1 M, while the post-treatment process of the mixture was the same as that of the colloidal synthesis approach. The schematic illustration for the fabrication of Ag_3PO_4/CNFs heterostructures is given in Figure 1. For the convenience of description, the Ag_3PO_4/CNFs heterostructures were herein called Ag_3PO_4/CNFs-1 and Ag_3PO_4/CNFs-2 synthesized from the colloidal synthesis approach and precipitation synthesis approach, respectively.

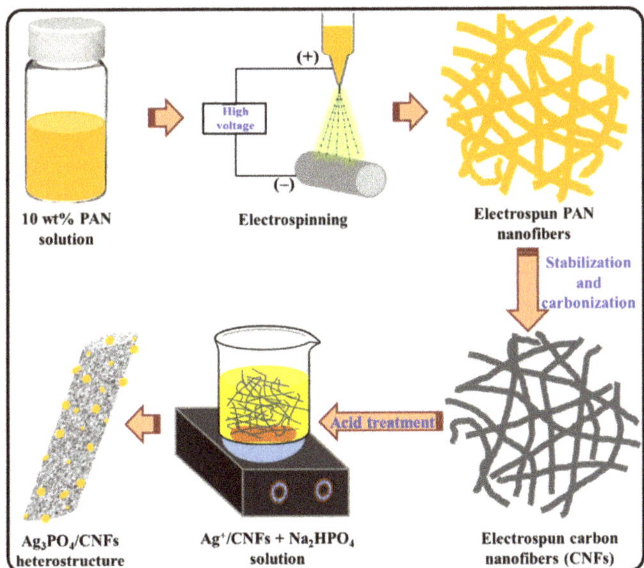

Figure 1. Schematic illustration for the fabrication of Ag_3PO_4/CNFs heterostructures.

2.4. Characterization

A field emission scanning electron microscope (FESEM, GeminiSEM 500, Carl Zeiss Microscopy GmbH, 73,447 Oberkochen, Germany) and transmission electron microscope (TEM/HR-TEM; JEM-2200FS, JEOL Ltd., Akishima, Tokyo, Japan) equipped with energy dispersive X-ray spectroscopy (EDS) were used to characterize the morphology and distribution of nanoparticles on the surface CNFs. EDS being attached to FESEM was used to analyze the composition of samples. X-ray diffraction (XRD) measurement was carried out to characterize the phase and crystallinity of as-prepared samples using an X-ray diffractometer (XRD, Empyrean, PANAlytical, Eindhoven 5651 GH, the Netherlands) with Cu Kα (λ = 1.540 Å) radiation over Bragg angles ranging from 10° to 80°. Fourier transform infrared (FTIR) spectroscopy was performed to characterize the bonding configurations of CNFs with nanoparticles by using Fourier-transform infrared (FT-IR, FT/IR-4200, Jasco, international Co., Ltd., Hachioji, Tokyo, 193-0835, Japan). UV-vis diffuse reflectance spectra (DRS) were measured using a UV-vis spectrophotometer (UV-2600 240 EN, SHIMADZU CORPORATION, Kyoto, Japan).

2.5. Photocatalytic Test

Photocatalytic performances of different photocatalysts were examined by observing the degradation of MB dye solution. For comparison, standard photocatalyst such as Evonic P25 TiO_2 was also employed in this study. All the experiments were performed at room temperature using a solar simulator (DYX300P, DYE TECH Co., Seoul, Korea) equipped with an internal xenon lamp and provided with a UV cut-off filter. Typically, suspension of each photocatalyst was prepared by dispersing 150 mg of photocatalyst in 50 mL of dye solution with initial concentration of 10 ppm and magnetically stirred in the dark for 30 min to attain adsorption/desorption equilibrium between dye and photocatalyst. The suspension was then irradiated with visible light (λ > 420 nm) produced from a 200 W xenon lamp under constant magnetic stirring. Aliquots were taken at 10 min time interval and the concentration of dye solution was measured using a UV-Vis spectrophotometer (HP 8453 UV-Vis spectroscopy system, Hudson, MA, USA).

3. Results and Discussion

The surface morphology of the precursor PAN nanofibers, CNFs and Ag_3PO_4/CNFs heterostructures were observed in FESEM images (Figure 2). As shown in Figure 2a, electrospun precursor PAN nanofibers exhibited uniform and continuous nanofibers with random orientation forming an interwoven 3-D network without the presence of any beads. The average diameter of these precursor nanofibers was found to be 320 nm. However, after carbonization of precursor PAN nanofibers, it could be clearly seen that CNFs still kept their nanofibrous morphology with a relatively smooth surface and random orientation without the presence of any secondary nanostructures. The average diameter of the CNFs was found to have decreased to 240 nm (Figure 2b). This decrease in fiber diameter can be attributed to the shrinkage in volume of PAN nanofibers due to weight loss during carbonization process. This is because during carbonization evaporation of residual solvent, removal of unwanted elements, densification of carbon, and evolution of the various gases such as H_2O, N_2, HCN, etc. from the precursor PAN nanofibers takes place, resulting in a decrease in fiber diameter [38,39].

After fabrication of heterostructures, both the samples remained with nanofibrous morphology. However, the surface of CNFs was no longer smooth due to growth of secondary Ag_3PO_4 nanoparticles (Figure 2c,d). High-magnification FESEM image of Ag_3PO_4/CNFs-1 heterostructure (inset of Figure 2c) depicted the uniform distribution spherical Ag_3PO_4 nanoparticles with an average size of 20 nm grown on the surface of CNFs without any evident aggregation offering a high level of exposure of the nanoparticle surface; however, the nanoparticles are not visible in low-magnification images (Figure 2c). On the other hand, in the Ag_3PO_4/CNFs-2 heterostructure, the Ag_3PO_4 nanoparticles with a bigger size were not evenly distributed across the surface of CNFs and found in agglomerated form, causing lesser exposure of their effective surface (Figure 2d). Figure 2e,f are the EDS spectra of CNFs and Ag_3PO_4/CNFs-1 heterostructure, respectively. It was indicated that C and O elements existed in pure CNFs, while C, O, Ag, and P elements existed in Ag_3PO_4/CNFs-1 heterostructure, confirming the successful fabrication of Ag_3PO_4/CNFs heterostructures.

Figure 3 represents the elemental mapping of the Ag_3PO_4/CNFs-1 heterostructure; as shown in the figure, carbon (Figure 3b) and oxygen (Figure 3e) have high density as they are incorporated in CNFs; however, silver (Figure 3c) and phosphorus (Figure 3d) have relatively low density. Additionally, from the elemental mapping, one can suggest the formation of Ag_3PO_4 nanoparticles on the CNFs surface.

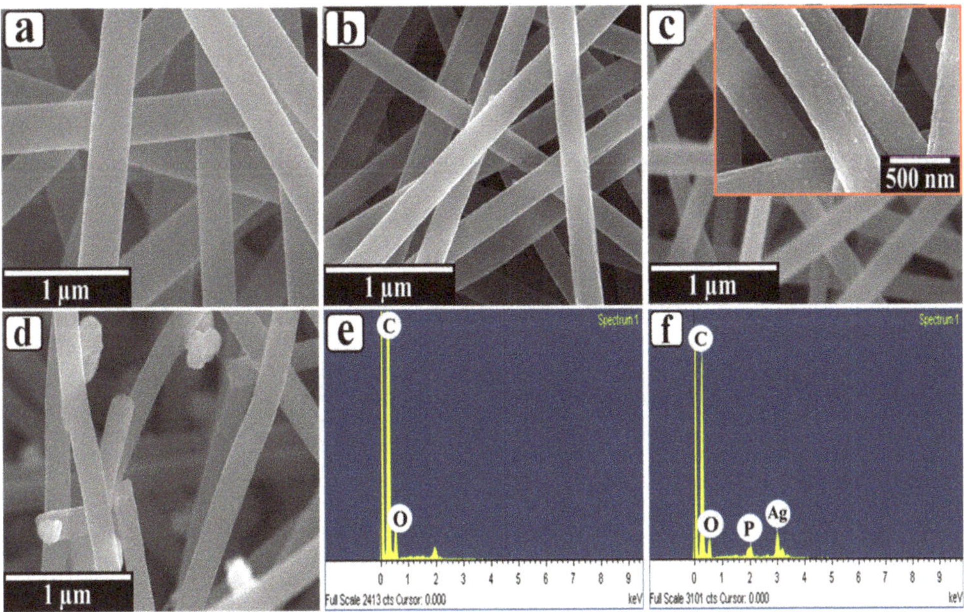

Figure 2. FESEM images; (**a**) precursor PAN nanofibers, (**b**) CNFs, (**c**) Ag_3PO_4/CNFs-1 heterostructure, and (**d**) Ag_3PO_4/CNFs-2 heterostructure. Panel (**e**,**f**) represent FESEM-EDS of CNFs, and Ag_3PO_4/CNFs-1 heterostructure, respectively. Inset; high-magnification FESEM image of Ag_3PO_4/CNFs-1 heterostructure.

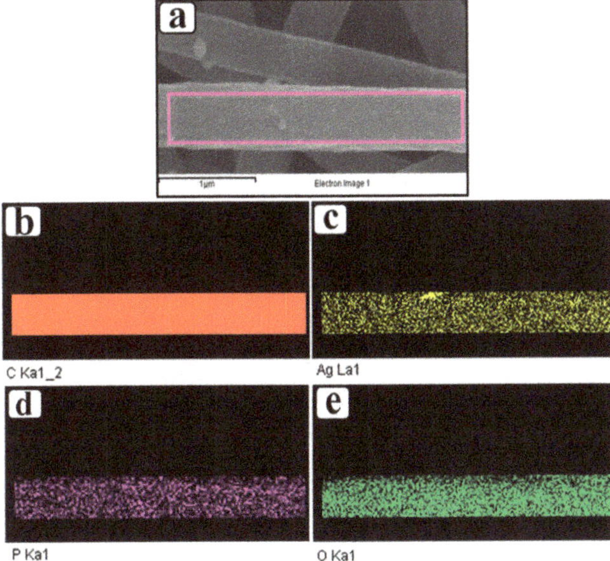

Figure 3. Elemental mapping analysis; (**a**) Ag_3PO_4/CNFs-1 heterostructure, (**b**) carbon, (**c**) silver, (**d**) phosphorus, and (**e**) oxygen elements.

XRD analysis of different samples (Ag_3PO_4 powder, CNFs, Ag_3PO_4/CNFs-1 heterostructure, and Ag_3PO_4/CNFs-2 heterostructure) was carried out to characterize their composition and crystal structure (Figure 4). The broad peak centered at around 25° was attributed to the (002) plane of the carbon structure in the CNFs, Ag_3PO_4/CNFs-2, and Ag_3PO_4/CNFs-2 (Figure 4b–d). Compared to the CNFs, the diffraction peaks at 2θ of 20.879°, 29.700°, 33.300°, 36.581°, 47.799°, 52.682°, 55.019°, 57.280°, 61.658°, and 71.898° in Ag_3PO_4 powder, Ag_3PO_4/CNFs-1, and Ag_3PO_4/CNFs-2 were attributed to the crystal planes of (110), (200), (210), (211), (310), (222), (320), (321), (400), (421) and (332) of Ag_3PO_4 (JCPDS card No: 74-0911), respectively (Figure 4a, c, d) [40]. Furthermore, no peaks for any other impurities were observed. Therefore, these results suggested that both the heterostructures were composed of CNFs and Ag_3PO_4 nanoparticles with high crystallinity and purity, which was also justified by FESEM-EDS.

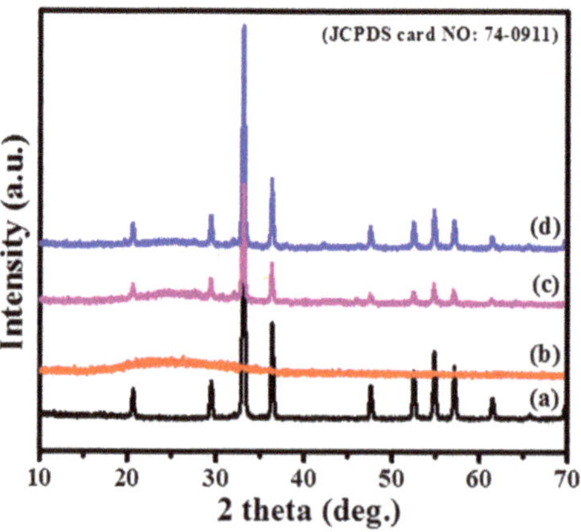

Figure 4. XRD patterns; (**a**) Ag_3PO_4 powder, (**b**) CNFs, (**c**) Ag_3PO_4/CNFs-1 heterostructure, and (**d**) Ag_3PO_4/CNFs-2 heterostructure.

The fiber morphology and degree of dispersion of nanoparticles on the surface of CNFs were observed by HR/TEM analysis (Figure 5). Figure 5a represents a typical TEM image of CNF with nanofibrous morphology possessing a relatively smooth surface as described in FESEM analysis. It is noteworthy that the morphology of CNFs obtained after acid treatment was also consistent and well-retained (inset, Figure 5a). As compared to the CNFs (Figure 5a), the heterostructure (Ag_3PO_4/CNFs-1) exhibited numerous Ag_3PO_4 nanoparticles uniformly distributed and successfully loaded on the surface of CNFs (Figure 5b). The average size of Ag_3PO_4 nanoparticles was found to be 20 nm. More interestingly, the sonication process conducted during the sample preparation for TEM analysis did not cause the Ag_3PO_4 nanoparticles decrease of the CNFs. This indicated that Ag_3PO_4 nanoparticles were successfully grown on the surface of CNFs with strong attachment. Moreover, highly magnified TEM images of the Ag_3PO_4/CNFs-1 heterostructure obtained from the marked area in Figure 5b is shown in Figure 5 c in order to display the attachment of colloidal nanoparticles with the CNFs surface. At the same time, the perfectly crystalline structure of Ag_3PO_4 nanoparticle was observed by HR-TEM image (Figure 5d) recorded from the marked area in Figure 5c. As shown in the figure, the inter planer spacing of 0.299 nm and 0.268 nm corresponding to (200) and (210) crystallographic planes of Ag_3PO_4 nanoparticle were clearly observed revealing its excellent crystallinity. Additionally, the heterojunction displayed that Ag_3PO_4 nanoparticles were attached on the surface of CNF having disordered stacking features indicating its low crystallinity.

In order to analyze the interaction between CNFs and inorganic phase, FTIR analysis was performed; the results are displayed in Figure 6. The IR absorption spectra of Ag_3PO_4 powder (Figure 6a) were compared with that of acid treated CNFs (Figure 6b) and Ag_3PO_4/CNFs-1 heterostructure (Figure 6c). As depicted in the figure, the broad absorption band centered at about 3200 cm^{-1} for Ag_3PO_4 powder or Ag_3PO_4/CNFs heterostructure and a strong absorption band at about 1660 cm^{-1} for Ag_3PO_4 powder were attributed to the stretching vibration of O-H and bending vibration of H-O-H of water molecules, respectively. Furthermore, for Ag_3PO_4 powder, two absorption bands centered at about 1055 cm^{-1} and 1400 cm^{-1} could be assigned to the molecular vibrations of PO_4^{3-} [41] and the stretch of the doubly bonded oxygen P=O [42]. For CNFs, a strong broad band was observed at about 3400–3500 cm^{-1}, which might be attributed to the bending vibration of absorbed water molecule and stretching vibration of -OH groups [43]. Similarly, two bands obtained at about 1200 cm^{-1} and 1580 cm^{-1} were due to the C-C stretching vibration [44,45] and carboxylic stretching COO^- vibration [46]. Hence, FTIR analysis clearly showed that the CNFs were -COOH functionalized; however, their peak intensities in the Ag_3PO_4/CNFs heterostructures were found to be decreased after the growth of Ag_3PO_4 nanoparticles on CNFs surface. Additionally, in the heterostructures, the bands corresponding to PO_4^{3-} and P=O were observed with a decrease in their peak intensities, suggesting that the carboxylic groups might play a significant role for the nucleation sites during ion exchange reaction.

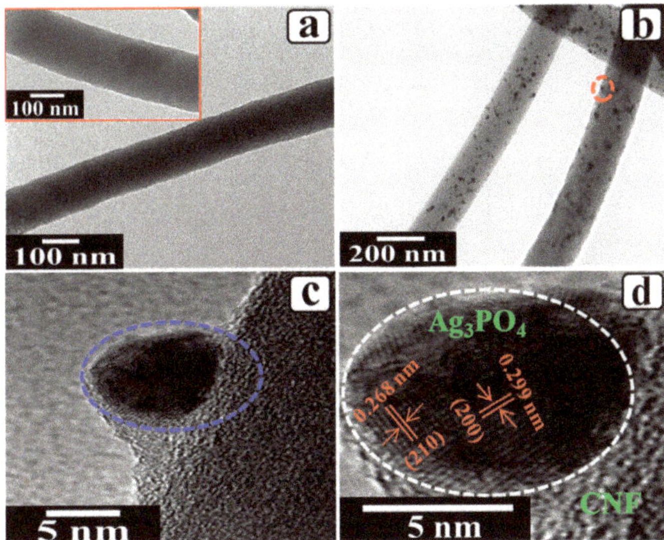

Figure 5. TEM images; (**a**) CNF, (**b**) Ag_3PO_4/CNFs-1heterostructure, (**c**) magnified TEM image of heterojunction region of red circled area for (**b**), and (**d**) HR-TEM image of blue circled area for (**c**). Inset, TEM image of acid-treated CNF.

Since the catalytic performances of any photocatalyst depends on its optical absorption properties, UV-vis diffusive reflectance spectra of Ag_3PO_4 powder and Ag_3PO_4/CNFs heterostructures were analyzed to investigate their optical properties; the results are presented in Figure 7. The pure Ag_3PO_4 powder could absorb light with a wavelength at around 520 nm (Figure 7a) and exhibit strong absorption in the visible range as reported before [47]. In the case of heterostructures, Ag_3PO_4/CNFs-1 (Figure 7b) also exhibited excellent absorption in the visible region with a wavelength at around 520 nm, while Ag_3PO_4/CNFs-2 (Figure 7c) could show relatively weak absorption in the visible region. Thus, these results indicated that Ag_3PO_4/CNFs-1 sample could absorb more photon and might prove itself as a more favorable photocatalyst with enhanced activity.

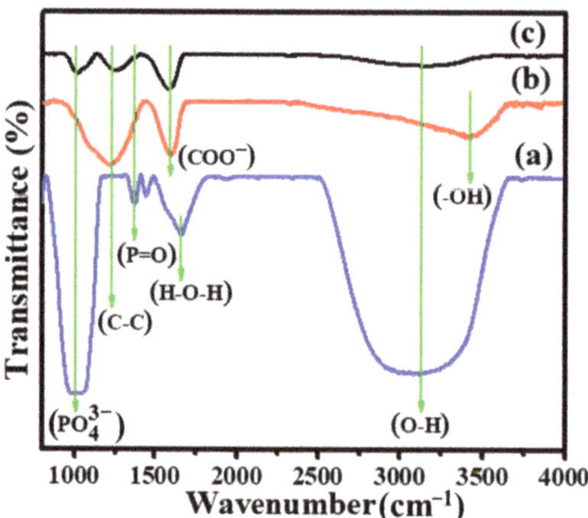

Figure 6. FTIR spectra; (**a**) Ag_3PO_4 powder, (**b**) acid-treated CNFs, and (**c**) Ag_3PO_4/CNFs-1 heterostructure.

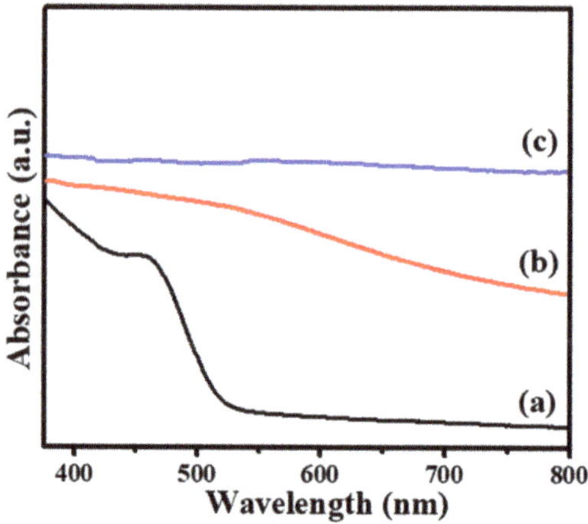

Figure 7. Uv-vis diffuse reflectance spectra; (**a**) Ag_3PO_4 powder, (**b**) Ag_3PO_4/CNFs-2 heterostructure, and (**c**) Ag_3PO_4/CNFs-1 heterostructure.

Results of photocatalytic performances exhibited by different photocatalysts towards the photodegradation of MB dye solution under simulated solar light irradiation are shown in Figure 8a. In order to evaluate the absorption property of electrospun CNFs, an experiment was carried out under similar conditions using bare CNFs, which showed negligible absorption of MB. The degradation is represented as the variation of (C/C_0) with irradiation time, where C_0 and C are the concentrations of dye solution at the initial time and at time t, respectively. When using P25 as the standard photocatalyst, only ~38% of MB dye solution was degraded in 50 min, indicating a low photocatalytic activity of P25. Similarly, the degradation results showed that less than 95% of MB dye solution was degraded within 50 min utilizing Ag_3PO_4 powder and Ag_3PO_4/CNFs-2 heterostructure. Contrastingly, more than 98% of dye solution was degraded within 40 min utilizing Ag_3PO_4/CNFs-1 heterostructure. Here, we believe that the higher photocatalytic efficiency of Ag_3PO_4/CNFs-1 heterostructure than that of Ag_3PO_4/CNFs-2 heterostructure is attributed to the strong uniformity and smaller Ag_3PO_4 particles dispersed on CNFs offering high level of exposure. Additionally, the lower photocatalytic activity of Ag_3PO_4 powder may refer to its low solubility and photocorrosion [48]. Furthermore, the enhanced photocatalytic activity of

Ag_3PO_4/CNFs heterostructures can be assigned to the synergistic effect between Ag_3PO_4 and CNFs due to their interfacial adhesion, which leads to the migration of excited electron from conduction band (CB) of Ag_3PO_4 to CNFs. As mentioned earlier [28,29], CNFs exhibit good conductivity and can act as an electron sink to prevent the recombination of photo-generated charge carriers [30]. Hence, this transfer phenomenon of electrons from CB of Ag_3PO_4 to CNFs could protect Ag_3PO_4 from photocorrosion, as shown in Formula (1) [25].

$$Ag^{+} + e^{-} \longrightarrow Ag \quad (1)$$

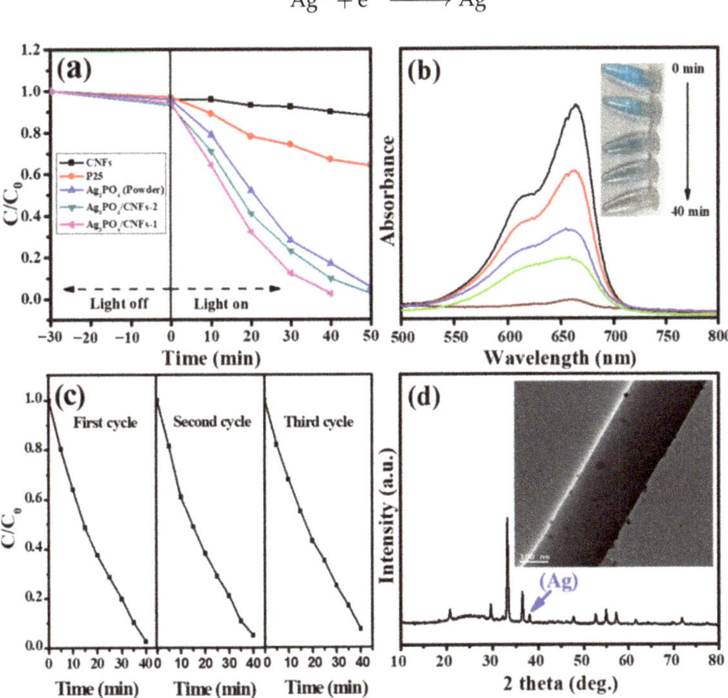

Figure 8. (**a**) Photocatalytic performances of CNFs, P25, Ag_3PO_4 powder, and Ag_3PO_4/CNFs heterostructures towards the photodegradation of MB solution; (**b**) absorbance variation of MB solution utilizing Ag_3PO_4/CNFs-1heterostructure; (**c**) cyclic photocatalytic performances of Ag_3PO_4/CNFs-1heterostructure, and (**d**) XRD pattern of used Ag_3PO_4/CNFs-1heterostructure. Insets; (**b**) digital photo of change in color of MB solution corresponding to degradation times and (**d**) TEM image of nanofiber of used Ag_3PO_4/CNFs-1 heterostructure.

The absorbance variations of MB solution utilizing Ag_3PO_4/CNFs-1 heterostructure under visible light irradiation at different times are shown in Figure 8b. The absorbance peak corresponding to MB at 665 nm is diminished gradually with the increase in irradiation time. At the same time, the color of MB dye solution also gradually diminished in presence of Ag_3PO_4/CNFs-1 heterostructure, as shown in inset, Figure 8b. Furthermore, the maximum absorption wave lengths of MB were not shifted, which indicated that the benzene/heterocyclic ring was decomposed rather than the simple decoloration process [49]. Moreover, the stability of the photocatalyst was evaluated by conducting a cycle test for the degradation of MB solution. For this purpose, used sample was separated by centrifugation and dried at room temperature then applied again for the degradation process under similar conditions. From the results (Figure 8c), it was found that the Ag_3PO_4/CNFs-1 heterostructure could work as a stable and efficient visible light photocatalyst up to the third cycle; however, there was a decrease in the activity during the recycling reaction, which might be attributed to the loss of photocatalyst during cycling experiments.

Moreover, to elucidate the structural integrity of photocatalyst, XRD analysis and TEM characterization of used Ag_3PO_4/CNFs-1 heterostructure were performed. Figure 8d depicts the XRD pattern of the Ag_3PO_4/CNFs-1 heterostructure after the third time cyclic test. As shown in the XRD pattern, an additional minor diffraction peak of metallic Ag appeared along with the corresponding diffraction peaks of Ag_3PO_4 and CNFs indicating the partial reduction of Ag_3PO_4 during the cyclic process. Additionally, the TEM image (inset, Figure 8d) of nanofiber of used Ag_3PO_4/CNFs-1 heterostructure showed its integrity even after three cycle tests without a significant loss of nanoparticles.

From the above discussions and results, it can be supposed that the CNFs play a key role in enhancing the photocatalytic activity and stability of Ag_3PO_4. However, the exact mechanism for the photocatalytic degradation of MB dye solution by Ag_3PO_4/CNFs heterostructures still needs further study; we have proposed a possible mechanism as shown in the schematic illustration (Figure 9) based on the above discussion and previous reports [13,50]. Under visible light irradiation, electrons (e^-) from the valance band (VB) of Ag_3PO_4 become excited due to the conduction band (CB) generating electron–hole (e^-–h^+) pairs (photogenerated charge carriers). During the photochemical reaction, the excited electrons in the CB of Ag_3PO_4 can ultimately migrate to the CNFs and react with adsorbed oxygen molecules to form $O_2^{\bullet -}$ radicals. At the same time, isolated h^+ can be utilized in the degradation of MB dye solution by direct oxidation, or it can react with water to produce $OH^\bullet$ radicals. Thus, the obtained reactive oxygen species (ROS) $O_2^{\bullet -}$ and $OH^\bullet$ radicals as well as holes (h^+) ultimately oxidize MB dye solution. The possible photochemical reactions taking place can be summarized by Formula (2)–(7) as follows:

$$Ag_3PO_4 \xrightarrow{h\upsilon} h^+(VB) + e^-(CB) \quad (2)$$

$$Ag_3PO_4(e^-_{CB}) + O_2 \longrightarrow Ag_3PO_4 + O_2^{\bullet -} \quad (3)$$

$$Ag_3PO_4(h^+_{VB}) + OH^- \longrightarrow Ag_3PO_4 + OH^\bullet \quad (4)$$

$$Ag_3PO_4(e^-_{CB}) + CNFs \longrightarrow Ag_3PO_4 + CNFs\ (e^-) \quad (5)$$

$$CNFs\ (e^-) + O_2 \longrightarrow CNFs + O_2^{\bullet -} \quad (6)$$

$$h^+/OH^\bullet/O_2^{\bullet -} + \text{dye solution} \longrightarrow \text{degradation} \quad (7)$$

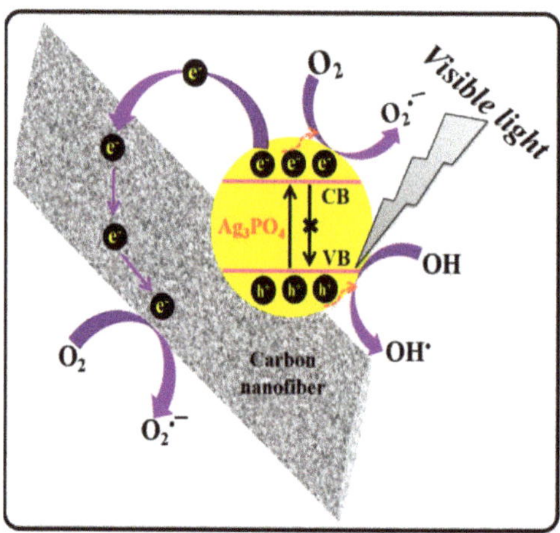

Figure 9. Schematic illustration for the photodegradation of MB solution using proposed Ag_3PO_4/CNFs heterostructure.

4. Conclusions

In summary, with the aid of electrospinning and the carbonization process, two types of Ag_3PO_4/CNFs heterostructures are fabricated. FESEM and TEM analyses revealed the uniform distribution of small crystalline Ag_3PO_4 nanoparticles on the surface of CNFs in the Ag_3PO_4/CNFs heterostructure obtained by the colloidal synthesis approach (Ag_3PO_4/CNFs-1). Contrastingly, bigger-sized Ag_3PO_4 nanoparticles were found to be distributed on the surface of CNFs with remarkable agglomeration in the Ag_3PO_4/CNFs heterostructure obtained by the precipitation synthesis approach (Ag_3PO_4/CNFs-2). Photocatalytic investigation of both the formulations towards the degradation of MB solution under visible light irradiation suggested that Ag_3PO_4/CNFs-1 is more advantageous over Ag_3PO_4/CNFs-2. Additionally, Ag_3PO_4/CNFs heterostructures could be easily separated from solution after use by sedimentation caused due to the high length-to-diameter ratio of CNFs. Finally, it is believed that the Ag_3PO_4/CNFs-1 heterostructure possessing enhanced photocatalytic performances will promote its practical application to remove organic pollutants from wastewater.

Author Contributions: For this research articles, the individual contribution of author was as follows: conceptualization and data analysis, G.P.; methodology, G.P.; writing—original draft preparation, G.P.; writing—reviewing and editing, G.P. and M.P.; resource and supervision, M.P.; funding acquisition, M.P. All authors have read and agreed to the published version of the manuscript.

Funding: This research was supported by the Traditional Culture Convergence Research Program through the National Research Foundation of Korea (NRF) funded by the Ministry of Science, ICT and Future Planning (2018M3C1B5052283). This work was also supported by the National Research Foundation of Korea (NRF) grant funded by the Korea government (MSIT) (No. NRF2019R1A2C1004467).

Institutional Review Board Statement: Not applicable.

Informed Consent Statement: Not applicable.

Conflicts of Interest: The authors declare no conflict of interest.

References

1. Hoffmann, M.R.; Martin, S.T.; Choi, W.; Bahnemannt, D.W. Environmental Applications of Semiconductor Photocatalysis. *Chem. Rev.* **1995**, *95*, 69–96. [CrossRef]
2. Fujishima, A.; Honda, K. Electrochemical photolysis of water at a semiconductor electrode. *Nature* **1972**, *238*, 37–38. [CrossRef]
3. Livraghi, S.; Paganini, M.C.; Giamello, E.; Selloni, A.; Valentin, C.D.; Pacchioni, G. Origin of photoactivity of nitrogen-doped titanium dioxide under visible light. *J. Am. Chem. Soc.* **2006**, *128*, 15666–15671. [CrossRef]
4. Yu, J.G.; Yu, X.X.; Huang, B.B.; Zhang, X.Y.; Dai, Y. Hydrothermal synthesis and visible-light photocatalytic activity of novel cage-like ferric oxide hollow spheres. *Cryst. Growth. Des.* **2009**, *9*, 1474–1480. [CrossRef]
5. Li, Q.; Guo, B.; Yu, J.; Ran, J.; Zhang, B.; Yan, H.; Gong, J.R. Highly efficient visible-light-driven photocatalytic hydrogen production of CdS-cluster-decorated graphene nanosheets. *J. A. Chem. Soc.* **2011**, *133*, 10878–10884. [CrossRef]
6. Panthi, G.; Yousef, A.; Barakat, N.A.M.; Khalil, K.A.; Akhter, S.; Choi, Y.; Kim, H.Y. Mn_2O_3/TiO_2 nanofibers with broad-spectrum antibiotics effect and photocatalytic activity for preliminary stage of water desalination. *Ceram. Int.* **2013**, *39*, 2239–2246. [CrossRef]
7. Zhao, D.; Chen, C.C.; Wang, Y.; Ma, W.H.; Zhao, J.C.; Rajh, T.; Zang, L. Enhanced photocatalytic degradation of dye pollutants under visible irradiation on Al (III)-modified TiO_2: Structure, interaction, and interfacial electron transfer. *Environ. Sci. Technol.* **2008**, *42*, 308–314. [CrossRef] [PubMed]
8. Panthi, G.; Kwon, O.H.; Kuk, Y.S.; Gyawali, K.R.; Park, Y.W.; Park, M. Ternary Composite of Co-Doped CdSe@electrospun Carbon Nanofibers: A Novel Reusable Visible LightDriven Photocatalyst with Enhanced Performance. *Catalysts* **2020**, *10*, 348. [CrossRef]
9. Wang, Z.; Ci, X.B.; Dai, H.J.; Yin, L.; Shi, H.X. One-step synthesis of highly active Ti-containing Cr-modified MCM-48 mesoporous material and the photocatalytic performance for decomposition of H_2S under visible light. *Appl. Surf. Sci.* **2012**, *258*, 8258–8263. [CrossRef]
10. Tang, J.T.; Gong, W.; Cai, T.J.; Xie, T.; Deng, C.; Peng, Z.H.; Deng, Q. Novel visible light responsive Ag@(Ag_2S/Ag_3PO_4) photocatalysts: Synergistic effect between Ag and Ag_2S for their enhanced photocatalytic activity. *RSC. Adv.* **2013**, *3*, 2543–2547. [CrossRef]
11. Yi, Z.; Ye, J.; Kikugawa, N.; Kako, T.; Ouyang, S.; Williams, H.S.; Yang, H.; Cao, J.; Luo, W.; Li, Z. An orthophosphate semiconductor with photooxidation properties under visible-light irradiation. *Nat. Mater.* **2010**, *9*, 559–564. [CrossRef]

12. Wan, J.; Liu, E.; Fan, J.; Hu, X.; Sun, L.; Tang, C.; Yin, Y.; Li, H.; Hu, Y. In-situ synthesis of plasmonic Ag/Ag_3PO_4 tetrahedron with exposed {111} facets for high visible-light photocatalytic activity and stability. *Ceram. Int.* **2015**, *41*, 6933–6940. [CrossRef]
13. Panthi, G.; Park, S.J.; Chae, S.H.; Kim, T.W.; Chung, H.J.; Hong, S.T.; Park, M.; Kim, H.Y. Immobilization of Ag3PO4 nanoparticles on electrospun PAN nanofibers via surface oximation: Bifunctional composite membrane with enhanced photocatalytic and antimicrobial activities. *J. Ind. Eng. Chem.* **2017**, *45*, 277–286. [CrossRef]
14. Panthi, G.; Barakat, N.A.M.; Park, M.; Kim, H.K.; Park, S.J. Fabrication of PdS/ZnS NPs doped PVAc hybrid electrospun nanofibers: Effective and reusable catalyst for dye photodegradation. *J. Ind. Eng. Chem.* **2015**, *21*, 298–302. [CrossRef]
15. Yang, X.F.; Cui, H.Y.; Li, Y.; Qin, J.L.; Zhang, R.X.; Tang, H. Fabrication of Ag_3PO_4-Graphene Composites with Highly Efficient and Stable Visible Light Photocatalytic Performance. *ACS Catal.* **2013**, *3*, 363–369. [CrossRef]
16. Tang, J.T.; Liu, Y.H.; Li, H.Z.; Tan, Z.; Li, D.T. A novel Ag_3AsO_4 visible-light-responsive photocatalyst: Facile synthesis and exceptional photocatalytic performance. *Chem. Commun.* **2013**, *49*, 5498–5500. [CrossRef] [PubMed]
17. Panthi, G.; Gyawali, K.R.; Park, M. Towards the Enhancement in Photocatalytic Performance of Ag3PO4 Nanoparticles through Sulfate Doping and Anchoring on Electrospun Nanofibers. *Nanomaterials* **2020**, *10*, 929. [CrossRef] [PubMed]
18. Xu, J.W.; Gao, Z.D.; Han, K.; Liu, Y.; Song, Y.Y. Synthesis of Magnetically Separable $Ag_3PO_4/TiO_2/Fe_3O_4$ Heterostructure with Enhanced Photocatalytic Performance under Visible Light for Photoinactivation of Bacteria. *ACS Appl. Mater. Interfaces* **2014**, *6*, 15122–15131. [CrossRef]
19. Yao, W.; Zhang, B.; Huang, C.; Ma, C.; Song, X.; Xu, Q. Synthesis and characterization of high efficiency and stable Ag_3PO_4/TiO_2 visible light photocatalyst for the degradation of methylene blue and rhodamine B solutions. *J. Mater. Chem.* **2012**, *22*, 4050–4055. [CrossRef]
20. Dong, C.; Wu, K.L.; Li, M.R.; Liu, L.; Wei, X.W. Synthesis of Ag_3PO_4-ZnO nanorod composites with high visible-light photocatalytic activity. *Catal. Commun.* **2014**, *46*, 32–35. [CrossRef]
21. Yang, X.; Qin, J.; Jiang, Y.; Li, R.; Li, Y.; Tang, H. Bifunctional TiO_2/Ag_3PO_4/graphene composites with superior visible light photocatalytic performance and synergistic inactivation of bacteria. *RSC Adv.* **2014**, *4*, 18627–18636. [CrossRef]
22. Liu, L.; Liu, J.; Sun, D.D. Graphene oxide enwrapped Ag_3PO_4 composite: Towards a highly efficient and stable visible-light-induced photocatalyst for water purification. *Catal. Sci. Technol.* **2012**, *2*, 2525–2532. [CrossRef]
23. Liang, Q.; Shi, Y.; Ma, W.; Li, Z.; Yang, X. Enhanced photocatalytic activity and structural stability by hybridizing Ag_3PO_4 nanospheres with graphene oxide sheets. *Phys. Chem. Chem. Phys.* **2012**, *14*, 15657–15665. [CrossRef]
24. Wang, Z.; Yin, L.; Zhang, M.; Zhou, G.; Fei, H.; Shi, H.; Dai, H. Synthesis and characterization of Ag_3PO_4/multiwalled carbon nanotube composite photocatalyst with enhanced photocatalytic activity and stability under visible light. *J Mater. Sci.* **2014**, *49*, 1585–1593. [CrossRef]
25. Xu, H.; Wang, C.; Song, Y.; Zhu, J.; Xu, Y.; Yan, J.; Song, Y.X.; Li, H. CNT/Ag_3PO_4 composites with highly enhanced visible light photocatalytic activity and stability. *Chem. Eng. J.* **2014**, *241*, 35–42. [CrossRef]
26. Ma, J.; Zou, J.; Li, L.; Yao, C.; Kong, Y.; Cui, B.; Zhu, R.; Li, D. Nanocomposite of attapulgite-Ag_3PO_4 for Orange II photodegradation. *Appl. Catal. B Environ.* **2014**, *144*, 36–40. [CrossRef]
27. Chen, P.; Zhang, L.; Wu, Q.; Yao, W. Novel Synthesis of Ag3PO4/CNFs/Silica-fiber hybrid composite as an efficient photocatalyst. *J. Taiwan Inst. Chem. Eng.* **2016**, *63*, 506–511. [CrossRef]
28. Unalan, H.; Wei, D.; Suzuki, K.; Dalal, S.; Hiralal, P.; Matsumoto, H.; Imaizumi, S.; Minagawa, M.; Tanioka, A.; Flewitt, A.; et al. Photoelectrochemical cell using dye sensitized zinc oxide nanowires grown on carbon fibers. *Appl. Phys. Lett.* **2008**, *93*, 133116–133118. [CrossRef]
29. Mu, J.; Shao, C.; Guo, Z.; Zhang, Z.; Zhang, M.; Zhang, P.; Chen, B.; Liu, Y. High Photocatalytic Activity of ZnO-Carbon Nanofiber Heteroarchitectures. *ACS Appl. Mater Interfaces* **2011**, *3*, 590–596. [CrossRef]
30. Mu, J.; Shao, C.; Guo, Z.; Zhang, M.; Zhang, Z.; Zhang, P.; Chen, B.; Liu, Y.N. In_2O_3 nanocubes/carbon nanofibers heterostructures with high visible light photocatalytic activity. *J. Mater. Chem.* **2012**, *22*, 1786–1793. [CrossRef]
31. Lee, J.S.; Kwon, O.S.; Park, S.J.; Park, E.Y.; You, S.A.; Yoon, H.; Jang, J. Fabrication of Ultrafine Metal-Oxide-Decorated Carbon Nanofibers for DMMP Sensor Application. *ACS Nano* **2011**, *5*, 7992–8001. [CrossRef]
32. Lakshminarayanan, P.V.; Toghiani, H.; Pittman, C.U., Jr. Nitric acid oxidation of vapor grown carbon nanofibers. *Carbon* **2004**, *42*, 2433–2442. [CrossRef]
33. Seo, M.K.; Park, S.J. Influence of air-oxidation on electric double layer capacitances of multi-walled carbon nanotube electrodes. *Curr. Appl. Phys.* **2010**, *10*, 241–244. [CrossRef]
34. Okajima, K.; Ohta, K.; Sudoh, M. Capacitance behavior of activated carbon fibers with oxygen-plasma treatment. *Electrochim. Acta* **2005**, *50*, 2227–2231. [CrossRef]
35. Oh, H.S.; Kim, K.; Ko, Y.J.; Kim, H. Effect of chemical oxidation of CNFs on the electrochemical carbon corrosion in polymer electrolyte membrane fuel cells. *Int. J. Hydrog. Energy* **2010**, *35*, 701–708. [CrossRef]
36. Yu, H.; Jin, Y.; Peng, F. Kinetically Controlled Side-Wall Functionalization of Carbon Nanotubes by Nitric Acid Oxidation. *J. Phys. Chem. C* **2008**, *112*, 6758–6763. [CrossRef]
37. Khan, A.; Qamar, M.; Muneer, M. Synthesis of highly active visible-light-driven colloidal silver orthophosphate. *Chem. Phys. Lett.* **2012**, *519–520*, 54–58. [CrossRef]
38. Lafdi, K.; Wright, M.A. Carbon Fibers. In *Handbook of Composites*; Peters, S.T., Ed.; Chapman & Hall: London, UK, 1998.

39. Zhou, Z.; Lai, C.; Zhang, L.; Qian, Y.; Hou, H.; Reneker, D.H. Development of carbon nanofibers from aligned electrospun polyacrylonitrile nanofiber bundles and characterization of their microstructural, electrical, and mechanical properties. *Polymer* **2009**, *50*, 2999–3006. [CrossRef]
40. Liu, B.; Li, Z.; Xu, S.; Han, D.; Lu, D. Enhanced visible-light photocatalytic activities of Ag_3PO_4/MWCNT nanocomposites fabricated by facile in situ precipitation method. *J. Alloys Compd.* **2014**, *596*, 19–24. [CrossRef]
41. Miller, L.M.; Vairavamurthy, V.; Chance, M.R.; Mendelsohn, R.; Paschalis, E.P.; Betts, F.; Boskey, A.L. In situ analysis of mineral content and crystallinity in bone using infrared micro spectroscopy of the v4 PO_4^{3-} vibration. *Biochim. Biophys. Acta Gen. Subj.* **2001**, *1527*, 11–19. [CrossRef]
42. Moustafa, Y.M.; El-Egili, K. Infrared studies on the structure of sodium phosphate glasses. *J. Non-Cryst. Solids* **1998**, *240*, 144–153. [CrossRef]
43. Tang, T.; Shi, Z.; Yin, J. Poly(benzimidazole) functionalized multi-walled carbon nanotubes/100% acidified poly(hydroxyaminoether) composites: Synthesis, characterization and properties. *Mater. Chem. Phys.* **2011**, *129*, 356–364. [CrossRef]
44. Qin, Y.H.; Yang, H.H.; Zhang, X.S.; Li, P.; Zhou, X.G.; Niu, L.; Yuan, W.K. Electrophoretic deposition of network like carbon nanofibers as a palladium catalyst support for ethanol ixidation in alkaline media. *Carbon* **2010**, *48*, 48–3323. [CrossRef]
45. Mawhinney, D.; Naumenko, V.; Kuznetsova, A.; Yates, J.; Liu, J.R. Smalley, Infrared Spectral Evidence for the Etching of Carbon Nanotubes: Ozone Oxidation at 298 K. *J. Am. Chem. Soc.* **2000**, *122*, 2383–2384. [CrossRef]
46. Zhu, C.; Guo, S.; Wang, P.; Xing, L.; Fang, Y.; Zhai, Y.; Dong, S. One-pot, water-phase approach to high-quality graphene/TiO_2 composite nanosheets. *Chem. Commun.* **2010**, *46*, 7148–7150. [CrossRef] [PubMed]
47. Panthi, G.; Ranjit, R.; Kim, H.Y.; Mulmi, D.D. Size dependent optical and antibacterial properties of Ag_3PO_4 synthesized by facile precipitation and colloidal approach in aqueous solution. *Optik* **2018**, *156*, 60–68. [CrossRef]
48. Lv, Y.; Huang, K.; Zhang, W.; Yang, B.; Chi, F.; Ran, S.; Liu, X. One step synthesis of Ag/Ag_3PO_4/$BiPO_4$ double-heterostructured nanocomposites with enhanced visible-light photocatalytic activity and stability. *Ceram. Int.* **2015**, *40*, 8087–8092. [CrossRef]
49. Panthi, G.; Park, M.; Park, S.J.; Kim, H.K. PAN electrospun nanofibers reinforced with Ag_2CO_3 nanoparticles: Highly efficient visible light photocatalyst for photodegradation of Organic Contaminants in waste water. *Macromol. Res.* **2015**, *23*, 149–155. [CrossRef]
50. Dong, P.Y.; Wang, Y.H.; Cao, B.C.; Xin, S.Y.; Guo, L.N.; Zhang, J.; Li, F.H. Ag_3PO_4/reduced graphite oxide sheets nanocomposites with highly enhanced visible light photocatalytic activity and stability. *Appl. Catal. B Environ.* **2013**, *132–133*, 45–53. [CrossRef]

MDPI

Article

Structural and Optical Properties of $Bi_{12}NiO_{19}$ Sillenite Crystals: Application for the Removal of Basic Blue 41 from Wastewater

Billal Brahimi [1], Hamza Kenfoud [2], Yasmine Benrighi [2] and Oussama Baaloudj [2,*]

1 Laboratory of Transfer Phenomena, Faculty of Mechanical Engineering and Process Engineering, University of Science and Technology Houari Boumediene (USTHB), BP 32, 16111 Algiers, Algeria; billalbrahimigpi@gmail.com

2 Laboratory of Reaction Engineering, Faculty of Mechanical Engineering and Process Engineering, University of Science and Technology Houari Boumediene (USTHB), BP 32, 16111 Algiers, Algeria; hamza.kenfoud.93@gmail.com (H.K.); benrighiyasmine@gmail.com (Y.B.)

* Correspondence: obaaloudj@gmail.com; Tel.: +213-661-899-266

Abstract: This article covers the structural and optical property analysis of the sillenite $Bi_{12}NiO_{19}$ (BNO) in order to characterize a new catalyst that could be used for environmental applications. BNO crystals were produced by the combustion method using Polyvinylpyrrolidone as a combustion reagent. Different approaches were used to characterize the resulting catalyst. Starting with X-ray diffraction (XRD), the structure was refined from XRD data using the Rietveld method and then the structural form of this sillenite was illustrated for the first time. This catalyst has a space group of I23 with a lattice parameter of a = 10.24 Å. In addition, the special surface area (SSA) of BNO was determined by the Brunauer-Emmett-Teller (BET) method. It was found in the range between 14.56 and 20.56 $cm^2 \cdot g^{-1}$. Then, the morphology of the nanoparticles was visualized by Scanning Electron Microscope (SEM). For the optical properties of BNO, UV-VIS diffusion reflectance spectroscopy (DRS) was used, and a 2.1 eV optical bandgap was discovered. This sillenite's narrow bandgap makes it an effective catalyst for environmental applications. The photocatalytic performance of the synthesized $Bi_{12}NiO_{19}$ was examined for the degradation of Basic blue 41. The degradation efficiency of BB41 achieved 98% within just 180 min at pH ~9 and with a catalyst dose of 1 g/L under visible irradiation. The relevant reaction mechanism and pathways were also proposed in this work.

Keywords: sillenite $Bi_{12}NiO_{19}$; Rietveld method; optical properties; photodegradation; BB41 dye

Citation: Brahimi, B.; Kenfoud, H.; Benrighi, Y.; Baaloudj, O. Structural and Optical Properties of $Bi_{12}NiO_{19}$ Sillenite Crystals: Application for the Removal of Basic Blue 41 from Wastewater. *Photochem* **2021**, *1*, 319–329. https://doi.org/10.3390/photochem1030020

Academic Editor: Vincenzo Vaiano

Received: 17 August 2021
Accepted: 16 September 2021
Published: 27 September 2021

Publisher's Note: MDPI stays neutral with regard to jurisdictional claims in published maps and institutional affiliations.

1. Introduction

Traditional methods of water treatment cannot effectively eliminate pollutants from wastewaters and could cause great harm to the environment [1–4]. On the other hand, it has been shown that photocatalysis is a promising approach for the degradation of non-biodegradable compounds in water [5,6]. It is based on a photocatalyst that can be activated by light such as sunlight and using this energy to remove various types of pollutants [7]. Among these pollutants, basic blue 41 (BB41) has been identified as one of the most problematic dyes. It is present in industrial effluents and commonly used in acrylic, nylon, silk, cotton, and wool dyeing [8–10]. This dye is potentially fatal to living organisms. It is effective as a strainer for identifying avian leukocytes, blood, and bone marrow cells. It can also induce short periods of fast or difficult breathing when inhaled, and it can also cause nausea, vomiting, excessive perspiration, mental disorientation, and methemoglobinemia when consumed through the mouth [11]. Therefore, its detection and elimination are challenging goals. Based on our previous studies, photocatalysis-using catalysts have successfully removed both organic and inorganic pollutants in wastewater [12,13]. A novel catalyst with high photocatalytic activity and a narrow bandgap must be developed and tested in order to achieve this goal [14].

The increase in research in recent years for new material categories allowed new studies on structural and optical properties [15,16]. Among these materials—despite being relatively new—sillenites have attracted the attention of many researchers due to their unusual crystal formations, their peculiar electronics, their interesting photochromic, photorefractive, electro-optic, piezoelectric, dielectric properties, and promising optical activity [17–19]. They are used in various industrial applications, such as image amplification, phase conjugation, real-time and multiwavelength holography, optical memories for data storage, optical communications, signal processing and a variety of photocatalytic applications [20,21]. Sillenite crystals with a general formula of $Bi_{12}MO_{20x}$ (BMO) have a body-centered cubic crystal structure crystallized in the I23 space group [22]. Their overall structure is described by the atom of bismuth surrounded by seven oxygen atoms that share corners with other similar Bi polyhedrons and with MO_4 which represents a tetravalent ion or a combination of ions that exist both at the cube center and on the corners in the BMO structure [23,24]. There are a significant number of new sillenites that are used as photo-catalysts in previous research, such as $Bi_{12}TiO_{20}$ [25], $Bi_{12}GeO_{20}$ [21], $Bi_{12}PbO_{19}$[22], $Bi_{12}CoO_{20}$ [26], $Bi_{12}SiO_{20}$ [21], $Bi_{12}MnO_{20}$ [17], $Bi_{12}FeO_{20}$ [27] and $Bi_{12}ZnO_{20}$ [28]. Among the sillenite crystals, bismuth nickelate $Bi_{12}NiO_{19}$ (BNO) has not been used as a photocatalyst yet, although it has recently gained considerable attention because of its tiny bandgap, its high photoconductivity, ease of separation of photogenerated electron-hole pairs and ease of recycling [29–31]. BNO is a single-phase multiferroic, which co-exists with ferroelasticity and has a magnetic character. It is also a lead-free and environmentally friendly material, which makes it very promising for photocatalytic applications. Due to its novelty in the photocatalyst field and its interesting proprieties, we selected BNO as a photocatalyst for this work.

We report in this study the synthesis of $Bi_{12}NiO_{19}$ sillenite by the sol-gel method using polyvinylpyrrolidone (PVP) as a combustion reagent. Firstly, the phase of the crystals was identified by X-ray diffraction (XRD); then, the structure and lattice constants of the phase were refined using the Rietveld method. The special surface area (SSA) of BNO was determined by the Brunauer–Emmett–Teller (BET) method. Then, the morphology of the nanoparticles was investigated by Scanning Electron Microscope (SEM). After that, the BNO's optical properties were investigated using UV-VIS diffusion reflectance spectroscopy (DRS), and the obtained bandgap was discussed. The photocatalytic activity of the sillenite $Bi_{12}NiO_{19}$ was tested for photodegradation of basic blue 41 dye.

2. Materials and Methods

2.1. Chemicals

Chemicals used in the present study were: nickel nitrate hexahydrate $[Ni(NO_3)_2{\cdot}6H_2O]$ (98% Biochem), bismuth nitrate pentahydrate $[Bi(NO_3)_3{\cdot}5H_2O]$ (98.5% Chem-Lab), Polyvinylpyrrolidone PVP K30 (Pharmalliance pharmaceutical company, Ouled Fayet, Algeria), ethanol (Biochem), nitric acid, HCl and NaOH (Sigma Aldrich; St. Louis, MO, USA). The basic blue 41 (>98% purity) was provided by Aldrich. Distilled water was used as a solvent. Without further purification, all chemicals were used as obtained.

2.2. Synthesis of the Sillenite $Bi_{12}NiO_{19}$

The $Bi_{12}NiO_{19}$ material (sillenite type) was synthesized by mixing nitrates of bismuth pentahydrate $[Bi(NO_3)_3{\cdot}5H_2O]$, and nickel nitrate hexahydrate $[Ni(NO_3)_2{\cdot}6H_2O]$ was dissolved in water using stoichiometric amounts (12:1 ratio). In order to improve the solubility of the solutions, nitric acid was added. The reaction was carried out according to the following equation:

$$12\,[Bi(NO_3)_3{\cdot}5H_2O] + [Ni(NO_3)_2{\cdot}6H_2O] \rightarrow Bi_{12}NiO_{19} + 38\,NO_2 + 66\,H_2O + 19/2\,O_2 \quad (1)$$

After total solubilization, PVP K30 as a complexing agent was added to the reaction solution with a concentration of 15% *w*/*w* to obtain its complexing role [32]. The obtained solution was dehydrated by evaporation on a hot plate until it turned to a gel, then burned to form Xerogel, an amorphous powder, and after that denitrified at 600 °C for 3 h. Before calcination, the obtained powder was homogenized by grinding in an agate mortar; it was calcined then in an air oven for 6 h at 800 °C. The calcination step was done to increase the crystallinity and to remove all carbonated waste left after the combustion reaction. All the synthesis process was summarized and illustrated in Figure 1. Then, the sample was subjected to phase identification, structural characterization and optical study.

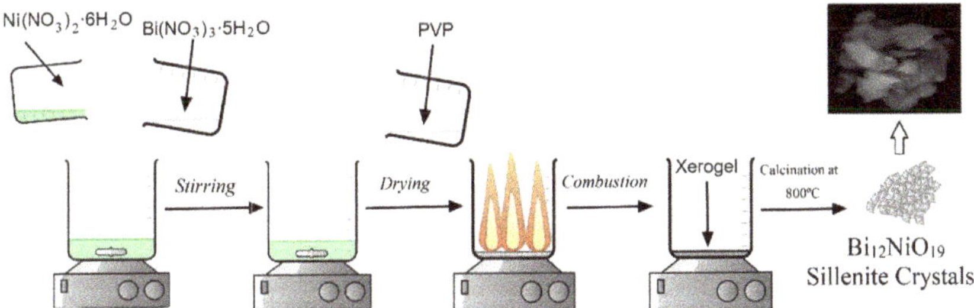

Figure 1. Process for preparing $Bi_{12}NiO_{19}$.

2.3. Characterization

X-ray diffraction (XRD) was conducted using a Phillips PW 1730. With the MAUD software (version 2.9.3), the Rietveld technique was used to measure structural properties, refinement and crystallinity. VESTA (version 3.4.0) was used to create the structure illustration. With an electron scanning microscope FEI Quanta 650, crystal pictures were captured. The sample's UV-visible diffuse reflection (DRS) was measured using a Cary 5000 UV-Vis spectrophotometer.

2.4. Photocatalysis Test

The photocatalytic degradation of Basic Blue 41 was carried out in a double-walled reactor with a magnetic stirring and cooling system. A 1 g/L dose of catalyst was suspended in a solution of BB41 aqueous solution with a concentration of 15 mg/L and pH ~9, which was found to be optimal conditions. The pH was modified by adding small amounts of HCl and NaOH. The adsorption experiment was performed in the absence of irradiation for 120 min before the photocatalysis test to reach the adsorption equilibrium. After the adsorption equilibrium, the reactor was exposed to visible irradiation using a tungsten lamp supplied by Osram (200 W). The temperature of the solution was maintained at a nearly constant 25 °C during the photocatalytic experiments using a thermostatic bath and a double-walled reactor as a cooling system. The concentration of BB41 was followed by measuring the absorbance in the wavelength (610 nm) with a UV–Vis spectrophotometer (OPTIZEN, UV-3220UV). The degradation efficiency was calculated from the relation:

$$\text{Degradation efficiency } \% = \frac{abs_{ad} - abs}{abs_{ad}} \times 100 \quad (2)$$

where abs_{ad} and abs are the initial absorbance of BB41 and absorbance after time t, respectively.

3. Results

3.1. Characterization of the Sillenite $Bi_{12}NiO_{19}$

3.1.1. Phase Identification and Structural Investigation

To explore the formation of sillenite crystals $Bi_{12}NiO_{19}$, X-ray diffraction (XRD) was used (Figure 2). All diffraction peaks are attributed to the sillenite phase of $Bi_{12}NiO_{19}$ (JCPDS card, PDF No. 43-0448) [33]. This indicates good crystallization at 800 °C. The XRD data and I23 cubic structure were used to conduct Rietveld refinement, experimental findings and theoretical data determined by Maud are shown in Figure 2 as points and red lines, respectively. According to the results, the findings estimated using the Rietveld technique matched well with the experimental X-ray diffraction pattern. The Rietveld refinement results gave identical results compared to other known methods of structural properties. [34,35]. Figure 1 also showed the purity and good crystallinity of our sillenite phase through the great congruence between the results of the experiment and the theoretical data. It was determined that the refined values represented a cubic form with a space group of I23 and a lattice parameter of a = 10.24 Å. The Rietveld refined parameters such as reliability factors R_p, R_{exp}, R_{wp} and Sig with the cell parameter (a) and atomic position (x,y,z) are presented in Table 1.

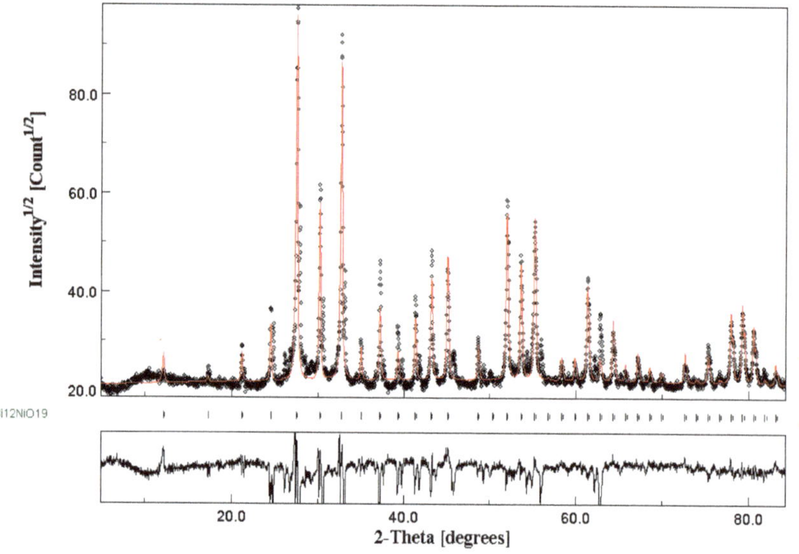

Figure 2. XRD diffractogram of $Bi_{12}NiO_{19}$ with Reitveld refinement.

Table 1. Structural and lattice parameters.

Phase		**$Bi_{12}NiO_{19}$**			
Groupe Space		I 2 3			
a (Å)		10.244759			
Atoms	Atom	x	y	z	Biso
	Ni1	0.000000	0.000000	0.000000	0.077
	Ni2	0.8216535	0.68130636	0.9814812	0.077
	Bi1	0.000000	0.000000	0.000000	0.923
	Bi2	0.9814812	0.68130636	0.8216535	0.923
	O1	0.83216524	0.6798176	0.459461	1
	O2	0.729963	0.729963	0.729963	1
	O3	0.123646714	0.123646714	0.123646714	0.75
V (Å^3)			1083.4283		
D (nm)			59.46		
R Factors	Rb		9.5436		
	Rexp		4.0007		
	Rwp		14.3421		
	Sig		3.58		

The structural representation of the sillenite BNO was illustrated in Figure 3 by Vesta using the structural parameters in Table 1. The atoms′ colors for Ni, Bi and O are green, yellow and red, respectively. As can be seen, the phase is a cubic structure (space group I23). The atoms in the crystal structure show shared occupancy between bismuth and nickel while bismuth atoms are dominant at around 92% due to the ratio of atoms being 12:1 for bismuth and nickel.

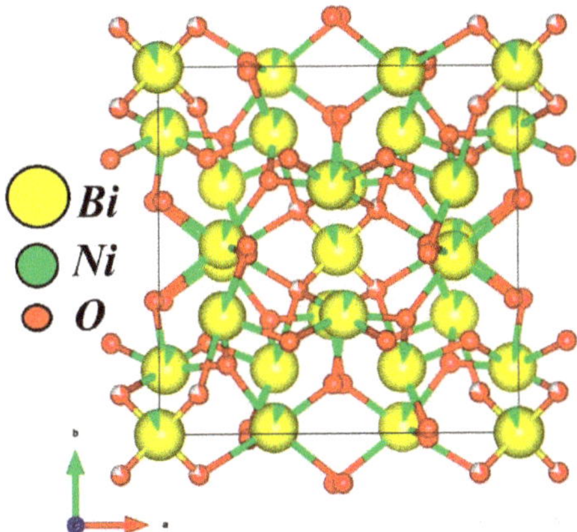

Figure 3. The structural representation of the crystal $Bi_{12}NiO_{19}$.

The crystallite size, X-ray density, and special surface area were calculated from the following equations [36]:

$$D = \frac{K\lambda}{\beta \cos(\theta)} \tag{3}$$

$$\rho = \frac{ZM}{N_A V} \tag{4}$$

$$S = \frac{6 \times 10^3}{D \times \rho} \tag{5}$$

where D is the regular phase crystallite, K is the Scherener constant, β is the full width at the highest half of the phase, θ is the angle Braggs, N_A is Avogadro's number, M is the Molecular mass of BNO (2870.45 g·mol^{-1}), ρ is the density of X-ray, Z is the count of model units in a cell (Z = 2 for sillenites [26]) and V is the unit cell volume $(1083.428335\ A\ 3)$ and S is the specific surface area.

The crystallite size was calculated from the main peaks of the XRD diffractogram, and it was found in the range between 33.20 and 33.20 nm. The density of X-rays was found to be 8.79 g·cm^{-3}. The special surface area (SSA) of the particular BNO was estimated by the Brunauer–Emmett–Teller (BET) method in Equation (4), and it was found in the range between 14.56 and 20.56 cm^2·g^{-1}.

3.1.2. Morphology Investigation

In order to analyze the morphology of the BNO crystals, a Scanning Electron Microscopy (SEM) was used. Figure 4 shows typical SEM images of the BNO crystals. A small agglomeration can be seen due to the ultrafine nature of the sample [37]. This leads to a non-uniform distribution of crystals of different shapes and a noticeable porosity in the sample [28]. The viscosity of the suspension in the synthesis route plays an important role in the development of the porous structure [1,38]. The viscosity of suspension during synthesis causes the porous structure to get compacted. The sol-gel method is well-known for its great viscosity because it forms a gel during the synthesis. Due to this, we have significant porosity in our sample.

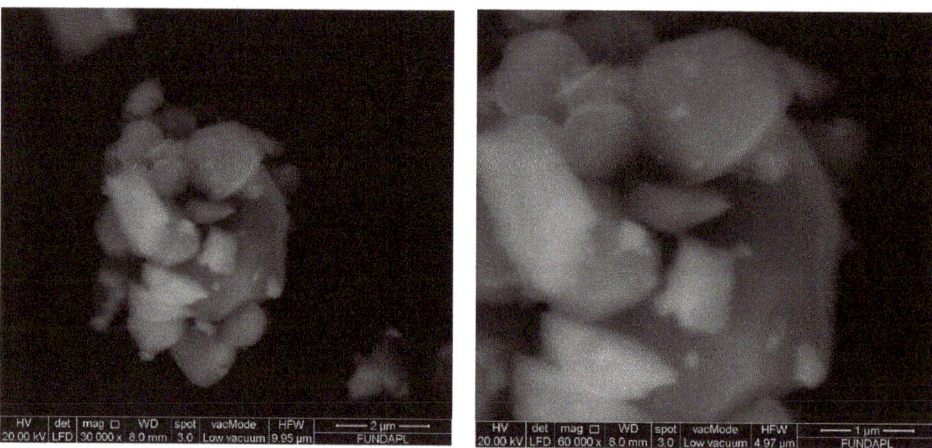

Figure 4. SEM images of the crystals $Bi_{12}NiO_{19}$.

3.1.3. Optical Study

To determine the catalyst's performance, knowledge about the gap is highly necessary. The optical properties of BNO crystals were investigated using UV-vis diffuse reflectance spectroscopy (DRS). The bandgap energy (E_g) is determined by plotting the Tauc plot (Figure 5), which is the dependence of the absorption coefficient (α) on the photon energy (hv). It is expressed in the following relationship [37,39]:

$$(\alpha h\nu)^{\frac{1}{n}} = K(h\nu - E_g) \quad (6)$$

where α is the absorption coefficient, K is a proportionality constant, and the exponent n= 2 or 1/2 refers to the nature of the transition. The direct bandgap was estimated by the interception of the linear plot $(\alpha h\nu)^2$ and the hv axis. The band-gap energy (E_g) of NBO crystals was about 2.1 ± 0.1 eV which is a smaller bandgap than of TiO_2 and ZnO (3.2 eV) [40,41]. This shows that BNO has a significant absorption level for both UV and visible light from wavelength 200 to 800 nm where the fraction of the light irradiance is converted into electrical and/or chemical energy. This tends to be more effective than the photocatalysts ZnO and TiO_2 and leads in visible light irradiation to an increase in the formation of pairs of electron holes. This sillenite's narrow bandgap makes it a new promising catalyst for photocatalysis applications for environmental aquatic pollution.

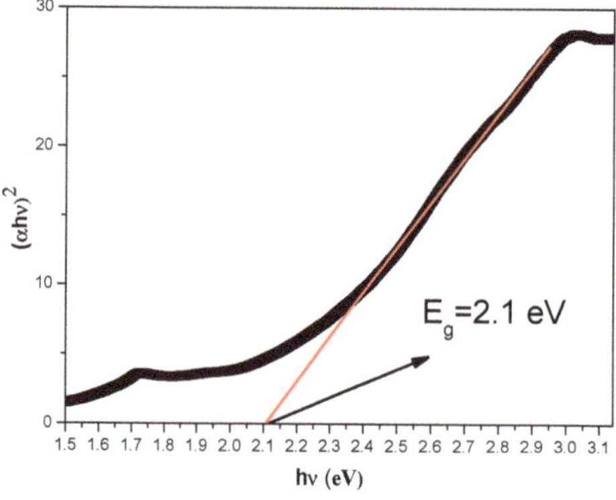

Figure 5. UV–Vis diffuse reflectance spectrum of the crystals $Bi_{12}NiO_{19}$.

3.2. *Photocatalytic Activity*

To test the photocatalytic activity of this catalyst, BB41 was selected as a topical example of organic pollutants. The photolysis effect of irradiation must be taken into account. An irradiation test for the elimination of the effects of photolysis was carried without the Catalyst BNO. Photolysis showed only a small percentage of BB41 degradation, not even above 5% percent. The effects of photolysis could consequently be overlooked. After that, the test was performed in the presence of the catalyst BNO. Before illumination of the photocatalytic reactor, it is important to eliminate the adsorption effect in a dark condition, the adsorption has not a significant removal for BB41.

After that, photocatalysis tests were started by the effect of pH, as the pH played a major role in the degradation. The tests were done in different pH mediums with a concentration of 15 mg/L and a catalyst dose of 1 g/L for 3 h, the results are shown in Figure 6. As can be seen, pH 9 was the optimal pH condition, because BB41 is a cationic dye, its photodegradation is favored in the basic medium, and the photodegradation performance of BNO for BB41 is boosted in the basic condition.

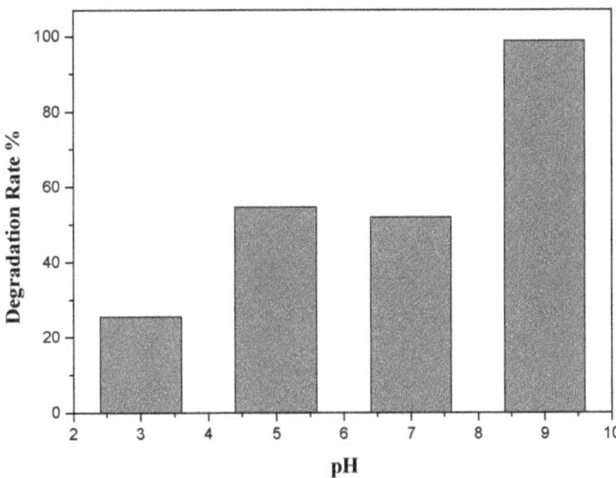

Figure 6. Effect of pH on the degradation of BB41.

After selecting the optimal pH, a test with a kinetic study was carried out under optimal conditions. The results are illustrated in Figure 7; Figure 7a shows the evolution of BB41 degradation as a function of time, where Figure 7b shows the associated UV-vis spectra for each point and the absorption peaks of BB41 at 610 nm can be observed decreasing with time. As can be observed, the degradation efficiency of BB41 achieved 98% within just 180 min at pH ~9 and 25 °C. This result can be explained by the gap energy of the catalyst (2.1 eV), which offers a higher absorption in both UV and visible areas from 200 to 800 mm [42].

It was already demonstrated in our previous works [13,14] that the degradation of the dye as an organic compound in the photocatalytic process is mainly due to the reactive oxidative species (ROS) such as superoxide radicals ($O_2^{\bullet-}$) and hydroxyl radicals ($^{\bullet}OH$), where the electrons of $Bi_{12}NiO_{19}$ conduction band are degraded the Basic Blue 41 by reducing the absorbed O_2 to the super-radical anion $O_2^{\bullet-}$. The oxidation occurs concomitantly by radicals $^{\bullet}OH$ through a valence band that reacts with H_2O [43]. In order to investigate the active species for the degradation of BB41, isopropanol and benzoquinone were chosen as scavenger agents to capture $^{\bullet}OH$ and $O_2^{\bullet-}$, respectively. Figure 8 shows the degradation efficiency with and without the presence of scavengers.

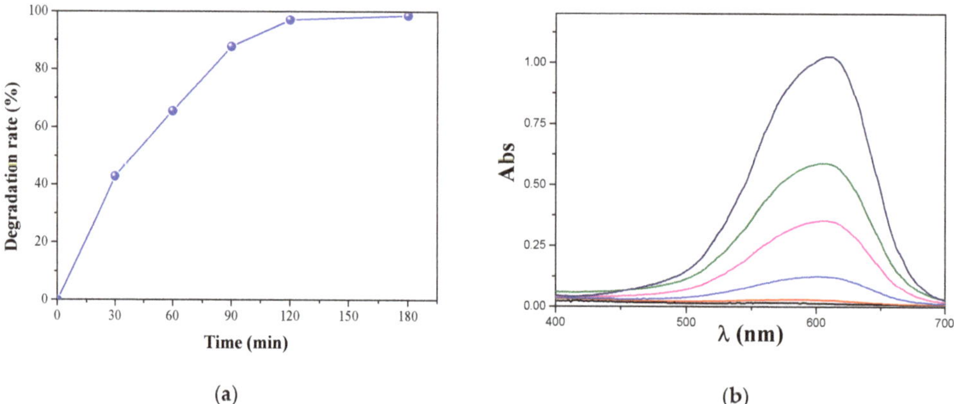

(a) (b)

Figure 7. (**a**) Photocatalytic efficiency on BB 41 as a function of time [C_o = 15 mg/L, dose =1 g/L, pH ~ 9 and T = 25 °C]. (**b**) UV–visible spectra.

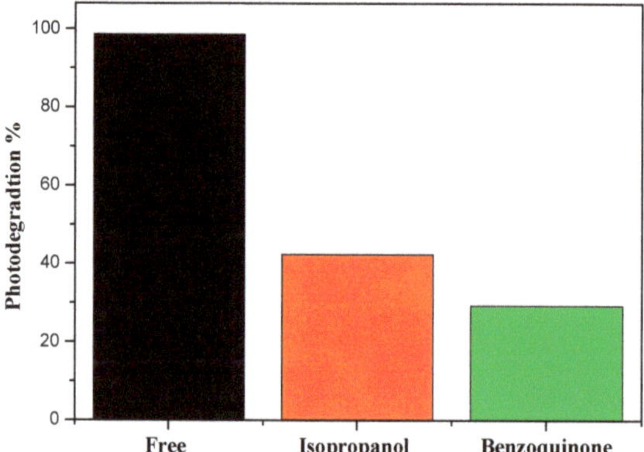

Figure 8. Effect of scavengers on the degradation of BB41.

As can be seen, adding benzoquinone and isopropanol to the solution reduced the BB41 degradation efficiency, indicating that both $O_2^{\bullet -}$ and $^{\bullet}OH$ were important active species in the photocatalytic activity. The photocatalytic reactions that occur at the solid/liquid interface can be bonded according to the following reaction mechanism:

$$Bi_{12}NiO_{19} + hv \rightarrow e^-_{\ CB} + h^+_{\ VB} \quad (7)$$

$$h^+_{\ VB} + H_2O \rightarrow {}^{\bullet}OH + H^+ \quad (8)$$

$$e^-_{\ CB} + O_2 \rightarrow O_2^{\bullet -} \quad (9)$$

$$Dye + {}^{\bullet}OH \ / \ O_2^{\bullet -} \rightarrow \text{Degradation products} \quad (10)$$

4. Conclusions

The main objective of this work was devoted to the study of the structural and optical properties of $Bi_{12}NiO_{19}$ and its application as a photocatalyst for the disposal of the basic blue 41 dye, used in the textile industry, under visible irradiation. BNO nanoparticles were synthesized using the combustion method using Polyvinylpyrrolidone as a combustion reagent. Different approaches were used to characterize the resulting catalyst. Starting with X-ray diffraction (XRD), the Rietveld method was used to refine the structure based on XRD data. which showed the purity and good crystallinity of our sillenite phase. Then, the structural form of this sillenite was illustrated for the first time. This catalyst has a space group of I23 with a lattice parameter of (a = 10.24 Å) for this catalyst. The special surface area (SSA) of BNO was determined by the Brunauer–Emmett–Teller (BET) method, it was found in the range between 14.56 and 20.56 $cm^2 \cdot g^{-1}$. Then, the morphology of the nanoparticles was visualized by Scanning Electron Microscope (SEM). Finally, the optical properties of BNO were determined by UV-VIS diffusion reflectance spectroscopy (DRS), a 2.1 eV optical bandgap was discovered. This narrow bandgap and the good crystallinity allow this sillenite to be a promising and effective catalyst for photocatalytic applications in the environmental field such as the treatment of polluted water under visible light radiation. That was confirmed by the application of this sillenite by performing the decomposition of the basic blue 41 dye under visible irradiation where a total degradation was obtained at pH ~9 and 25 °C in less than 180 min. For future research, we will test the effects of this sillenite on the degradation and reduction of organic and inorganic pollutants, as they are both hazardous contaminants to the environment.

Author Contributions: Conceptualization, O.B. and H.K.; methodology, O.B. and Y.B.; software, O.B.; validation, O.B. and B.B.; formal analysis, B.B.; investigation, Y.B. and O.B.; resources, B.B.; data curation, O.B.; writing—original draft preparation, O.B., H.K., B.B. and Y.B.; writing—review and editing, O.B.; visualization, O.B.; supervision, O.B.; All authors have read and agreed to the published version of the manuscript.

Funding: This research received no external funding.

Institutional Review Board Statement: Not applicable.

Informed Consent Statement: Not applicable.

Data Availability Statement: Not applicable.

Acknowledgments: This work was financially supported by the Faculties of Mechanical Engineering and Process Engineering.

Conflicts of Interest: The authors declare that they have no conflict of interest.

References

1. Baaloudj, O.; Nasrallah, N.; Kebir, M.; Khezami, L.; Amrane, A.; Assadi, A.A. A comparative study of ceramic nanoparticles synthesized for antibiotic removal: Catalysis characterization and photocatalytic performance modeling. *Environ. Sci. Pollut. Res.* **2020**, *28*, 13900–13912. [CrossRef] [PubMed]
2. Achour, S.; Amokrane, S.; Chegrouche, S.; Nibou, D.; Baaloudj, O. Artificial neural network modeling of the hexavalent uranium sorption onto chemically activated bentonite. *Res. Chem. Intermed.* **2021**, *24*, 1–8. [CrossRef]
3. Asadi-Ghalhari, M.; Mostafaloo, R.; Ghafouri, N.; Kishipour, A.; Usei, S.; Baaloudj, O. Removal of Cefixime from aqueous solutions via proxy electrocoagulation: Modeling and optimization by response surface methodology. *React. Kinet. Mech. Catal.* **2021**, *27*, 1–13. [CrossRef]
4. Bourkeb, K.; Baaloudj, O. Facile electrodeposition of ZnO on graphitic substrate for photocatalytic application: Degradation of antibiotic in a continuous stirred-tank reactor. *J. Solid State Electrochem.* **2021**, *6*, 1–8. [CrossRef]
5. Benrighi, Y.; Nasrallah, N.; Chaabane, T.; Sivasankar, V.; Darchen, A.; Baaloudj, O. Photocatalytic performances of $ZnCr_2O_4$ nanoparticles for cephalosporins removal: Structural, optical and electrochemical properties. *Opt. Mater.* **2021**, *115*, 111035. [CrossRef]
6. Brahimi, B.; Mekatel, E.; Mellal, M.; Trari, M. Synthesis of the hexaferrite semiconductor $SrFe_{12}O_{19}$ and its application in the photodegradation of Basic Red 46. *J. Mater. Sci. Mater. Electron.* **2021**, *32*, 11780–11790. [CrossRef]
7. Thi Mai Tho, N.; The Huy, B.; Nha Khanh, D.N.; Quoc Thang, N.; Thi Phuong Dieu, N.; Dai Duong, B.; Thi Kim Phuong, N. Mechanism of Visible-Light Photocatalytic Mineralization of Indigo Carmine Using $ZnBi_2O_4$-Bi_2S_3 Composites. *ChemistrySelect* **2018**, *3*, 9986–9994. [CrossRef]
8. Mahmoodi, N.M. Magnetic ferrite nanoparticle-alginate composite: Synthesis, characterization and binary system dye removal. *J. Taiwan Inst. Chem. Eng.* **2013**, *44*, 322–330. [CrossRef]
9. Mahmoodi, N.M. Surface modification of magnetic nanoparticle and dye removal from ternary systems. *J. Ind. Eng. Chem.* **2015**, *162*, 251–259. [CrossRef]
10. Mahmoodi, N.M.; Abdi, J.; Oveisi, M.; Alinia Asli, M.; Vossoughi, M. Metal-organic framework (MIL-100 (Fe)): Synthesis, detailed photocatalytic dye degradation ability in colored textile wastewater and recycling. *Mater. Res. Bull.* **2018**, *100*, 357–366. [CrossRef]
11. Mahmoodi, N.M.; Keshavarzi, S.; Ghezelbash, M. Synthesis of nanoparticle and modelling of its photocatalytic dye degradation ability from colored wastewater. *J. Environ. Chem. Eng.* **2017**, *5*, 3684–3689. [CrossRef]

12. Kenfoud, H.; Nasrallah, N.; Baaloudj, O.; Meziani, D.; Chaabane, T.; Trari, M. Photocatalytic reduction of Cr(VI) onto the spinel $CaFe_2O_4$ nanoparticles. *Optik* **2020**, *223*, 165610. [CrossRef]
13. Baaloudj, O.; Nasrallah, N.; Kebir, M.; Guedioura, B.; Amrane, A.; Nguyen-Tri, P.; Nanda, S.; Assadi, A.A. Artificial neural network modeling of cefixime photodegradation by synthesized $CoBi_2O_4$ nanoparticles. *Environ. Sci. Pollut. Res.* **2020**, *28*, 15436–15452. [CrossRef] [PubMed]
14. Baaloudj, O.; Assadi, I.; Nasrallah, N.; El, A.; Khezami, L. Simultaneous removal of antibiotics and inactivation of antibiotic-resistant bacteria by photocatalysis: A review. *J. Water Process Eng.* **2021**, *42*, 102089. [CrossRef]
15. Oliveira, O.G.; Mincache, A.J.; Dias, G.S.; Santos, I.A.; Guo, R.; Bhalla, A.S.; Cótica, L.F. Study of the crystal and electronic structures of $(Bi1-xNdx)FeO_3$ compositions using Rietveld refinements and the maximum entropy method. *Ferroelectrics* **2019**, *545*, 167–174. [CrossRef]
16. Im, W.B.; Page, K.; Denbaars, S.P.; Seshadri, R. Probing local structure in the yellow phosphor $LaSr_2\ AlO_5{:}Ce^{3+}$, by the maximum entropy method and pair distribution function analysis. *J. Mater. Chem.* **2009**, *19*, 8761–8766. [CrossRef]
17. Wu, X.; Li, M.; Li, J.; Zhang, G.; Yin, S. A sillenite-type $Bi_{12}MnO_{20}$ photocatalyst: UV, visible and infrared lights responsive photocatalytic properties induced by the hybridization of Mn 3d and O 2p orbitals. *Appl. Catal. B Environ.* **2017**, *219*, 132–141. [CrossRef]
18. Lin, X.; Huang, F.; Wang, W.; Xia, Y.; Wang, Y.; Liu, M.; Shi, J. Photocatalytic activity of a sillenite-type material $Bi_{25}GaO_{39}$. *Catal. Commun.* **2008**, *9*, 572–576. [CrossRef]
19. Zhang, H.; Lü, M.; Liu, S.; Xiu, Z.; Zhou, G.; Zhou, Y.; Qiu, Z.; Zhang, A.; Ma, Q. Preparation and photocatalytic properties of sillenite $Bi_{12}TiO_{20}$ films. *Surf. Coat. Technol.* **2008**, *202*, 4930–4934. [CrossRef]
20. Lima, A.F.; Lalic, M.V. First-principles study of the $BiMO_4$ antisite defect in the $Bi_{12}MO_{20}$ (M = Si, Ge, Ti) sillenite compounds. *J. Phys. Condens. Matter.* **2013**, *25*, 495505. [CrossRef]
21. Hou, D.; Hu, X.; Wen, Y.; Shan, B.; Hu, P.; Xiong, X.; Qiao, Y.; Huang, Y. Electrospun sillenite $Bi_{12}MO_{20}$ (M = Ti, Ge, Si) nanofibers: General synthesis, band structure, and photocatalytic activity. *Phys. Chem. Chem. Phys.* **2013**, *15*, 20698–20705. [CrossRef]
22. Yao, W.F.; Wang, H.; Xu, X.H.; Zhou, J.T.; Yang, X.N.; Zhang, Y.; Shang, S.X.; Wang, M. Sillenites materials as novel photocatalysts for methyl orange decomposition. *Chem. Phys. Lett.* **2003**, *377*, 501–506. [CrossRef]
23. Yao, W.F.; Xu, X.H.; Zhou, J.T.; Yang, X.N.; Zhang, Y.; Shang, S.X.; Wang, H.; Huang, B.B. Photocatalytic property of sillenite $Bi_{24}AlO_{39}$ crystals. *J. Mol. Catal. A Chem.* **2004**, *212*, 323–328. [CrossRef]
24. Valant, M.; Suvorov, D. Processing and Dielectric Properties of Sillenite Compounds $Bi_{12}MO_{20}$-δ (M: Si, Ge, Ti, Pb, Mn, $B_{1/2}P_{1/2}$). *ChemInform* **2010**, *33*, 2900–2904. [CrossRef]
25. Noh, T.H.; Hwang, S.W.; Kim, J.U.; Yu, H.K.; Seo, H.; Ahn, B.; Kim, D.W.; Cho, I.S. Optical properties and visible light-induced photocatalytic activity of bismuth sillenites ($Bi_{12}XO_{20}$, X = Si, Ge, Ti). *Ceram. Int.* **2017**, *43*, 12102–12108. [CrossRef]
26. Kenfoud, H.; Baaloudj, O.; Nasrallah, N.; Bagtache, R.; Assadi, A.A.; Trari, M. Structural and electrochemical characterizations of $Bi_{12}CoO_{20}$ sillenite crystals: Degradation and reduction of organic and inorganic pollutants. *J. Mater. Sci. Mater. Electron.* **2021**, *32*, 16411–16420. [CrossRef]
27. Vavilapalli, D.S.; Melvin, A.A.; Bellarmine, F.; Mannam, R.; Velaga, S.; Poswal, H.K.; Dixit, A.; Ramachandra Rao, M.S.; Singh, S. Growth of sillenite Bi^+FeO^+ single crystals: Structural, thermal, optical, photocatalytic features and first principle calculations. *Sci. Rep.* **2020**, *10*, 22052. [CrossRef] [PubMed]
28. Baaloudj, O.; Nasrallah, N.; Assadi, A.A. Facile synthesis, structural and optical characterizations of $Bi_{12}ZnO_{20}$ sillenite crystals: Application for Cefuroxime removal from wastewater. *Mater. Lett.* **2021**, *304*, 130658. [CrossRef]
29. Zhoua, P.; Xiaoa, F.; Jia, L. $Bi_{12}NiO_{19}$ micro-sheets grown on graphene oxide: Temperature-dependent facile synthesis and excellent electrochemical behavior for supercapacitor electrode. *J. Electroanal. Chem.* **2021**, *884*, 115075. [CrossRef]
30. Pei, L.Z.; Wei, T.; Lin, N.; Zhang, H. Synthesis of bismuth nickelate nanorods and electrochemical detection of tartaric acid using nanorods modified electrode. *J. Alloys Compd.* **2016**, *663*, 677–685. [CrossRef]
31. Rajamoorthy, M.; Geetha, D.; Sathiya Priya, A. Synthesis of Cobalt-Doped $Bi_{12}NiO_{19}$: Structural, Morphological, Dielectric and Magnetic Properties. *Arab. J. Sci. Eng.* **2021**, *46*, 737–744. [CrossRef]
32. Giannopoulou, I.; Saïs, F.; Thomopoulos, R. Handbook-of-pharmaceutical-excipients-6th-edition. *Rev. Nouv. Technol. Inf.* **2015**, *E28*, 257–262.
33. Ma, Y.; Qiu, F.L.; Wei, T.; Lin, F.F.; Yan, L.; Wu, H.; Zhang, Y.; Pei, L.Z.; Fan, C.G. Facile Synthesis of Polyaniline/Bismuth Nickelate Nanorod Composites for Sensitive Tartaric Acid Detection. *Surf. Eng. Appl. Electrochem.* **2019**, *55*, 335–341. [CrossRef]
34. Yé, Z.G.; Crottaz, O.; Vaudano, F.; Kubel, F.; Tissot, P.; Schmid, H. Single crystal growth, structure refinement, ferroelastic domains and phase transitions of the hausmannite $CuCr_2O_4$. *Ferroelectrics* **2011**, *162*, 103–118. [CrossRef]
35. Dollase, W.A.; O'Neill, H.S.C. The spinels $CuCr_2O_4$ and $CuRh_2O_4$. *Acta Crystallogr. Sect. C Cryst. Struct. Commun.* **1997**, *53*, 657–659. [CrossRef]
36. Karuppasamy, P.; Senthil Pandian, M.; Ramasamy, P.; Verma, S. Crystal growth, structural, optical, thermal, mechanical, laser damage threshold and electrical properties of triphenylphosphine oxide 4-nitrophenol (TP4N) single crystals for nonlinear optical applications. *Opt. Mater.* **2018**, *79*, 152–171. [CrossRef]
37. Kenfoud, H.; Nasrallah, N.; Baaloudj, O.; Derridj, F.; Trari, M. Enhanced photocatalytic reduction of Cr(VI) by the novel hetero-system $BaFe_2O_4/SnO_2$. *J. Phys. Chem. Solids* **2022**, *160*, 110315. [CrossRef]

38. Tripathy, S.; Saini, D.S.; Bhattacharya, D. Synthesis and fabrication of $MgAl_2O_4$ ceramic foam via a simple, low-cost and eco-friendly method. *J. Asian Ceram. Soc.* **2016**, *4*, 149–154. [CrossRef]
39. Makuła, P.; Pacia, M.; Macyk, W. How to Correctly Determine the Band Gap Energy of Modified Semiconductor Photocatalysts Based on UV-Vis Spectra. *J. Phys. Chem. Lett.* **2018**, *9*, 6814–6817. [CrossRef]
40. Serpone, N. Is the band gap of pristine TiO_2 narrowed by anion- and cation-doping of titanium dioxide in second-generation photocatalysts? *J. Phys. Chem. B* **2006**, *110*, 24287–24293. [CrossRef]
41. Miki-yoshida, M. Optical Band Gap Estimation of ZnO Nanorods a E = BˆE − Eg h. *Mat. Res.* **2016**, *19*, 33–38. [CrossRef]
42. Baaloudj, O.; Assadi, A.A.; Azizi, M.; Kenfoud, H.; Trari, M.; Amrane, A.; Assadi, A.A.; Nasrallah, N. Synthesis and Characterization of $ZnBi_2O_4$ Nanoparticles: Photocatalytic Performance for Antibiotic Removal under Different Light Sources. *Appl. Sci.* **2021**, *11*, 3975. [CrossRef]
43. Kenfoud, H.; Nasrallah, N.; Meziani, D.; Trari, M. Photoelectrochemical study of the spinel $CaFe_2O_4$ nanostructure: Application to Basic Blue 41 oxidation under solar light. *J. Solid State Electrochem.* **2021**, *25*, 1815–1823. [CrossRef]

MDPI

Article

Metal-Free g-C_3N_4/Nanodiamond Heterostructures for Enhanced Photocatalytic Pollutant Removal and Bacteria Photoinactivation

Natalya Kublik [1,2], Luiz E. Gomes [3], Luiz F. Plaça [1], Thalita H. N. Lima [4], Thais F. Abelha [4], Julio A. P. Ferencz [1], Anderson R. L. Caires [4] and Heberton Wender [1,*]

1 Nano & Photon Research Group, Laboratory of Nanomaterials and Applied Nanotechnology (LNNA), Institute of Physics, Federal University of Mato Grosso do Sul, Campo Grande 79070-900, Brazil; natalyaemkublik@gmail.com (N.K.); luiz.placa@ufms.br (L.F.P.); julio.ferencz@ufms.br (J.A.P.F.)
2 School of Manufacturing and System Networks, Ira A. Fulton Schools of Engineering, Arizona State University, Mesa, AZ 85212, USA
3 Centro de Tecnologias Estratégicas do Nordeste (CETENE), Av. Prof. Luiz Freire, 01, Recife 50740-540, Brazil; luizedugo@gmail.com
4 Optics and Photonics Group, Institute of Physics, Federal University of Mato Grosso do Sul, Campo Grande 79070-900, Brazil; thalita.lima212@gmail.com (T.H.N.L.); thaisbee@gmail.com (T.F.A.); anderson.caires@ufms.br (A.R.L.C.)
* Correspondence: heberton.wender@ufms.br

Citation: Kublik, N.; Gomes, L.E.; Plaça, L.F.; Lima, T.H.N.; Abelha, T.F.; Ferencz, J.A.P.; Caires, A.R.L.; Wender, H. Metal-Free g-C_3N_4/Nanodiamond Heterostructures for Enhanced Photocatalytic Pollutant Removal and Bacteria Photoinactivation. *Photochem* **2021**, *1*, 302–318. https://doi.org/10.3390/photochem1020019

Academic Editor: Vincenzo Vaiano

Received: 10 July 2021
Accepted: 10 September 2021
Published: 14 September 2021

Abstract: Heterogeneous photocatalysis has emerged as a promising alternative for both micropollutant removal and bacterial inactivation under solar irradiation. Among a variety of photocatalysts explored in the literature, graphite carbon nitride (g-C_3N_4) is a metal-free semiconductor with acceptable chemical stability, low toxicity, and excellent cost-effectiveness. To minimize its high charge recombination rate and increase the photocatalyst adsorption capacity whilst keeping the metal-free photocatalyst system idea, we proposed the heterojunction formation of g-C_3N_4 with diamond nanocrystals (DNCs), also known as nanodiamonds. Samples containing different amounts of DNCs were assessed as photocatalysts for pollutant removal from water and as light-activated antibacterial agents against *Staphylococcus sureus*. The sample containing 28.3 wt.% of DNCs presented the best photocatalytic efficiency against methylene blue, removing 71% of the initial dye concentration after 120 min, with a pseudo-first-order kinetic and a constant rate of 0.0104 min^{-1}, which is nearly twice the value of pure g-C_3N_4 (0.0059 min^{-1}). The best metal-free photocatalyst was able to promote an enhanced reduction in bacterial growth under illumination, demonstrating its capability of photocatalytic inactivation of *Staphylococcus aureus*. The enhanced photocatalytic activity was discussed and attributed to (i) the increased adsorption capacity promoted by the presence of DNCs; (ii) the reduced charge recombination rate due to a type-II heterojunction formation; (iii) the enhanced light absorption effectiveness; and (iv) the better charge transfer resistance. These results show that g-C_3N_4/DNC are low-cost and metal-free photoactive catalysts for wastewater treatment and inactivation of bacteria.

Keywords: photodegradation; heterostructure; diamond nanocrystals; bacterial photoinactivation

1. Introduction

The conversion of solar energy into chemical energy using a semiconductor material has been extensively investigated to generate clean fuels and degradation of persistent organic pollutants (POPs) through environmentally friendly processes [1]. For efficient solar energy conversion, semiconductor photocatalysts are required to effectively absorb light in the visible range to achieve practical applications, present satisfactory stability under experimental conditions, and substantially possess suitable energy band positions and the ability to avoid photogenerated charge recombinations [2–4].

In a photocatalytic reaction, the conduction band (CB) and valence band (VB) energy position of the semiconducting photocatalyst must present more negative and more positive values, respectively, compared to the reduction and oxidation potentials of the target molecules [4]. Examples include, but are not limited to, CO_2 reduction [5], H_2 and O_2 evolution by water splitting [3], and degradation of organic pollutants [6]. In the latter, in addition to the direct reduction or oxidation of the target molecules by the electrons and holes generated at the CB and BV, respectively, reactive oxygen species (ROS) such as superoxide anions ($O_2^{\bullet-}$) and hydroxyl radicals ($HO^{\bullet}$) are commonly involved in the pollutant removal and, in this case, the electron and hole potentials have to be suitable for their formation [4,7–9]. Additionally, some other relevant and desirable characteristics for a semiconductor material consist of low cost, recyclability, simple and scalable synthesis, and environmentally friendly composition [10,11].

Graphitic carbon nitride (g-C_3N_4) is a metal-free semiconductor that meets some of the above requirements since it presents good physicochemical stability and cost-effectiveness, negligible toxicity, and suitable CB energy as well as is easily prepared from a variety of precursors [12–14]. In regard to its energy band structure, g-C_3N_4 presents VB energies ranging from 1.40 to 1.59 eV (vs. NHE, at pH = 7), CB from −1.3 to −1.13 eV (vs. NHE, at pH = 7), and bandgap (E_g) energies of about 2.7 eV [15]. Its relative mild-gap provides only partial visible-light absorption, which along with its low separation efficiency of excited electron–hole pairs and deficient active sites, consists of the most relevant challenges concerning its optical activity [15]. These limitations, however, can be overcome by combining g-C_3N_4 with other suitable materials and thus fabricating heterojunctions [2,16–19].

Diamond nanocrystals (DNCs) are reported to have activation sites for both reduction and oxidation reactions, due to their high carrier mobility [20], as well as due to their potential intragap states caused by surface defects, unsaturated bonding, or impurities [21]. Based on its electronic properties, pristine DNC should not present significant photocatalytic activity under visible-light irradiation because of its high E_g value, of approximately 5.5 eV for bulk diamond and 3.3 eV for DNC powders [22], but the presence of impurities or defects may create donor or acceptor levels in its bandgap (intragap), facilitating the absorption of photons under visible-light irradiation conditions [23]. A coupling may occur between diamond and nondiamond C electrons in such a way that bandgap energies of less than 3.0 eV may be observed for the DNC [24]. The intragap states have been observed (but not commented on) in previous works [25], and treated as VB and CB of the DNC. However, DNC presents little to no cytotoxicity [26] and has been recognized as an outstanding material in photocatalysis along with graphene-based composites [24,27].

Relatively few works have applied DNC for photocatalytic degradation of pollutants [16,28,29], and even fewer have explored g-C_3N_4/DNC heterojunctions for wastewater purification and H_2 production [16,25,30]. Due to the various loose ends in the discussion on the chemical, physical, and photocatalytic properties of this system, further studies are highly desired. In the study carried out by Haleem et al. [30], 0.4 wt.% of DNC was added to g-C_3N_4 and the photocatalyst showed ~50% increased photocatalytic H_2 evolution compared to pristine g-C_3N_4, which was attributed to an enhanced charge transfer, low recombination rate, and higher surface area of the hybrid material. In a later work, the g-C_3N_4/DNC heterojunction prepared with 10 wt.% DNC showed the best photoactivity for H_2 evolution under visible-light irradiation [25]. In this case, the enhanced photoactivity was attributed to two factors: (i) the more localized charge carrier generation due to light-trapping from scattering properties of the DNCs, and (ii) the appropriated energy level matching of both materials to form a type-II heterojunction. Concerning a different preparation method and larger amounts of DNC in the heterojunction [16], a recent work evaluated the photocatalytic efficiency against the degradation of the methylene blue (MB) dye. The best photocatalytic degradation rate and dye adsorption capacity were observed in the sample containing 33 wt.% of DNC, which was associated with increased light absorption and greater surface area. Therefore, the g-C_3N_4/DNC is a metal-free composite material that has the potential to act as a photocatalyst in different photoactivated reactions.

Amongst the useful applications of g-C_3N_4 and DNC, their antibacterial capabilities have also been individually studied; a fact that is motivated by the urgent demand for new materials capable of tackling pathogenic bacteria [31–33]. Under light illumination, g-C_3N_4 alone and in association with metals/semiconductors inactivated microorganisms through the generation of ROS [31]. Besides having photochemotherapy capabilities, g-C_3N_4 nanosheets also displayed cytotoxicity by inducing the rupture of the cell membrane of bacteria upon direct mechanical contact [34]. A similar physical mechanism of action is proposed for DNC-based antimicrobial agents [33]. These materials have gained increased biomedical interest due to their favorable biocompatibility; however, cytotoxicity against both Gram-positive and Gram-negative bacteria was displayed by modifying their surface composition [35]. Nanodiamonds could also inhibit biofilm formation and disrupt biofilms, which is particularly important considering that microorganism aggregates are harder to tackle than planktonic dispersions [36]. In addition, the combination of DNC with photosensitizers enabled the photodynamic antimicrobial chemotherapy against planktonic *S. aureus* [37,38] and its biofilms [37].

In this context, we prepared metal-free photocatalysts by making a heterojunction between g-C_3N_4 and as-obtained DNC under the in situ urea decomposition. The heterojunctions were prepared, characterized, and evaluated in terms of their photocatalytic activities for the photodegradation of MB and *Staphylococcus aureus* photoinactivation. The mechanism of improved photocatalytic efficiency under illumination was thoroughly revised, revealing it to be a combination of increased adsorption capacity, better light absorption effectiveness, enhanced charge transfer resistance, and decreased charge recombination rate by a type-II heterojunction formation. These results show that g-C_3N_4/DNC are efficient low-cost and metal-free photoactive catalysts for wastewater treatment and photoinactivation of bacteria.

2. Materials and Methods

2.1. Pristine Samples Preparation

The g-C_3N_4 samples were obtained through urea decomposition (Alphatec—São Paulo, Brazil, 99.9%) in an alumina crucible with a loose lid (see details in Figure S1, Supplementary Materials), according to previous reports with modifications [39,40]. After cooling naturally, the obtained content was macerated into a yellow powder and stored at room temperature. Non-detonated DNC samples were purchased from NaBond with a purity of 99.95% and used without further purification.

2.2. Synthesis of the g-C_3N_4/DNC Heterojunctions

For the heterostructures formation, a suitable amount of urea (Table S1) was solubilized in 25 mL of deionized water in a crucible with a loose lid for 15 min, under magnetic stirring. The amount of urea was calculated based on the mean yield of its conversion to g-C_3N_4 that in our experiments was 4.4%. Then, a certain mass of DNC (Table S1) was added to the mixture and sonicated (UltraCleaner 1400 A—model unique, 40 kHz—315 W RMS) for 30 min. For the DNC, the initial mass was calculated considering the rate of mass loss obtained at 550 °C by thermogravimetric analysis (TGA). The solution was dried at 120 °C for 6 h using a heating rate of 5 °C min^{-1}. Next, the crucible was heated at 550 °C for 2 h using a rate of 3 °C min^{-1} under an air atmosphere. The obtained porous powder was ground and stored for further analysis. The flowchart in Scheme 1 illustrates the described methodology. Samples were named as g-C_3N_4/DNC-2, g-C_3N_4/DNC-11, and g-C_3N_4/DNC-28, according to their approximated experimental DNC *w*/*w*%, respectively, found by TGA analysis (see TGA results) and considering their calculated synthesis yield of 3.0%, 2.3%, and 1.9%, respectively (Table S1).

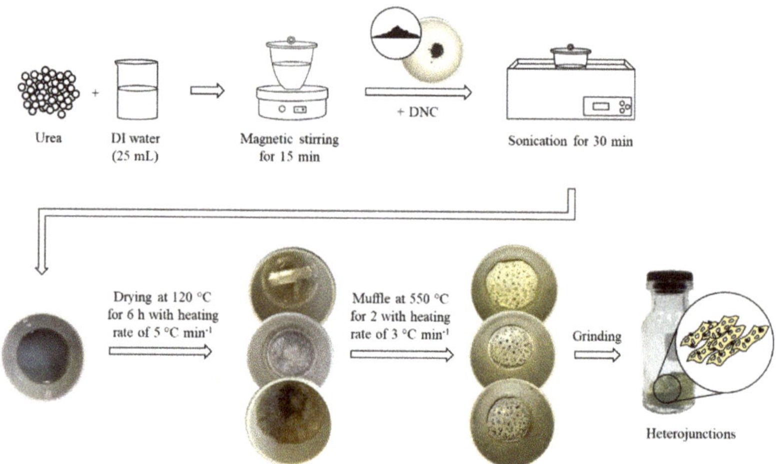

Scheme 1. The standard synthesis procedure of g-C_3N_4/DNC heterojunctions.

2.3. Characterization

Thermogravimetric (TGA) and differential scanning calorimetry (DSC) analysis were performed with a Netzsch STA 449 F3 Jupiter equipment under N_2 atmosphere, from 25 °C to 900 °C, with a 10 °C min^{-1} rate and 1 °C interval between measurements. The morphological features of the samples were observed by scanning electron microscopy (SEM) by using a JEOL JSM-6380 LV microscope and transmission electron microscopy (TEM) by using an FEI MORGANI 268 D operated at 80 kV. SEM measurements were performed by placing small amounts of the powder samples on the top of a carbon tape and TEM by dispersing the samples into isopropyl alcohol followed by deposition on carbon-coated Cu grids. Crystallinity was investigated by X-ray diffraction (XRD) using a Shimadzu XRD-6100 diffractometer. For the XRD analysis, the diffractometer was adjusted to 40.0 kV—30.0 mA, and 2θ scanning was in the 5°–85° range. Obtained diffractograms with Co-Kα X-ray source (λ_{Co} = 1.792850 Å) were adapted to typical Cu-Kα source typical conditions (λ_{Cu} = 1.544390 Å) to compare the peaks with the ones from the literature. Fourier transform infrared spectroscopy (FTIR) was carried out with an attenuated total reflection (ATR) accessory in a Perkin Elmer Spectrum 100 equipment. For dynamic light scattering (DLS) and Zeta potential (ZP) measurements (Malvern, Zetasizer Nano ZS), approximately 0.5 mg of each sample were sonicated in DI water for 10 min, and three measurements with 12 runs each were recorded at room temperature. Diffuse ultraviolet-visible reflectance spectroscopy (DRS-UV–Vis) with a Ø60 mm integrating sphere and Tauc plots was applied to determine the other optical bandgap. Brunauer–Emmett–Teller (BET) specific surface area measurements were examined by N_2 adsorption–desorption with an ASAP 2420 Micromeritics instrument. Steady-state photoluminescence (PL) studies were performed using an FS-2 fluorescence spectrometer (Scinco, Seoul, Korea) and data were collected at room temperature using right-angle geometry (90° excitation/emission geometry) and a quartz cuvette with four polished faces and a 10 mm optical path length. The PL spectra were obtained using a 150 W Xe-arc lamp as an excitation source; Czerny–Turner monochromators for selecting excitation and emission wavelengths; and an R928 photomultiplier for collecting the photoluminescence signal. PL emission spectra were obtained under excitation at 371 nm.

Electrochemical impedance spectroscopy (EIS) and Mott–Schottky (M–S) analysis were performed using a CorrTest potentiostat/galvanostat in a standard three-electrode electrochemical cell, with samples as the working electrode, Ag/AgCl as the reference electrode, and a platinum wire as the counter electrode, all immersed in a 0.1 M Na_2SO_4 electrolyte (pH 6). The photoelectrode preparation was carried out by sonicating 4 mg of the as-prepared sample powders in 2 mL of H_2O and then drop-casting 200 μL of catalyst ink directly onto a limited area of 1 cm^{-2} of pre-cleaned fluorine-doped tin oxide (FTO) glass substrate. After drying, 20 μL of Nafion solution (0.3 wt.%) was dropped on the film surface and left to dry spontaneously. EIS measurements were conducted with frequencies varying from 40 mHz to 100 kHz, the amplitude of 10 mV, and applied potential set as 0 V vs. the open circuit potential (OCP), under dark conditions. The Mott–Schottky (M–S) analyses were carried out by ranging the applied potential from −0.2 to −1.2 V at 1000 and 2000 Hz. Potentials versus the Ag/AgCl reference electrode ($E_{Ag/AgCl}$) were converted to RHE (E_{RHE}) by using the Nernst equation, where $E_{RHE} = E_{Ag/AgCl} + (0.059 \times pH) + 0.197$ V.

2.4. Photocatalytic Degradation Study

The photocatalytic properties of the samples were investigated against the photodegradation of methylene blue (MB) dye as a cationic model molecule to simulate pollutants removal from water. For each experiment, 25 mg of the photocatalyst was added to 25 mL of MB solution (30 mg L^{-1}) and left under dark conditions for 40 min (20 min under sonication and 20 min under magnetic stirring) to achieve dye adsorption equilibrium. The photoreactor was a borosilicate beaker placed below an Abet-Tech solar simulator equipped with a 150 W Xe lamp. The photon flux was calibrated to 200 mW cm^{-2} using a reference cell, as described elsewhere [9,41]. Once under irradiation, the solution was kept under magnetic stirring and the concentration of the remaining MB was monitored through UV−Vis spectrophotometry at 664 nm [42] every 20 min up to 160 min. Adsorption of the MB dye was investigated by C/C_i curves, where C is the concentration of MB at the time "t" and C_i is the initial concentration of MB. The photocatalytic effect was accounted for by the C/C_0 curves, where C_0 is the concentration of MB when the light was turned on.

2.5. Photocatalytic Inactivation of Bacteria

The photocatalytic bactericidal effect was tested against a *Staphylococcus aureus* strain (ATCC 25923). The bacterial suspensions were prepared with 40 μL of the bacterial strain added to 4 mL of Müller−Hinton Broth and incubated at 37 °C for 24 h. The catalyst at 125 mg L^{-1} was dispersed in 2 mL of a physiological saline (0.9% NaCl) solution containing the bacterial inoculum at 1.5×10^8 CFU/mL. After that, the samples were shaken at 120 rpm for 60 min. After the incubation, the samples were separated into two groups: non-illuminated and illuminated. The illuminated samples were placed in a 96-well plate (250 μL/well) and submitted to visible-light irradiation provide by RGB light-emitting diodes (LEDs). The samples were illuminated at 18 mW cm^{-2} for 1 h. Then, a serial dilution was performed until 1:32 for both illuminated and non-illuminated samples. Finally, the colony-forming units (CFU) were counted after 24 h of incubation at 37 °C.

3. Results and Discussions

Metal-free photocatalysts are of great interest for solar energy conversion due to their low cost of production. Herein, we prepared heterojunction composites between g-C_3N_4 prepared from urea decomposition and as-obtained DNC through an in situ methodology. To obtain the experimental percentage of DNC in the g-C_3N_4/DNC heterostructures, as well as the thermostability of the composites, TGA experiments were carried out (Figure 1). Small weight losses (~2%) are observed from room temperature up to 200 °C for both the pristine g-C_3N_4 and g-C_3N_4/DNC composites which are attributed to desorption of physically adsorbed and intercalated water [43]. From 200 to 500 °C, it is clear that the samples presented almost negligible weight loss, and for higher temperatures, the decomposition of g-C_3N_4 could be evidenced, which was completed at around 770 °C, in agreement with previous reports [44]. For the DNCs, the residual mass observed was as high as 94.4% at 794 °C, similar to that obtained by Piña-Salazar et al. [45]. This small mass loss of the DNC refers to the release of gases adsorbed on its surface, such as CO, CO_2, and H_2O. Based on that, the temperature of 794 °C was used as a reference point for obtaining the residual masses (as indicated in the graph) and consequently the wt.% of DNC in the composites, since at this temperature the residual mass percentage of the g-C_3N_4 is negligible. The amount (wt.%) of DNC in the photocatalysts was then estimated to be 1.6, 11.1, and 28.3% for samples g-C_3N_4/DNC-2, g-C_3N_4/DNC-11, and g-C_3N_4/DNC-28, respectively. This quantification considers the fact that the calculated synthesis yield decreased with respect to the increase in the initial DNC mass in the heterojunctions (Table S1). Therefore, there is a significant difference between the nominal amount of DNC introduced during the in situ synthesis and the experimentally obtained one, and this difference comes mainly from the reduced synthesis yield of g-C_3N_4 due to the presence of the DNC particles. To explain the synthesis yield decrease, we hypothesize that DNCs behaves as nanometer spots of heat transfer due to their highest thermal conductivity among existing materials [46], locally elevating the temperature of the surrounding intermediated products and accelerating the urea decomposition. As a result, the system ended with a lower content of g-C_3N_4 during the in situ heterojunction formation compared to the case without DNC. In addition, the DSC curve of the pristine g-C_3N_4 exhibited an endothermic peak centered at 720 °C, which shifted to 710 °C for the sample g-C_3N_4/DNC-28 (Figure S2 of the supplementary material). It shows that the g-C_3N_4 decomposition in the heterojunctions started and was accomplished earlier compared to the pristine g-C_3N_4. In addition, it is well-known that the yield of synthesis is strongly dependent on the temperature and time of reaction, and few increases in the temperature may result in a significantly reduced mass of g-C_3N_4 during the precursor decomposition [15]. To further validate our hypothesis, we prepared the heterojunctions by only physically mixing the DNC with the already prepared g-C_3N_4. The TGA results revealed a DNC experimental content remarkably close to the expected value (Figure S3), showing that in fact DNCs do not, in fact, cause direct decomposition of g-C_3N_4 but alters the kinetics and yield of urea decomposition to form the g-C_3N_4.

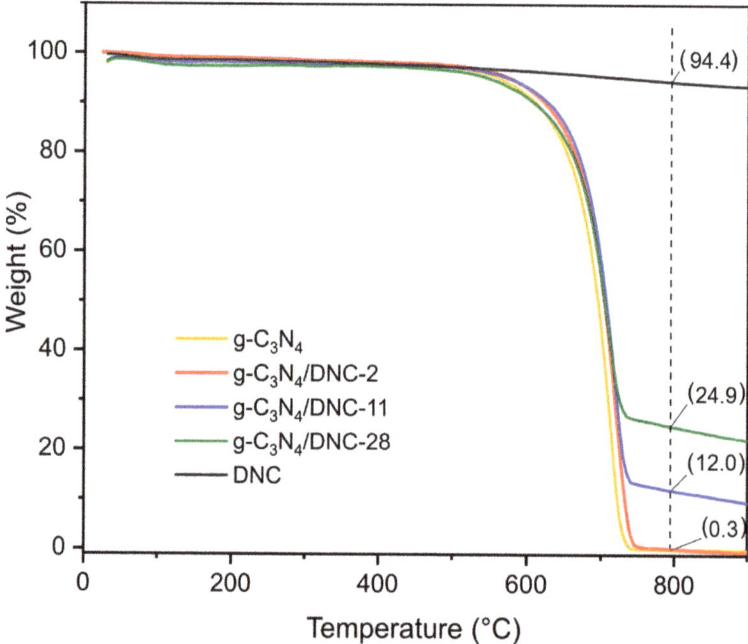

Figure 1. TG curves of pristine g-C_3N_4 and the g-C_3N_4/DNC heterojunctions.

Figure 2a,b, respectively, show a representative SEM and TEM image of the pure g-C_3N_4 sample that confirms that the morphology is composed of nanometer-thick sheets, as expected for a 2D material. Figure 2c shows the SEM image of the DNC sample, revealing the formation of individual nanoparticles of about 100–200 nm, without formation of large aggregates. DLS results confirmed the formation of non-agglomerated particles in solution, with mean diameters of 125.4 ± 48.9 nm (see Figure S4). SEM and TEM analysis were also used to investigate the morphology of the g-C_3N_4/DNC heterojunctions (Figure 2d–h). As a result, g-C_3N_4 nanosheets functioned as supports for the DNC nanoparticles, preventing the formation of large crystal clusters and, consequently, enhancing the physical interaction between both materials. For the lower concentration of DNC (g-C_3N_4/DNC-2, Figure 2d), the sample morphology is like that of the g-C_3N_4 pristine sample, as expected. Additionally, a slight fragmentation of the g-C_3N_4 structure occurred when increasing the amount of DNC (Figure 2e,f). The representative TEM images of the g-C_3N_4/DNC-28 heterojunction confirm that DNC particles are dispersed and anchored over the nanosheet structures (Figure 2g–h).

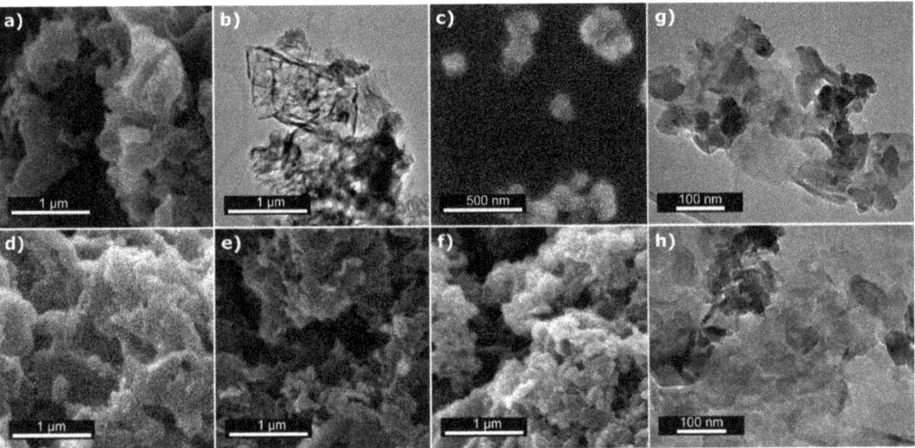

Figure 2. SEM (**a**) and TEM image (**b**) of the g-C_3N_4 photocatalyst. SEM images of the DNC pristine sample (**c**) and of the g-C_3N_4/DNC heterojunctions with (**d**) 1.6 wt.%, (**e**) 11.1 wt.%, and (**f**) 28.3 wt.% of DNC. Representative TEM images (**g**,**h**) of g-C_3N_4/DNC-28 heterojunctions.

Figure 3 shows the diffraction patterns of the pristine and heterojunctions samples. The XRD pattern of pristine g-C_3N_4 shows two peaks located at 12.8° and 27.4° that can be indexed as the (100) and (002) diffraction planes, respectively, of a graphite-like structure (JCPDS 87-1526). In addition, the latter corresponds to an interlayer distance of d = 0.326 nm, which matches the value commonly reported for graphite-like carbon nitride materials [47]. The former peak is, on the other hand, a signature of tri-*s*-triazine-based g-C_3N_4 formation [15,48]. For the DNC pristine sample, the XRD pattern contains one peak located at 44.3° that can be indexed as the (111) diffraction planes of diamond cubic structure (JCPDS 03-065-0537) [49]. The XRD patterns of the heterojunctions contain only the three peaks reported above, without a signal of secondary impurity phases. Additionally, with the increase in the concentration of DNC, there is an increase in the intensity of the diffraction peak located at 44.3° which is consistent with the expected higher concentration of DNC in the samples. A close look at the (002) planes of g-C_3N_4 reveals an angular shift of the diffraction peak from 27.5° to 27.9° for the heterojunctions which is a consequence of a decrease in the distances between the adjacent g-C_3N_4 layers [15]. Similar behavior has also been reported when the annealing of urea was conducted at different and increasing temperatures, from 450 to 600 °C, where an increase in the crystal stability of g-C_3N_4 was hypothesized [39]. This assumption is also supported also by the reduced full width at half maximum (FWHM) of the (002) peaks as the DNC was introduced, revealing that the g-C_3N_4 is more crystalline. Therefore, in agreement with TGA findings, the DNCs increased the crystal stability of the g-C_3N_4 in the heterojunction by the in situ formation of tight-packed nanosheets during the decomposition of urea. This result also confirms the hypothesis of localized spots of heat transfer around the DNC that may decrease the amount of g-C_3N_4 by facilitating its decomposition at temperatures slightly higher than 550 °C.

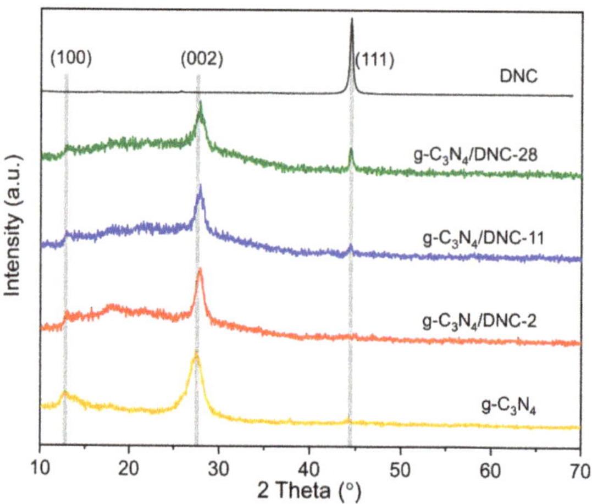

Figure 3. XRD patterns of pristine g-C_3N_4 and DNC, and the g-C_3N_4/DNC heterojunctions.

FTIR spectroscopy revealed chemical features of pristine DNC, g-C_3N_4, and the g-C_3N_4/DNC heterojunctions (Figure S5). It is possible to see that the spectra of pristine g-C_3N_4 and the heterojunctions are very similar and reflect the overall characteristics reported for g-C_3N_4 systems [15]. They show a sharp band at 810 cm^{-1} that is characteristic of this material and is attributed to the breathing mode of tri-*s*-triazine units [15,39]. The band at 890 cm^{-1} is ascribed to the out-of-plane C–H bond in aromatic domains [40] and, as also observed before, the spectra additionally present several strong bands in the 1200–1650 cm^{-1} region that are ascribed to the stretching mode of aromatic CN heterocycles [50]. These peaks together constitute a fingerprint of the g-C_3N_4 formation. Finally, the broadband observed between 3400 and 3000 cm^{-1} is ascribed to stretching vibrations of OH (3500–2500 cm^{-1}) and NH (3400–3300 cm^{-1}) groups, coming from water molecules physically adsorbed and/or the bridging C–NH–C units, respectively [15]. As expected, the pristine DNC presents less intense absorption bands since it contains fewer functional groups in both the bulk and surface chemical composition, showing a broad peak in the 3500–3000 cm^{-1} range, as observed for g-C_3N_4, as well as three main characteristic additional peaks at 1085, 1625, and 1707 cm^{-1}. The bands at 1085 and 1625 cm^{-1} can be attributed to the stretching vibration of C–O, and O–H, respectively. The broad band at 1700–1800 cm^{-1} is assigned to the C=O stretching vibration from the surface of DNC. It is also possible to observe that the DNC spectrum shows some minor features at 966, 915, and 810 cm^{-1} that together with the previously described peaks, are a fingerprint of the DNC. Therefore, the DNC surface contains functional groups that may help strengthen the physicochemical interaction with organic molecules or bacterial cells near the photocatalyst surface and enhance the overall composite photoactivity.

Figure 4a presents the UV–Vis–DRS curves of the samples. The g-C_3N_4 exhibited an optical reflectance edge at approximately 425 nm that corresponds to a direct bandgap of 2.91 eV (see Tauc plot in Figure 4b). Additionally, the increase in DNC concentration did not induce significant changes in the absorption edge (all samples presented bandgaps of ~2.9–3.0 eV) but resulted in an evident decrease in the total reflectance within the entire visible light range. This is a result of increased light absorption and/or light scattering effect caused by the addition of DNC. Su et al. observed similar results and mainly attributed it to the effective light scattering of NDs [25]. However, it is noteworthy that as the DNC and g-C_3N_4 refractive indexes are, respectively ~2.4 and <1.5, the DNC would promote the incident light propagation through refraction, increasing its optical path at the junction interface [51,52] and, thus, the probability of photon absorption by the g-C_3N_4 in the heterojunctions. In fact, the optical properties of the DNC show thought-provoking aspects. The first is that two linear regions could be found in Tauc plots (Figure 4c), which represent two distinct bandgap energies of 2.64 and 4.83 eV. The latter is well-recognized for the bulk diamond structure and the former possibly comes from an intragap state due to (i) structural or surface defects or (ii) quantum confinement effects [23,53]. Therefore, the DNC studied herein, although to a lesser extent, may harvest visible light photons, especially considering electron jumps from its intragap energy level, and scattering is not a negligible factor.

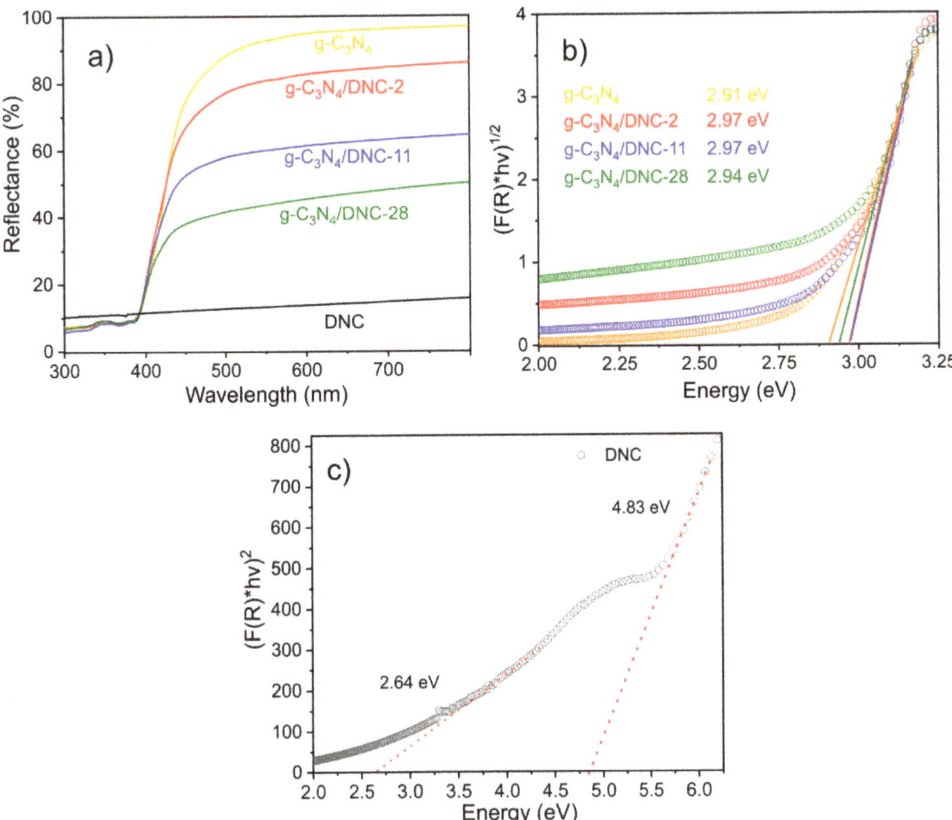

Figure 4. Diffuse reflectance spectra of pristine g-C_3N_4 and DNC, and the g-C_3N_4/DNC heterojunctions (**a**). Tauc plot for the indirect bang gap determination of pristine g-C_3N_4 and the prepared heterojunctions (**b**), and Tauc plot for the direct bang gap determination of the DNC (**c**).

To investigate the photocatalytic activity of the samples, experiments for photodegradation of the MB dye in water were conducted under simulated solar irradiation. Firstly, the MB dye adsorption efficiency without irradiation was studied for 40 min (Figure 5a), revealing that increasing DNC content enhances the adsorption efficiency of the MB in the photocatalysts, and pure DNC took a slightly longer time to achieve equilibrium in the dark (60 min, Figure S6). Similarly, the presence of DNC affected the zeta potential values, as the larger amount, the more electronegative were the heterojunctions: g-C_3N_4

(−17.2 mV), g-C_3N_4/DNC-2 (−15.5 mV), g-C_3N_4/DNC-11 (−18.2 mV), g-C_3N_4/DNC-28 (−27.6 mV), and DNC (−29.5 mV). The highly negative surface charge of the DNC favors the adsorption process through electrostatic interactions between cationic molecules and the negatively charged surface groups present in the photocatalysts. Specific surface area (S_{BET}) was determined according to the Brunauer−Emmett−Teller (BET) method using nitrogen adsorption isotherms, and the results show that the S_{BET} values of the pristine DNC and g-C_3N_4 are 83.6 and 60.9 m^2 g^{-1}, respectively. All studied heterojunctions showed higher S_{BET} values than the pristine materials but with values very close to each other, where g-C_3N_4/DNC-11 presented the largest surface area of 111.7 m^2 g^{-1}, followed by g-C_3N_4/DNC-28 (106.9 m^2 g^{-1}) and g-C_3N_4/DNC-2 (104.1 m^2 g^{-1}). Therefore, as it can be seen that the MB adsorption onto the photocatalysts does not follow the surface area trending perfectly. It is worth noting that the adsorption of the pollutant target molecule is an important but not the only step regarding its photocatalytic treatment. In this context, among the studied photocatalysts, the g-C_3N_4/DNC-28 showed a higher photoactivity for the removal of the MB under simulated solar irradiation at 200 mW cm^{-2} (Figure 5b). All photocatalytic results had a superior performance compared to the photolysis (the isolated light effect), showing that the photocatalyst material is crucial for MB degradation under light irradiation. The kinetic of the MB photocatalytic degradation follows a pseudo-first-order model with a constant rate of 0.0104 min^{-1} for the g-C_3N_4/DNC-28 sample, which is 76% higher than that obtained for pure g-C_3N_4 (0.0059 min^{-1}).

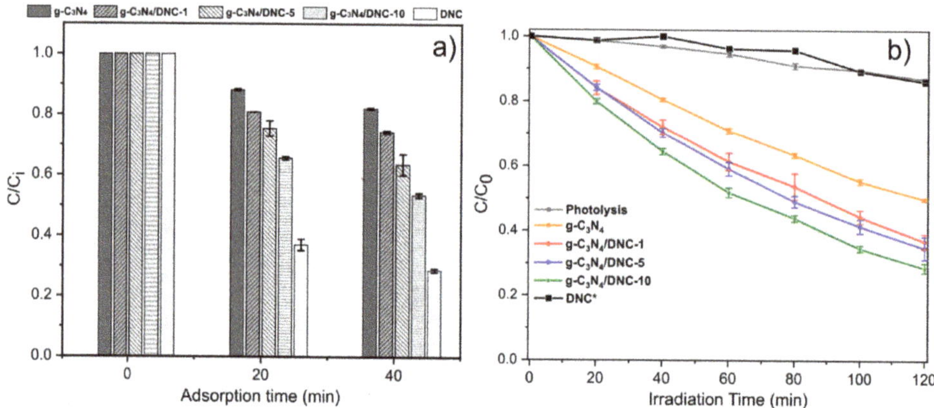

Figure 5. (**a**) C/C_i bar curves for MB adsorption under dark conditions (20 min under sonication followed by 20 min under magnetic stirring) at initial [MB] = 30 mg L^{-1}, 1 mg mL^{-1} of photocatalyst, and 25 mL liquid volume. (**b**) Photocatalytic removal of MB dye in water using pristine g-C_3N_4 and DNC, and g-C_3N_4/DNC heterojunction samples under simulated solar illumination (AM1.5G filter) at 200 mW cm^{-2}. Note: Figure 5b was produced immediately after the adsorption step in the dark (40 min for all samples, except DNC* that presented the highest adsorption ability where 60 min under dark was necessary to establish equilibrium—Figure S6), and represents the concentration of MB at the time "t" divided by the concentration of MB at the moment the light was turned on (C_0). The MB concentration was not restored to 30 mg L^{-1} and error bars represent the mean standard deviation of the triplicate experiments.

As the produced heterojunctions showed efficient results for the photocatalytic degradation of MB, we studied their ability for the photoinactivation of a model bacterium. Figure S7 shows representative images of Petri dishes with photocatalyst-containing *Staphylococcus aureus* colonies grown under illumination and dark conditions. Exposure to the heterojunction composites photocatalyst in the dark or under light and without the photocatalysts (photolysis) did not impair the *Staphylococcus aureus* growth, Figure 6. On the other hand, g-C_3N_4 induced a significant reduction in bacterial growth under illumination, demonstrating its capability of photocatalytic inactivation of *Staphylococcus aureus*. In contrast, DNC did not promote the photocatalytic inactivation of bacteria under the conditions explored herein, but the heterojunction containing 28 wt.% of DNC presented an improved photoinactivation capacity compared to the DNC-free sample ($g-C_3N_4$). This result may be due to the existing additional channel for oxidative radicals' production promoted by the heterojunction. Recent studies have demonstrated that $g-C_3N_4$ by itself can promote bacterial photocatalytic inactivation under visible-light illumination [14] and had its photoinactivation activity enhanced by preparing heterojunctions of $g-C_3N_4$ with different semiconductor nanomaterials [54,55].

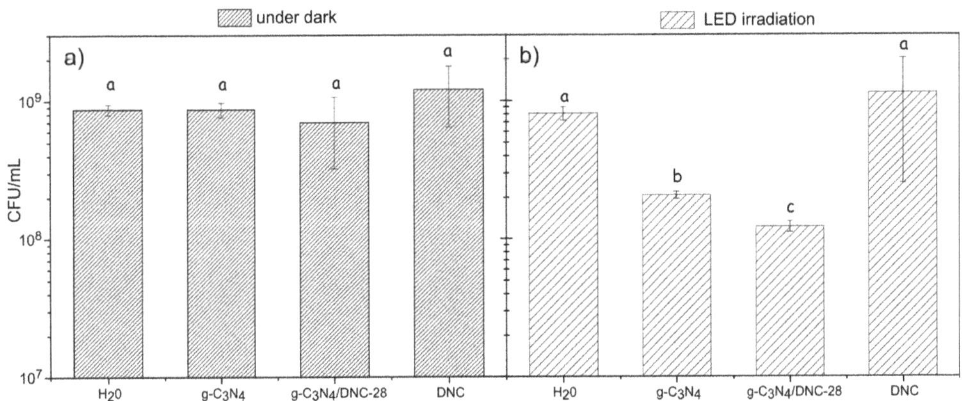

Figure 6. CFU mean values (±standard deviation) of *Staphylococcus aureus* when subjected to H_2O (negative control), and catalysts ($g-C_3N_4$, $g-C_3N_4$/DNC-28, and DNC) at 31 mg L^{-1} under dark (**a**) or LED irradiation (**b**). The duration of inactivation tests was 1 h and the irradiation source was RGB light-emitting diodes (LEDs) operated at 18 mW cm^{-2}. Means denoted by a different letter indicate significant differences between treatments ($p < 0.05$).

The photoluminescence (PL) technique can elucidate the role of DNC in the promotion or inhibition of e^-/h^+ recombination in $g-C_3N_4$. It is important to note that after photoexcitation of electrons to an excited state above or at the CB of a semiconductor, electron−hole pairs can recombine radiatively or non-radiatively, and these processes may occur through different routes with or without photon emissions, respectively [56]. Electron−hole recombination only contributes to PL band edge emission when photons are emitted with energies near the semiconductor bandgap, whilst nonradiative decays normally involve charge trapping at defect states within the bandgap [56,57]. This latter process involves the successful relaxation of charge carriers to the shallow surface trap states and the bottom of the CB−top of the VB in the case of holes. One should note that further radiative recombination coming from these trapped charges (donor−acceptor levels) may produce additional emission bands of lower energy than the band edge emission and may represent signs of intragap states in semiconductor materials. In addition, the intensity of emitted photons depends on the type/structure of materials [58] and the rate of e^-/h^+ recombination, where a lower PL emission intensity (at a fixed wavelength near the absorption band edge) may be indicative of lesser e^-/h^+ recombination [9,41,59,60]. Figure 7a presents the PL emission spectra obtained under excitation at 371 nm (excitation spectrum of pristine $g-C_3N_4$ is available in Figure S8) of the pristine $g-C_3N_4$ and the heterojunction samples. All samples show a broad PL emission that could be satisfactorily decomposed on three main Gaussian peaks (Figure 7b) centered at 438 nm (sigma $\approx$ 20), 466 nm (sigma $\approx$ 47), and 505 nm (sigma $\approx$ 47) which results from the recombination of e^-/h^+ from different pathways in $g-C_3N_4$ [30,61]. According to previous reports [62,63], these recombination paths come from the sp^3 C–N σ band, sp^2 C–N π band, and the lone pair (LP) state of the bridge N atom transitions, respectively, as depicted in the inset of Figure 7b. Additionally, the pure DNC did not show PL signal under these conditions, and as the amount of DNC increased, the overall PL intensity of the heterojunction was lowered, indicating that photogenerated electrons in the excited $g\text{-}C_3N_4$ are transferred to the DNC immediately after the photo-production, reducing, therefore, the recombination rate of charge carriers, as also reported before [25,29,30,61].

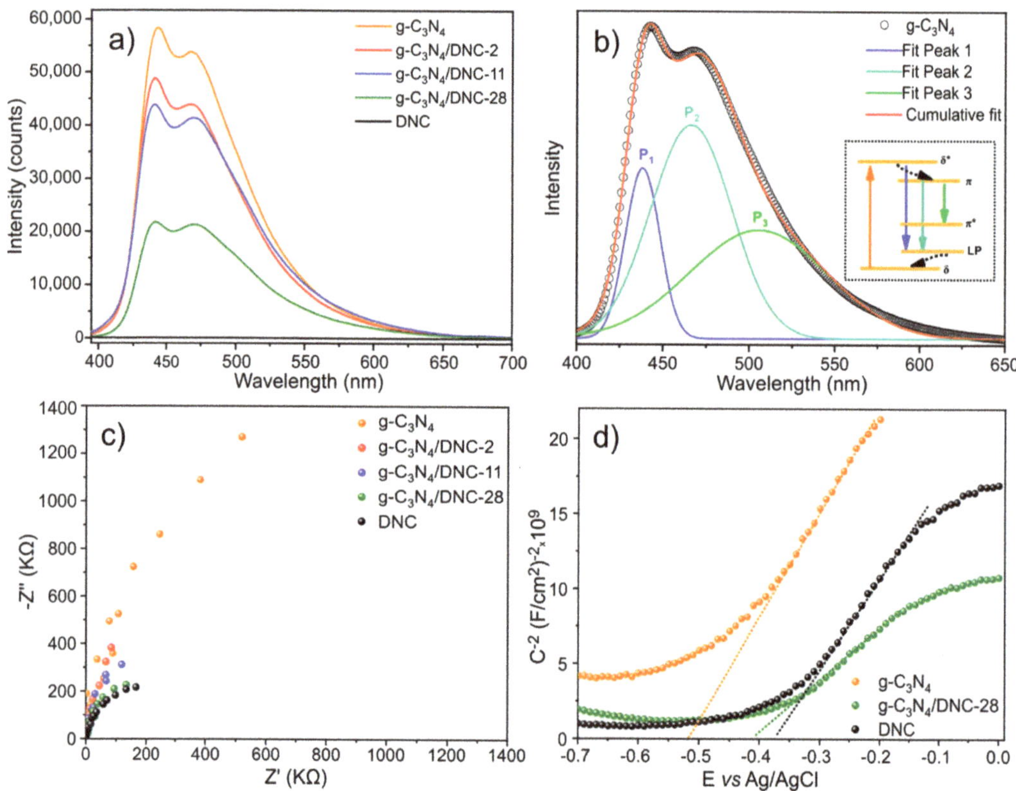

Figure 7. PL emission spectra under excitation at 371 nm (**a**); Gaussian peak fitting of $g\text{-}C_3N_4$ PL emission spectrum (**b**); EIS spectra under simulated solar light irradiation (**c**); and Mott–Schottky curves under dark conditions at 1000 Hz (**d**).

To further understand the charge transfer properties of the heterojunctions we performed EIS as a tool to study the charge transfer resistance at the interface formed between the semiconductor and the electrolyte and, indirectly, the separation of e^-/h^+ pairs at the electrode surface by monitoring the radii of the formed semicircles in the Nyquist plot (Figure 7c). It is known that the smaller the radius of the semicircle, the lower the resistance of charge transfer [64,65]. The EIS data showed a single semicircular arch for all the samples, and the radius of the semicircle diminished with the introduction of DNC in the compositions, thus indicating a lower charge transfer resistance when the diamond nanoparticles are present in the photoelectrode. Therefore, the EIS results corroborate the PL findings, indicating that DNC is working as an efficient cocatalyst on the surface of the $g\text{-}C_3N_4$, decreasing the recombination of the photogenerated charges and, consequently, improving the photocatalytic activity, as seen in the MB photodegradation and photoinactivation of *Staphylococcus aureus* bacteria.

For a better understanding of the photoactivity activity including the electronic properties of the photocatalysts, we conducted Mott-Schottky (M–S) analysis to determine the type of conductivity and the flab band potential (V_{fb}) of both pristine $g\text{-}C_3N_4$ and DNC, which is close to the CB in the case of n-type semiconductors. The M–S analysis consists of plotting $1/C^2$ against V, where C is the space charge layer capacitance, and V is the applied voltage [66]. The V_{fb} of the electrode material can be obtained through the curve intercept at the horizontal axis, Figure 7d. The M–S plots of pristine and the heterojunction samples showed positive slopes, demonstrating that the photocatalysts present n-type conductivity and electrons are the majority charge carriers. The V_{fb} of the electrodes is ~−0.52 and −0.36 V (vs. Ag/AgCl) for both pristine g-C_3N_4 and DNC, respectively, and changed to an intermediate value of −0.42 V (vs. Ag/AgCl) for the heterojunction with 10% wt. of DNC. As the V_{fb} of n-type semiconductor materials lies very close to their CB minimum (~0.1–0.2 eV) [66], we can assume CB minimum positions of g-C_3N_4 and DNC at −0.17 and −0.01 V (vs. RHE), respectively. Therefore, by using the bandgap values of the individual semiconductors, their respective VB maximum values could be estimated as 2.73 and 4.82 V (vs. RHE), and the band position alignment of the heterojunction is proposed as illustrated in Figure 8.

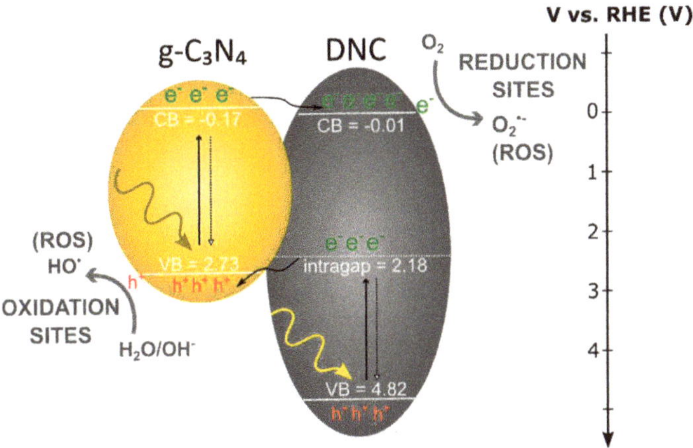

Figure 8. Mechanism of charge transfer and ROS formation in the g-C_3N_4/DNC nanosheets/NPs heterojunction.

The improved photoactivity of a heterojunction may come from the better separation of photogenerated charges and increased light absorption due to the dual-absorber semiconductor configuration [4,8]. Since more photons are absorbed, more free charges become available in the CB and VB of semiconductor materials. Depending on their electronic configurations, the photogenerated charges can physically separate, with electrons migrating to one of the semiconductors, and the holes to the other [4]. In this case, the recombination of electrons and holes decreases due to the created physical barrier, increasing the probability of redox reactions to occurs on the surface of the photocatalyst. On one hand, the photogenerated holes can drive direct oxidation reactions or react with H_2O or OH^- to produce hydroxyl radical hydroxyl, and on the other hand, the photogenerated electrons can drive direct reduction reactions or react with oxygen to form superoxide radicals.

In our case, a type-II g-C_3N_4/DNC heterojunction formed, in such a way that after photoexcitation of electrons to the CB of g-C_3N_4, those that did not straight recombine with the holes at the VB of g-C_3N_4 straight away could migrate to the CB of DNC due to its more positive potential. This latter surface works as an electron sink and triggers reduction reactions to directly attack the targeting adsorbed molecules or to generate superoxide radicals (Figure 8). On the other hand, after photoexcitation of electrons to the intragap states of DNC, they can recombine back to the VB of DNC or migrate to the VB of g-C_3N_4 due to its more negative potential, and recombine with holes present therein. However, it is important to note that the photoexcitation of DNC is not as efficient as for g-C_3N_4 under simulated solar light or LED irradiation and, therefore, the photoholes in the latter material are not completely consumed by recombination with electrons coming from the

DNC intragap states. Therefore, the oxidation reactions can still occur at the surface of the g-C_3N_4, with the reduction counterpart reactions happening at the DNC surface.

Therefore, the enhanced photocatalytic activity of the g-C_3N_4/DNC nanosheets heterojunction is a combination of some important factors. The first ones are the increased adsorption capacity of targeting molecules and enhanced light absorption effectiveness promoted by the presence of DNC. This is due to DNC's large surface area, available surface functional groups, and intragap states formation. Secondly, but no less important, is the decreased charge recombination rates due to the type-II heterojunction formation. Finally, the in situ formation of tight-packed g-C_3N_4 nanosheets during the decomposition of urea in the presence of DNC, correlated with the higher electrical conductivity of the latter, may facilitate charge transfer at the semiconductor/water interface. These results show that g-C_3N_4/DNC are efficient low-cost and metal-free photoactive catalysts for wastewater treatment and photoinactivation of bacteria.

4. Conclusions

This work demonstrated an effective and simple methodology to prepare type-II g-C_3N_4/DNC heterojunctions by adding DNC into urea decomposition in situ. Their structural, morphological, optical, and electronic properties were investigated in detail and the materials were applied as photocatalysts for the degradation of MB and photoinactivation of *Staphylococcus aureus*. The morphology of pure g-C_3N_4 was composed of nanometer-thick sheets, whilst DNC formed individual nanoparticles with a mean diameter of 125.4 ± 48.9 nm. Under the g-C_3N_4/DNC heterojunctions formation, the g-C_3N_4 nanosheets were slightly fragmented and functioned as supports for the DNC nanoparticles. The pristine g-C_3N_4 exhibited a direct bandgap of 2.91 eV and the DNC showed two distinct bandgap energies of 4.83 and 2.64 eV, the latter being a signal of intragap states due to structural/surface defects and/or quantum confinement effects. The obtained g-C_3N_4/DNC nanocomposite heterojunctions exhibited enhanced photocatalytic activity against MB degradation and *Staphylococcus aureus* inactivation under simulated solar light and LED irradiation, respectively. Due to its negative surface charge, the DNC enhanced the adsorption efficiency of the targeting molecules onto the photocatalyst by favoring electrostatic interactions between the cationic molecules and their negatively charged surface groups. The MB photocatalytic degradation followed a pseudo-first-order kinetic and the best photocatalyst (28 wt.% of DNC) showed a 76% higher degradation constant than pure g-C_3N_4. This sample also presented the highest *Staphylococcus aureus* photoinactivation activity. As the amount of DNC increased in the heterojunctions, the overall PL intensity was lowered, and the charge transfer resistance at the semiconductor/electrolyte interface followed an inverse trend, decreasing with increasing DNC concentration. This indicates that photogenerated electrons in the excited g-C_3N_4 are physically separated from the holes by transferring to the DNC nanoparticles immediately after photoexcitation, resulting in a reduced recombination rate of charge carriers. Both the g-C_3N_4 and DNC showed n-type semiconducting features and the mechanism of the decreased charge recombination was evidenced by a type-II heterojunction formation, with reduction and oxidation reactions occurring at the DNC and g-C_3N_4 surfaces, respectively. Overall, the improved photoactivity of the heterojunctions was due to better separation of photogenerated charges, increased light absorption, enhanced adsorption, and decreased charge transfer resistance. These results suggest that g-C_3N_4/DNC are efficient low-cost and metal-free photoactive catalysts for wastewater treatment and bacterial photoinactivation.

Supplementary Materials: The following are available online at https://www.mdpi.com/article/10.3390/photochem1020019/s1. Flowchart of g-C_3N_4 synthesis (Figure S1); DSC of g-C_3N_4 and g-C_3N_4/DNC-28 (Figure S2); TGA profiles for yield comparisons (Figure S3); DLS of the pristine DNC (Figure S4); Normalized FTIR spectra of pristine and prepared heterojunctions (Figure S5); Adsorption of MB under dark using the diamond nanocrystals (Figure S6); Growth of Staphylococcus aureus colonies in Petri dishes (Figure S7); Excitation PL spectrum of pristine g-C_3N_4 for emission fixed at 443 nm (Figure S8); Nominal content of DNC (wt%), masses used of both precursors in the synthesis, and the obtained experimental DNC content (Table S1); and Kinect constant (k) and R-squared values for first-order MB photodegradation reaction using prepared samples (Table S2).

Author Contributions: N.K.: investigation, methodology, and writing—original draft. L.E.G.: investigation, formal analysis, and writing—original draft. L.F.P.: methodology and formal analysis. T.H.N.L. and T.F.A.: formal analysis, resources, and writing—review and editing. J.A.P.F.: investigation, methodology, and writing—original draft. A.R.L.C.: conceptualization, formal analysis, resources, supervision, and writing—review and editing. H.W.: conceptualization, funding acquisition, project administration, supervision, and writing—review and editing. All authors have read and agreed to the published version of the manuscript.

Funding: This research was funded by the Brazilian funding agency CNPq (grant numbers: 427835/2016-0; 310066/2017-4; 302245/2021-9; and 313300/2020-8). The authors also acknowledge the financial support provided by the CAPES-PrInt funding program (grant numbers: 88887.353061/2019-00, 88881.311921/2018-01 and 88887.311920/2018-00) and the National Institute of Science and Technology of Basic Optics and Optics Applied to Life Science (grant number: 465360/2014-9). This study was supported by the Universidade Federal de Mato Grosso do Sul—UFMS/MEC—Brasil and financed in part by the Coordenação de Aperfeiçoamento de Pessoal de Nível Superior—Brasil (CAPES)—Finance Code 001.

Data Availability Statement: Data are available upon request from the authors.

Acknowledgments: The authors are grateful to the Brazilian funding agencies CNPq and CAPES, Universidade Federal de Mato Grosso do Sul—UFMS/MEC—Brasil, and the Laboratório de Pesquisa em Nanotecnologia (LANano)/CETENE.

Conflicts of Interest: The authors declare no conflict of interest.

References

1. Li, M.; Liu, Y.; Dong, L.; Shen, C.; Li, F.; Huang, M.; Ma, C.; Yang, B.; An, X.; Sand, W. Recent advances on photocatalytic fuel cell for environmental applications—The marriage of photocatalysis and fuel cells. *Sci. Total Environ.* **2019**, *668*, 966–978. [CrossRef] [PubMed]
2. Masih, D.; Ma, Y.; Rohani, S. Graphitic C_3N_4 based noble-metal-free photocatalyst systems: A review. *Appl. Catal. B Environ.* **2017**, *206*, 556–588. [CrossRef]
3. Du, H.; Liu, Y.-N.; Shen, C.-C.; Xu, A.-W. Nanoheterostructured photocatalysts for improving photocatalytic hydrogen production. *Chin. J. Catal.* **2017**, *38*, 1295–1306. [CrossRef]
4. Stelo, F.; Kublik, N.; Ullah, S.; Wender, H. Recent advances in Bi_2MoO_6 based Z-scheme heterojunctions for photocatalytic degradation of pollutants. *J. Alloys Compd.* **2020**, *829*, 154591. [CrossRef]
5. Ye, S.; Wang, R.; Wu, M.-Z.; Yuan, Y.-P. A review on g-C_3N_4 for photocatalytic water splitting and CO_2 reduction. *Appl. Surf. Sci.* **2015**, *358*, 15–27. [CrossRef]
6. Navarro, R.M.; Alvarez-Galvan, M.C.; de la Mano, J.A.V.; Al-Zahrani, S.; Fierro, J.L.G. A framework for visible-light water splitting. *Energy Environ. Sci.* **2010**, *3*, 1865–1882. [CrossRef]
7. Feil, A.F.; Wender, H.; Gonçalves, R.V. Photovoltaic, Photocatalytic Application, and water Splitting. *Nanocatal. Ion. Liq.* **2016**, 275–294. [CrossRef]
8. Gonçalves, R.V.; Wender, H.; Khan, S.; Melo, M.A. Photocatalytic Water Splitting by Suspended Semiconductor Particles. *Nanoenergy* **2017**, 107–140. [CrossRef]
9. Nogueira, A.C.; Gomes, L.E.; Ferencz, J.A.P.; Rodrigues, J.E.F.S.; Gonçalves, R.V.; Wender, H. Improved Visible Light Photoactivity of $CuBi_2O_4$/CuO Heterojunctions for Photodegradation of Methylene Blue and Metronidazole. *J. Phys. Chem. C* **2019**, *123*, 25680–25690. [CrossRef]
10. Homem, V.; Santos, L. Degradation and removal methods of antibiotics from aqueous matrices—A review. *J. Environ. Manag.* **2011**, *92*, 2304–2347. [CrossRef]
11. Julkapli, N.M.; Bagheri, S.; Hamid, S.B.A. Recent Advances in Heterogeneous Photocatalytic Decolorization of Synthetic Dyes. *Sci. World J.* **2014**, *2014*, 692307. [CrossRef] [PubMed]
12. Sagara, N.; Kamimura, S.; Tsubota, T.; Ohno, T. Photoelectrochemical CO_2 reduction by a p-type boron-doped g-C_3N_4 electrode under visible light. *Appl. Catal. B Environ.* **2016**, *192*, 193–198. [CrossRef]
13. Mao, J.; Peng, T.; Zhang, X.; Li, K.; Ye, L.; Zan, L. Effect of graphitic carbon nitride microstructures on the activity and selectivity of photocatalytic CO2 reduction under visible light. *Catal. Sci. Technol.* **2013**, *3*, 1253–1260. [CrossRef]
14. Zhao, H.; Yu, H.; Quan, X.; Chen, S.; Zhang, Y.; Zhao, H.; Wang, H. Fabrication of atomic single layer graphitic-C_3N_4 and its high performance of photocatalytic disinfection under visible light irradiation. *Appl. Catal. B Environ.* **2014**, *152-153*, 46–50. [CrossRef]
15. Wen, J.; Xie, J.; Chen, X.; Li, X. A review on g-C_3N_4—Based photocatalysts. *Appl. Surf. Sci.* **2017**, *391*, 72–123. [CrossRef]

16. Zhou, L.; Zhang, H.; Guo, X.; Sun, H.; Liu, S.; Tade, M.; Wang, S. Metal-free hybrids of graphitic carbon nitride and nanodiamonds for photoelectrochemical and photocatalytic applications. *J. Colloid Interface Sci.* **2017**, *493*, 275–280. [CrossRef]
17. Silva, G.S.T.; Carvalho, K.T.G.; Lopes, O.; Ribeiro, C. g-C_3N_4/Nb_2O_5 heterostructures tailored by sonochemical synthesis: Enhanced photocatalytic performance in oxidation of emerging pollutants driven by visible radiation. *Appl. Catal. B Environ.* **2017**, *216*, 70–79. [CrossRef]
18. Yilmaz, E.; Soylak, M. Facile and green solvothermal synthesis of palladium nanoparticle-nanodiamond-graphene oxide material with improved bifunctional catalytic properties. *J. Iran. Chem. Soc.* **2017**, *14*, 2503–2512. [CrossRef]
19. Tian, Y.; Ge, L.; Wang, K.; Chai, Y. Synthesis of novel MoS_2/g-C_3N_4 heterojunction photocatalysts with enhanced hydrogen evolution activity. *Mater. Charact.* **2014**, *87*, 70–73. [CrossRef]
20. Isberg, J.; Hammersberg, J.; Johansson, E.; Wikström, T.; Twitchen, D.J.; Whitehead, A.J.; Coe, S.E.; Scarsbrook, G.A. High Carrier Mobility in Single-Crystal Plasma-Deposited Diamond. *Science* **2002**, *297*, 1670–1672. [CrossRef]
21. Holt, K.B. Undoped diamond nanoparticles: Origins of surface redox chemistry. *Phys. Chem. Chem. Phys.* **2010**, *12*, 2048–2058. [CrossRef]
22. Hirai, H.; Terauchi, M.; Tanaka, M.; Kondo, K. Estimating band gap of amorphous diamond and nanocrystalline diamond powder by electron energy loss spectroscopy. *Diam. Relat. Mater.* **1999**, *8*, 1703–1706. [CrossRef]
23. Choudhury, S.; Kiendl, B.; Ren, J.; Gao, F.; Knittel, P.; Nebel, C.; Venerosy, A.; Girard, H.; Arnault, J.-C.; Krueger, A.; et al. Combining nanostructuration with boron doping to alter sub band gap acceptor states in diamond materials. *J. Mater. Chem. A* **2018**, *6*, 16645–16654. [CrossRef]
24. Lin, Z.; Xiao, J.; Li, L.; Liu, P.; Wang, C.; Yang, G. Nanodiamond-Embedded p-Type Copper(I) Oxide Nanocrystals for Broad-Spectrum Photocatalytic Hydrogen Evolution. *Adv. Energy Mater.* **2015**, *6*. [CrossRef]
25. Su, L.-X.; Huang, Q.-Z.; Lou, Q.; Liu, Z.-Y.; Sun, J.-L.; Zhang, Z.-T.; Qin, S.-R.; Li, X.; Zang, J.-H.; Dong, L.; et al. Effective light scattering and charge separation in nanodiamond@g-C_3N_4 for enhanced visible-light hydrogen evolution. *Carbon* **2018**, *139*, 164–171. [CrossRef]
26. Yu, S.-J.; Kang, M.-W.; Chang, H.-C.; Chen, K.-M.; Yu, Y.-C. Bright Fluorescent Nanodiamonds: No Photobleaching and Low Cytotoxicity. *J. Am. Chem. Soc.* **2005**, *127*, 17604–17605. [CrossRef] [PubMed]
27. Pastrana-Martínez, L.M.; Morales-Torres, S.; Carabineiro, S.A.C.; Buijnsters, J.; Faria, J.; Figueiredo, J.; Silva, A.M.T. Nanodiamond-TiO_2Composites for Heterogeneous Photocatalysis. *ChemPlusChem* **2013**, *78*, 801–807. [CrossRef] [PubMed]
28. Kim, K.-D.; Dey, N.K.; Seo, H.O.; Kim, Y.D.; Lim, D.C.; Lee, M. Photocatalytic decomposition of toluene by nanodiamond-supported TiO_2 prepared using atomic layer deposition. *Appl. Catal. A Gen.* **2011**, *408*, 148–155. [CrossRef]
29. Sampaio, M.J.; Pastrana-Martínez, L.M.; Silva, A.M.T.; Buijnsters, J.G.; Han, C.; Silva, C.G.; Carabineiro, S.A.C.; Dionysiou, D.D.; Faria, J. Nanodiamond–TiO_2 composites for photocatalytic degradation of microcystin-LA in aqueous solutions under simulated solar light. *RSC Adv.* **2015**, *5*, 58363–58370. [CrossRef]
30. Haleem, Y.A.; He, Q.; Liu, D.; Wang, C.; Xu, W.; Gan, W.; Zhou, Y.; Wu, C.; Ding, Y.; Song, L. Facile synthesis of mesoporous detonation nanodiamond-modified layers of graphitic carbon nitride as photocatalysts for the hydrogen evolution reaction. *RSC Adv.* **2017**, *7*, 15390–15396. [CrossRef]
31. Zhang, C.; Li, Y.; Shuai, D.; Shen, Y.; Xiong, W.; Wang, L. Graphitic carbon nitride (g-C_3N_4)-based photocatalysts for water disinfection and microbial control: A review. *Chemosphere* **2018**, *214*, 462–479. [CrossRef] [PubMed]
32. Norouzi, N.; Ong, Y.; Damle, V.G.; Najafi, M.B.H.; Schirhagl, R. Effect of medium and aggregation on antibacterial activity of nanodiamonds. *Mater. Sci. Eng. C* **2020**, *112*, 110930. [CrossRef] [PubMed]
33. Wang, C.; Makvandi, P.; Zare, E.N.; Tay, F.R.; Niu, L. Advances in Antimicrobial Organic and Inorganic Nanocompounds in Biomedicine. *Adv. Ther.* **2020**, *3*. [CrossRef]
34. Cui, H.; Gu, Z.; Chen, X.; Lin, L.; Wang, Z.; Dai, X.; Yang, Z.; Liu, L.; Zhou, R.; Dong, M. Stimulating antibacterial activities of graphitic carbon nitride nanosheets with plasma treatment. *Nanoscale* **2019**, *11*, 18416–18425. [CrossRef]
35. Wehling, J.; Dringen, R.; Zare, R.N.; Maas, M.; Rezwan, K. Bactericidal Activity of Partially Oxidized Nanodiamonds. *ACS Nano* **2014**, *8*, 6475–6483. [CrossRef] [PubMed]
36. Zhang, T.; Kalimuthu, S.; Rajasekar, V.; Xu, F.; Yiu, Y.C.; Hui, T.K.C.; Neelakantan, P.; Chu, Z. Biofilm inhibition in oral pathogens by nanodiamonds. *Biomater. Sci.* **2021**, *9*, 5127–5135. [CrossRef] [PubMed]
37. Openda, Y.I.; Ngoy, B.P.; Nyokong, T. Photodynamic Antimicrobial Action of Asymmetrical Porphyrins Functionalized Silver-Detonation Nanodiamonds Nanoplatforms for the Suppression of Staphylococcus aureus Planktonic Cells and Biofilms. *Front. Chem.* **2021**, *9*. [CrossRef] [PubMed]
38. Openda, Y.I.; Nyokong, T. Detonation nanodiamonds-phthalocyanine photosensitizers with enhanced photophysicochemical properties and effective photoantibacterial activity. *Photodiagn. Photodyn. Ther.* **2020**, *32*, 102072. [CrossRef]
39. Yang, Y.; Chen, J.; Mao, Z.; An, N.; Wang, D.; Fahlman, B.D. Ultrathin g-C_3N_4 nanosheets with an extended visible-light-responsive range for significant enhancement of photocatalysis. *RSC Adv.* **2017**, *7*, 2333–2341. [CrossRef]
40. Zheng, Y.; Zhang, Z.; Li, C. A comparison of graphitic carbon nitrides synthesized from different precursors through pyrolysis. mboxemphJ. Photochem. Photobiol. A Chem. **2017**, *332*, 32–44. [CrossRef]
41. Gomes, L.E.; da Silva, M.F.; Gonçalves, R.V.; Machado, G.; Alcantara, G.B.; Caires, A.R.L.; Wender, H. Synthesis and Visible-Light-Driven Photocatalytic Activity of Ta_4+ Self-Doped Gray Ta_2O_5 Nanoparticles. *J. Phys. Chem. C* **2018**, *122*, 6014–6025. [CrossRef]

42. Gomes, L.E.; Nogueira, A.C.; da Silva, M.F.; Plaça, L.F.; Maia, L.J.Q.; Gonçalves, R.V.; Ullah, S.; Khan, S.; Wender, H. Enhanced photocatalytic activity of $BiVO_4$/Pt/PtOx photocatalyst: The role of Pt oxidation state. *Appl. Surf. Sci.* **2021**, *567*, 150773. [CrossRef]
43. Li, K.; Gao, S.; Wang, Q.; Xu, H.; Wang, Z.; Huang, B.; Dai, Y.; Lu, J. In-Situ-Reduced Synthesis of Ti3+ Self-Doped TiO_2/g-C3N_4 Heterojunctions with High Photocatalytic Performance under LED Light Irradiation. *ACS Appl. Mater. Interfaces* **2015**, *7*, 9023–9030. [CrossRef] [PubMed]
44. Ismael, M.; Wu, Y.; Taffa, D.H.; Bottke, P.; Wark, M. Graphitic carbon nitride synthesized by simple pyrolysis: Role of precursor in photocatalytic hydrogen production. *New J. Chem.* **2019**, *43*, 6909–6920. [CrossRef]
45. Piña-Salazar, E.-Z.; Kukobat, R.; Futamura, R.; Hayashi, T.; Toshio, S.; Ōsawa, E.; Kaneko, K. Water-selective adsorption sites on detonation nanodiamonds. *Carbon* **2018**, *139*, 853–860. [CrossRef]
46. Ishida, N.; Kato, K.; Suzuki, N.; Fujimoto, K.; Hagio, T.; Ichino, R.; Kondo, T.; Yuasa, M.; Uetsuka, H.; Fujishima, A.; et al. Preparation of amino group functionalized diamond using photocatalyst and thermal conductivity of diamond/copper composite by electroplating. *Diam. Relat. Mater.* **2021**, *118*, 108509. [CrossRef]
47. Gao, J.; Zhou, Y.; Li, Z.; Yan, S.; Wang, N.; Zou, Z. High-yield synthesis of millimetre-long, semiconducting carbon nitride nanotubes with intense photoluminescence emission and reproducible photoconductivity. *Nanoscale* **2012**, *4*, 3687–3692. [CrossRef]
48. Wang, X.; Maeda, K.; Thomas, A.; Takanabe, K.; Xin, G.; Carlsson, J.M.; Domen, K.; Antonietti, M. A metal-free polymeric photocatalyst for hydrogen production from water under visible light. *Nat. Mater.* **2008**, *8*, 76–80. [CrossRef]
49. Pina-Salazar, E.-Z.; Urita, K.; Hayashi, T.; Futamura, R.; Vallejos-Burgos, F.; Włoch, J.; Kowalczyk, P.; Wiśniewski, M.; Sakai, T.; Moriguchi, I.; et al. Water Adsorption Property of Hierarchically Nanoporous Detonation Nanodiamonds. *Langmuir* **2017**, *33*, 11180–11188. [CrossRef]
50. Xiang, Q.; Yu, J.; Jaroniec, M. Preparation and Enhanced Visible-Light Photocatalytic H_2-Production Activity of Graphene/C_3N_4 Composites. *J. Phys. Chem. C* **2011**, *115*, 7355–7363. [CrossRef]
51. Nunn, N.; Torelli, M.; McGuire, G.; Shenderova, O. Nanodiamond: A high impact nanomaterial. *Curr. Opin. Solid State Mater. Sci.* **2017**, *21*, 1–9. [CrossRef]
52. Aharonovich, I.; Greentree, A.D.; Prawer, S. Diamond photonics. *Nat. Photonics* **2011**, *5*, 397–405. [CrossRef]
53. Hertkorn, J.; Fyta, M. Electronic features of vacancy, nitrogen, and phosphorus defects in nanodiamonds. *Electron. Struct.* **2019**, *1*, 025002. [CrossRef]
54. Li, J.; Yin, Y.; Liu, E.; Ma, Y.; Wan, J.; Fan, J.; Hu, X. In situ growing Bi_2MoO_6 on g-C_3N_4 nanosheets with enhanced photocatalytic hydrogen evolution and disinfection of bacteria under visible light irradiation. *J. Hazard. Mater.* **2017**, *321*, 183–192. [CrossRef] [PubMed]
55. Li, G.; Nie, X.; Chen, J.; Jiang, Q.; An, T.; Wong, P.K.; Zhang, H.; Zhao, H.; Yamashita, H. Enhanced visible-light-driven photocatalytic inactivation of Escherichia coli using g-C_3N_4/TiO_2 hybrid photocatalyst synthesized using a hydrothermal-calcination approach. *Water Res.* **2015**, *86*, 17–24. [CrossRef]
56. Tan, H.L.; Wen, X.; Amal, R.; Ng, Y.H. $BiVO_4$ {010} and {110} Relative Exposure Extent: Governing Factor of Surface Charge Population and Photocatalytic Activity. *J. Phys. Chem. Lett.* **2016**, *7*, 1400–1405. [CrossRef]
57. Klimov, V.; Bolívar, P.H.; Kurz, H. Ultrafast carrier dynamics in semiconductor quantum dots. *Phys. Rev. B* **1996**, *53*, 1463–1467. [CrossRef] [PubMed]
58. Lin, X.; Hou, J.; Jiang, S.; Lin, Z.; Wang, M.; Che, G. A Z-scheme visible-light-driven Ag/Ag_3PO_4/Bi_2MoO_6 photocatalyst: Synthesis and enhanced photocatalytic activity. *RSC Adv.* **2015**, *5*, 104815–104821. [CrossRef]
59. Zhang, L.; Dai, Z.; Zheng, G.; Yao, Z.; Mu, J. Superior visible light photocatalytic performance of reticular $BiVO_4$ synthesized via a modified sol–gel method. *RSC Adv.* **2018**, *8*, 10654–10664. [CrossRef]
60. Geng, Y.; Zhang, P.; Li, N.; Sun, Z. Synthesis of Co doped $BiVO_4$ with enhanced visible-light photocatalytic activities. *J. Alloys Compd.* **2015**, *651*, 744–748. [CrossRef]
61. Su, L.-X.; Liu, Z.-Y.; Ye, Y.-L.; Shen, C.-L.; Lou, Q.; Shan, C.-X. Heterostructured boron doped nanodiamonds@g-C3N4 nanocomposites with enhanced photocatalytic capability under visible light irradiation. *Int. J. Hydrogen Energy* **2019**, *44*, 19805–19815. [CrossRef]
62. Yuan, Y.; Zhang, L.; Xing, J.; Utama, M.I.B.; Lu, X.; Du, K.; Li, Y.; Hu, X.; Wang, S.; Genç, A.; et al. High-yield synthesis and optical properties of g-C_3N. *Nanoscale* **2015**, *7*, 12343–12350. [CrossRef] [PubMed]
63. Das, D.; Shinde, S.L.; Nanda, K.K. Temperature-Dependent Photoluminescence of g-C_3N_4: Implication for Temperature Sensing. *ACS Appl. Mater. Interfaces* **2016**, *8*, 2181–2186. [CrossRef] [PubMed]
64. Zhang, W.; Li, R.; Zhao, X.; Chen, Z.; Law, W.K.A.; Zhou, K. A Cobalt-Based Metal–Organic Framework as Cocatalyst on BiVO 4 Photoanode for Enhanced Photoelectrochemical Water Oxidation. *ChemSusChem* **2018**, *11*, 2710–2716. [CrossRef] [PubMed]
65. Liu, C.; Luo, H.; Xu, Y.; Wang, W.; Liang, Q.; Mitsuzaki, N.; Chen, Z. Cobalt–phosphate-modified Mo:$BiVO_4$ mesoporous photoelectrodes for enhanced photoelectrochemical water splitting. *J. Mater. Sci.* **2019**, *54*, 10670–10683. [CrossRef]
66. Ullah, S.; Fayeza; Khan, A.A.; Jan, A.; Aain, S.Q.; Neto, E.P.; Serge-Correales, Y.E.; Parveen, R.; Wender, H.; Rodrigues-Filho, U.P.; et al. Enhanced photoactivity of $BiVO_4$/Ag/Ag_2O Z-scheme photocatalyst for efficient environmental remediation under natural sunlight and low-cost LED illumination. *Colloids Surfaces A Physicochem. Eng. Asp.* **2020**, *600*, 124946. [CrossRef]

Article

Photocatalytic Decomposition of Nitrobenzene in Aqueous Solution by Ag/Cu_2O Assisted with Persulfate under Visible Light Irradiation

Wen-Shing Chen * and Jyun-Yang Chen

Department of Chemical and Materials Engineering, National Yunlin University of Science & Technology, 123 University Road, Section 3, Douliou 640, Yunlin, Taiwan; m10915009@yuntech.edu.tw
* Correspondence: chenwen@yuntech.edu.tw; Tel.: +88-6553-42601 (ext. 4624); Fax: +88-6553-12071

Abstract: The mineralization of nitrobenzene was executed using an innovative method, wherein Ag/Cu_2O semiconductors stimulated by visible light irradiation were supported with persulfate anions. Batch-wise experiments were performed for the evaluation of effects of silver metal contents impregnated, persulfate concentrations and Ag/Cu_2O dosages on the nitrobenzene removal efficiency. The physicochemical properties of fresh and reacted Ag/Cu_2O were illustrated by X-ray diffraction analyses, FE-SEM images, EDS Mapping analyses, UV–Vis diffuse reflectance spectra, transient photocurrent analyses and X-ray photoelectron spectra, respectively. Because of intense scavenging effects caused by benzene, 1-propanol and methanol individually, the predominant oxidant was considered to be sulfate radicals, originated from persulfate anions via the photocatalysis of Ag/Cu_2O. As regards oxidation pathways, nitrobenzene was initially transformed into hydroxycyclohexadienyl radicals, followed with the production of 2-nitrophenol, 3-nitrophenol or 4-nitrophenol. Afterwards, phenol compounds descended from denitration of nitrophenols were converted into hydroquinone and *p*-benzoquinone.

Keywords: nitrobenzene; Ag/Cu_2O; persulfate; sulfate radical

Citation: Chen, W.-S.; Chen, J.-Y. Photocatalytic Decomposition of Nitrobenzene in Aqueous Solution by Ag/Cu_2O Assisted with Persulfate under Visible Light Irradiation. *Photochem* **2021**, *1*, 220–236. https://doi.org/10.3390/photochem1020013

Received: 15 July 2021
Accepted: 16 August 2021
Published: 27 August 2021

1. Introduction

Nitrobenzene is commonly used for the manufacture of polyurethane by way of intermediates of aniline. It has been also applied in the following industries: plastics, pesticides, pharmaceuticals and explosives [1]. Due to high risks for mutagenicity and carcinogenicity suffered, effluent contaminated with nitrobenzene and related derivatives would cause strong damage to the aqueous circumstances [2,3]. Consequently, much effort has been focused on the development of economically and effectively treating manners for industrial wastewater.

Advanced oxidation processes have been extensively explored for the mineralization of nitrobenzene in wastewater due to its resistance to biodegradation resulted from the electron-withdrawing property of nitro groups [4]. Firstly, much research has been carried out on hydroxyl radical-based processes, such as Fenton's methods [5–7], Fenton-like manners [8–10], Fenton reagents with auxiliary ultrasound [11] and Fenton reagents coupled with fluidization flow patterns [12,13]. Secondly, as commercial titanium dioxide is under investigation, the significant enhancement of the nitrobenzene destruction efficiency took advantage of doping ferric oxides, which successfully prevent the combination of photon-induced electrons with holes [14]. In another aspect, ultraviolet absorbance band was obviously changed to the visible light range by virtue of impregnating ammonium and cerium nitrates simultaneously [15,16]. On the other hand, ozone supported with ultrasound was employed for nitrobenzene removal, wherein hydroxyl radicals were claimed to be predominant oxidizing agents [17,18]. The oxidation of nitrobenzene through the catalysis of ozone over aluminum silicate was performed to elucidate the influences of operating parameters [19,20]. Additionally, the electrochemical oxidation of nitrobenzene was conducted by the modified PbO_2 electrode plates, which were incorporated with titanium dioxide nanotubes, resulting in ordered arrangement and surface areas extended [21,22].

To date, sulfate-radical-related processes have been also dedicated to decomposing nitrobenzene in wastewater. The thermal activation of persulfate was effective for the disposal of effluents contaminated with nitrobenzene, wherein reaction pathways consist of 2-nitrophenol, 4-nitrophenol, 2,6-dinitrophenol and 2,4-dinitrophenol [23]. Except ferrous ions for the catalytic transformation of persulfate to sulfate radicals, zinc metal and magnetized iron metal also exhibited fruitful performance on nitrobenzene removal [24–26]. With a view to promoting nitrobenzene abatement, experiments were carried out utilizing persulfate anions irradiated with ultraviolet light [27]. Even though persulfate activated with photoelectrons descended from semiconductors excited by visible light, this has been barely studied in regard to nitrobenzene oxidation. As expected, the band gap energy of semiconductors meets the luminous energy of the incident light, leading to the generation of photo-induced electron–hole pairs. Persulfate anions would be transformed into reactive sulfate radicals via the activation of photo-induced electrons [28].

In this research, an innovative technique for the effective removal of nitrobenzene in wastewater was developed. Considerable sulfate radicals could be produced by way of persulfate via photoelectrons originated from Ag/Cu_2O irradiated with visible light, which are well known as semiconductors [29–32]. The influences of operating parameters on the nitrobenzene degradation behaviors were explored, such as persulfate concentrations and Ag/Cu_2O dosages. Nitrobenzene decomposition pathways catalyzed by the Ag/Cu_2O coupled with persulfate under visible light irradiation would be investigated in the meantime.

2. Experimental Methods

2.1. Testing of Photocatalytic Oxidation of Nitrobenzene by Ag/Cu_2O with Assistance of Persulfate

The experimental system containing the main equipment was referred to in our previous report [33]. The photocatalytic cell was a quartz cylinder fitted internally with a magnetic stirrer and cooling coils, in which the testing temperature was maintained through a thermostat (PIIN JIA Technol. Co. New Taipei City, Taiwan). Visible light irradiation was supplied from twelve lamps (8.6 W each) surrounding the cell with three chief peaks of 438 nm, 550 nm and 619 nm, respectively (Philips Corp. PL-S, Hanover, MD, US). Owing to consistence with practical concentrations of industrial wastewater, feedstock was prepared at 1.0 mM concentrations of nitrobenzene ($\geq$99.5%, Riedel-de Haen, Seelze, Germany) [34], being well agitated with proper weights of sodium persulfate ($\geq$99.5%, Fluka, Seelze, Germany) beforehand. The Ag/Cu_2O, manufactured from Cu_2O powder (SHOWA, Tokyo, Japan) by incipient impregnation with 1–5 wt% of silver nitrate ($\geq$99.5%, Fluka), respectively, and sequential 3 h calcination at 473 K by sieving with 400 mesh [35], was carefully loaded into the basket and fixed near the center of photocatalytic cell. For the duration of tests, samples were taken from the cell at constant time intervals, and sequentially quenched to the temperature of 273 $\pm$ 0.5 K to terminate nitrobenzene oxidation [36]. Aqueous samples were executed on total organic carbon analyses to evaluate residual organic compounds. The Ag/Cu_2O separated from oxidation tests were examined by an X-ray photoelectron spectrometer. In this work, experiments were performed in a series of persulfate concentrations of 35.0 to 70.0 mM to elucidate the sulfate radical effect at the pH values of 5 to 6. Photocatalytic tests under diverse dosages of Ag/Cu_2O (1.05 up to 1.50 g L^{-1}) were carried out for enhancement on nitrobenzene removal efficiency. All experiments were undertaken repeatedly for the affirmation of data reliability.

2.2. Total Organic Carbon (TOC) Analysis

Within the duration of photocatalytic testing by Ag/Cu_2O assisted with persulfate, wastewater was periodically sampled and measured, utilizing a TOC analyzer (GE Corp. Sievers InnovOx, Boston, MA, USA). The hydrocarbons involved were completely oxidized into carbon dioxide and quantified through nondispersive infrared (NDIR) analyses, wherein persulfate mineralization was carried out under supercritical water conditions. On the contrary, non-hydrocarbons were exhausted in the species of carbonic acid. In this work, the TOC concentrations reported were calibrated to the standard curve, fulfilled faithfully in the range (0–5.0 mM) utilizing standard chemicals of potassium hydrogen phthalate.

2.3. *Physicochemical Properties of Ag/Cu_2O*

The crystal compositions of fresh Ag/Cu_2O semiconductors were examined using an X-ray diffractometer (Rigaku, MiniFlex II, Tokyo, Japan) integrated with high-intensity monochromated CuKα radiation at the wavelength of 0.15418 nm, operating with the current of 30 mA and 40 kV volts over the 2θ range from 10 to 80 degrees. For the inspection of surface morphology and silver content impregnated on Ag/Cu_2O semiconductors, surface scanning was executed by a field-emission scanning electron microscope (JEOL, JSM-6500F) equipped with an energy dispersive X-ray spectroscope (JEOL, JED-2300). As far as light absorption band is concerned, the Ultraviolet–Visible diffuse reflectance spectra on Ag/Cu_2O were measured using a UV–Vis spectrometer (PerkinElmer, Lambda-850, Waltham, MA, USA), of which wavelength was among the range of 250 to 800 nm with reference to $BaSO_4$. The photocurrent measurements of the samples were carried out using a potentiostat (Zensor, ECAS-100, Etterbeek, Belgium) under continuous visible light irradiation (Philips, PL-S), wherein a Pt wire served as an auxiliary electrode, coated with Ag/Cu_2O and polymer electrolyte membrane, and was referenced to a saturated calomel electrode. Further, electronic states of fresh and reacted Ag/Cu_2O semiconductors were monitored by means of XPS spectra from an X-ray photoelectron spectrometer (Kratos Analytical Ltd. Axis Ultra, Manchester, UK), in which monochromated AlKα irradiation was used as a light source and the binding energy of samples was calibrated to 284.8 eV for C 1s core level of adventitious carbon.

2.4. *Gas Chromatography–Mass Spectrometry (GC-MS) Analysis*

A proper amount of wastewater was withdrawn from the photocatalytic cell after nitrobenzene oxidation testing for 30 min. The microextraction fiber spread with carboxen/polydimethylsiloxane (Supelco, Bellefonte, PA, USA) was added into aqueous solution for the effective adsorption of degradation intermediates. Then, the fiber was directly packed into a micro-needle which was immediately injected into the orifice of the gas chromatograph–mass spectrometer (Hewlett Packard, MASS 59864B/5973, Palo Alto, CA, USA). The metal capillary column for ingredient separation was used at the dimension of 30 m $\times$ 0.25 mm (Ultra ALLOY UA-5), wherein helium gas served as the carrier gas. The degradation intermediates resolved were trustworthy based on mass spectra obtained in comparison with those of standards.

2.5. *Scavenging Effects*

In order to disclose the main oxidants on the mineralization of nitrobenzene, testing was conducted in the presence of diverse scavengers simultaneously, such as methanol, 1-propanol and benzene, respectively [37,38]. The nitrobenzene degradation percentage was determined on the basis of the peak area (262 nm) shown in a UV–Vis spectrophotometer (PerkinElmer, Lambda 850) [8]. In the course of pretesting, benzene was verified as the most sharp scavenger. The benzene scavenging effect may represent sulfate radical yields at different experimental conditions. To evaluate sulfate radical yields, the decrement of nitrobenzene degradation percentage was examined upon the addition of suitable amounts of benzene scavenger into wastewater.

3. Results and Discussion

3.1. *Comparison of Photocatalytic Oxidation by Ag/Cu_2O Alone and Ag/Cu_2O Assisted with Persulfate Respectively*

Figure 1 demonstrates the time flow patterns of TOC removal efficiency executed by photocatalytic oxidation over Ag/Cu_2O alone and Ag/Cu_2O aided with persulfate, respectively. Apparently, the nitrobenzene removal rate caused by Ag/Cu_2O supported with persulfate process was much higher than those simply using persulfate anions, Cu_2O and Ag/Cu_2O alone. Noteworthily, Ag(1%)/Cu_2O integrated with persulfate displayed a synergistic performance in comparison with photocatalytic behaviors exhibited with the Ag(1%)/Cu_2O and persulfate individually. The observation could be ascribed to great enhancement on reactive sulfate radical yields. In fact, persulfate anions have been successfully transformed to sulfate radicals through the photocatalysis of Cu_2O [39]. It has been also recognized that Ag metal functions as an electron sink and strengthens charge separation, leading to the repression of the combination of photo-induced electrons and holes over Cu_2O [40,41]. As expected, a higher extent of Ag metal doped on the surface of Cu_2O gave rise to higher nitrobenzene removal efficiency (refer to Figure 2). The main reactions that occurred could be concluded as follows.

(1) $Ag/Cu_2O + h\upsilon \rightarrow h^+_{vb} + e^-_{cb}$
(2) $S_2O_8^{2-} + e^-_{cb} \rightarrow SO_4\bullet^- + SO_4^{2-}$
(3) $SO_4^{2-} + h^+_{vb} \rightarrow SO_4\bullet^-$
(4) $H_2O + h^+_{vb} \rightarrow HO\bullet + H^+$

where in e^-_{cb} represents photo-induced electrons in the conduction band and h^+_{vb} represents photo-induced holes in the valence band. Ag(5%)/Cu_2O was chosen as a candidate for next testing due to its better nitrobenzene degradation behavior.

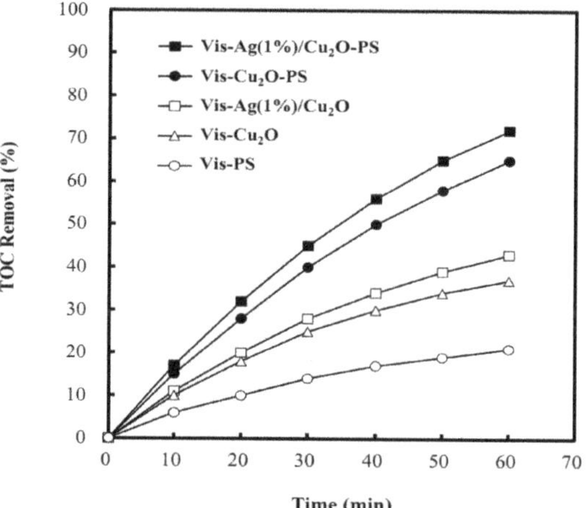

Figure 1. Time flow patterns of TOC removal efficiency by Cu_2O, persulfate, Ag(1 wt%)/Cu_2O, Cu_2O-persulfate and Ag(1 wt%)/Cu_2O-persulfate, respectively under the conditions of visible light power = 103.2 W, persulfate concentration = 50 mM, Cu_2O or Ag(1 wt%)/Cu_2O dosage = 1.20 g L^{-1} and T = 318 K.

XPS measurements were carried out to elucidate surface electronic states of Ag(5%)/Cu_2O. Figure 3 illustrates the Cu2p XPS spectra of fresh Ag(5%)/Cu_2O and reacted Ag(5%)/Cu_2O. As far as fresh Ag(5%)/Cu_2O semiconductor is concerned, two peaks centered at 933.0 and 952.0 eV were clearly found, which were appointed to the binding energy of $Cu^+2p_{(3/2)}$ and $Cu^+2p_{(1/2)}$, respectively [42–44]. With regard to reacted Ag(5%)/Cu_2O, four peaks centered at 933.5, 941.0, 953.0 and 961.0 eV were present, which were separately assigned to the binding energy of $Cu^+2p_{(3/2)}$, $Cu^{2+}2p_{(3/2)}$, $Cu^+2p_{(1/2)}$ and $Cu^{2+}2p_{(1/2)}$ [45,46]. Obviously, Cu^+ cations on the surface of reacted Ag(5%)/Cu_2O shifted to higher oxidation states in comparison with the fresh one, in consideration of the migration of photo-induced electrons to persulfate anions [39,47]. The observations manifest the above hypothesis that persulfate anions could be converted into sulfate radicals upon activation by photo-induced electrons. Instead, sulfate anions may be also transformed to sulfate radicals via photo-induced holes over Ag(5%)/Cu_2O [48]. It makes partial contributions for nitrobenzene oxidation.

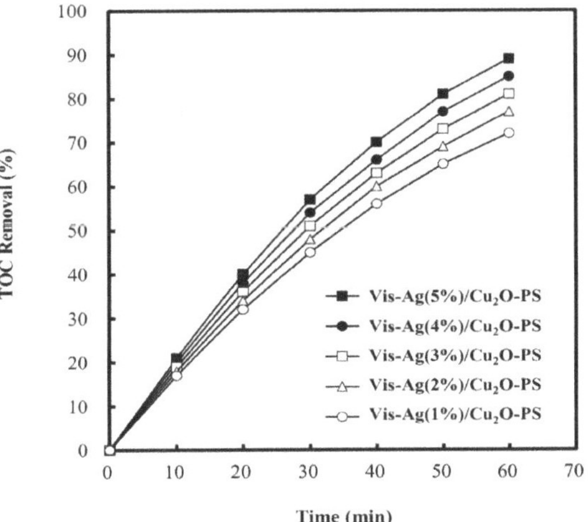

Figure 2. Effect of silver metal contents on the TOC removal efficiency under the conditions of visible light power = 103.2 W, persulfate concentration = 50 mM, Ag/Cu_2O dosage = 1.20 g L^{-1} and T = 318 K.

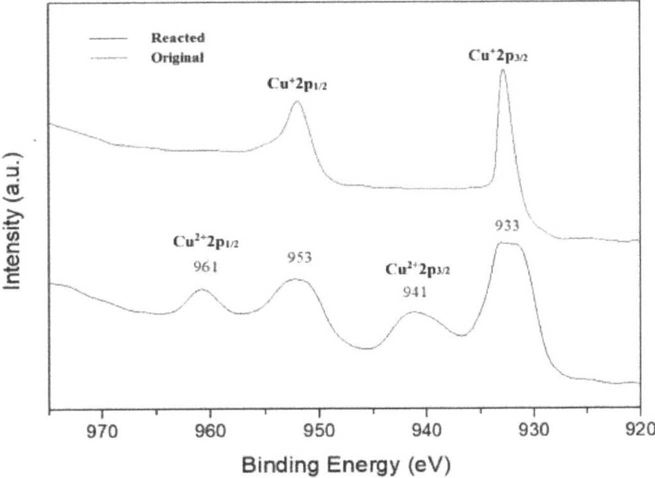

Figure 3. X-ray photoelectron spectra of Cu^+2p or $Cu^{2+}2p$ core level for original Ag(5 wt%)/Cu_2O and reacted Ag(5 wt%)/Cu_2O semiconductors.

3.2. Physicochemical Properties of Ag/Cu_2O

The X-ray diffraction patterns of Ag/Cu_2O semiconductors are displayed in Figure 4. Major peaks in the spectra were matched with crystal planes of Cu_2O, wherein the characteristic peak at 2θ value of 36.5° was ascribed to the (111) plane [48,49]. Conversely, the diffraction peak at 2θ values of 38.1° was appointed to the (111) plane of Ag metal [50]. It is evident that slight weight of Ag metal was doped on the surface of Cu_2O. Figure 5 presents field-emission SEM images of Ag/Cu_2O semiconductors. Obviously, the majority of the Cu_2O surface was smooth. In contrast, a few irregular-shaped sediments were found over

Ag/Cu_2O and more clumps of particles were observed upon increasing Ag extent. It is apparent that Ag metal was well spread on the surface of Cu_2O. The Energy-Dispersive X-ray Mapping analysis on Ag/Cu_2O is illustrated in Figure 6. Ag metal was well dispersed, and its contents measured agreed with those impregnated theoretically (see Table 1).

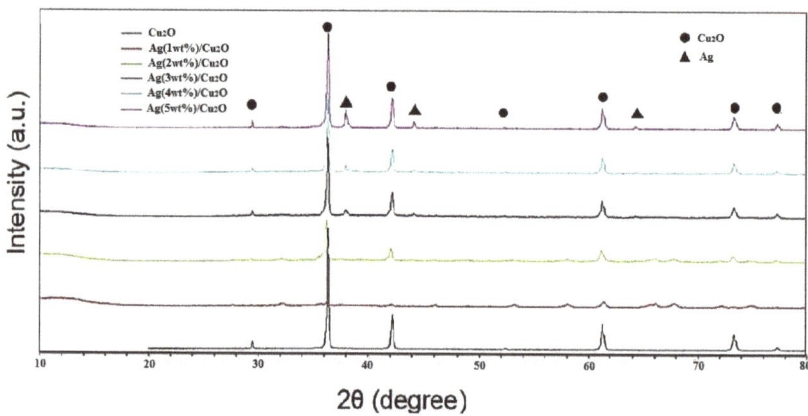

Figure 4. The XRD patterns of Cu_2O and Ag/Cu_2O semiconductors.

Figure 7 depicts the UV–Vis diffuse reflectance spectra for a series of Ag/Cu_2O semiconductors. The analogous spectra were observed between Ag/Cu_2O and Cu_2O at the absorbance wavelength between 400 and 530 nm, which fall into the visible light range. Particularly, the light absorbance intensity of Ag/Cu_2O was stronger than that of Cu_2O. It implies that Ag/Cu_2O semiconductors are more responsive to the visible light irradiation. This phenomenon could be mainly ascribed to Ag metal dopant, creating an electron sink and restraining a combination of photo-induced electrons with holes over Cu_2O [51,52]. Further, the Ag/Cu_2O band gap energy was resolved by means of a Tauc's equation $[(\alpha h\nu)^{1/n} = A(h\nu - E_g)]$, in which $h\nu$ stands for incident optical energy. The "n" parameter was set at the value of 1/2 according to the electronic transition state of Ag/Cu_2O semiconductors. The draft of $(\alpha h\nu)^2$ varied with incident optical energy; $(h\nu)$ was drawn to obtain the band gap energy by intercepting tangent lines to the X-axis [53–55]. Consequently, the Cu_2O band gap energy was determined to be 2.17 eV, consistent with that reported by Muthukumaran et al. [56]. For a series of Ag/Cu_2O semiconductors, the band gap energy was estimated to be 2.06, 1.92, 1.75, 1.55 and 1.43 eV, respectively, upon increasing Ag metal doping (refer to Table 2). The superior photocatalytic performance presented by $Ag(5\%)/Cu_2O$ could be reasonably attributed to a significant yield of photo-induced electrons, caused by a lower band gap energy. In other words, the optical energy of visible light could stimulate Ag/Cu_2O semiconductors for the generation of electron–hole pairs. Persulfate anions would be effectively converted into reactive sulfate radicals via activation of photo-induced electrons. Likewise, photo-induced holes may also transform sulfate anions into sulfate radicals. Figure 8 illustrates transient photocurrence of Cu_2O and $Ag(5\%)/Cu_2O$ excited under visible light irradiation. The photocurrent intensity of the latter was clearly higher than that of the former. It means that $Ag(5\%)/Cu_2O$ possessed the higher yield of photo-induced electron [40,41]. The results support the issue of the inhibition of a combination of photo-induced electrons and holes over Cu_2O through the impregnation of Ag metal.

Figure 5. FE-SEM images of the (**a**) Cu_2O, (**b**) Ag(1 wt%)/Cu_2O, (**c**) Ag(2 wt%)/Cu_2O, (**d**) Ag(3 wt%)/Cu_2O, (**e**) Ag(4 wt%)/Cu_2O and (**f**) Ag(5 wt%)/Cu_2O semiconductors.

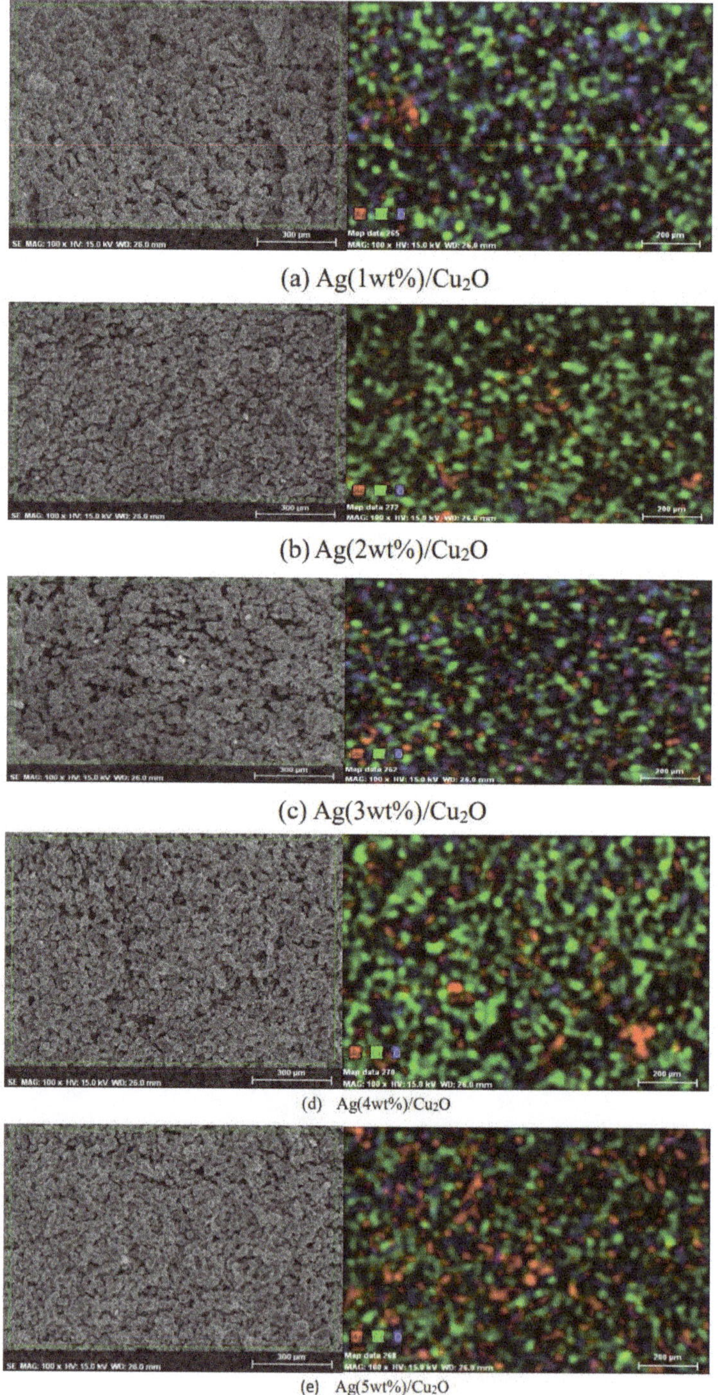

Figure 6. The EDS Mapping analyses on Ag/Cu_2O semiconductors: (**a**) Ag(1 wt%)/Cu_2O, (**b**) Ag(2 wt%)/Cu_2O, (**c**) Ag(3 wt%)/Cu_2O, (**d**) Ag(4 wt%)/Cu_2O and (**e**) Ag(5 wt%)/Cu_2O.

Table 1. The elemental compositions of semiconductors by EDS analyses.

Semiconductor	Ag(wt%)	Cu(wt%)	O(wt%)
Cu_2O	0.00	85.81	14.19
Ag(1 wt%)/Cu_2O	1.15	87.41	11.44
Ag(2 wt%)/Cu_2O	2.05	87.56	10.39
Ag(3 wt%)/Cu_2O	3.56	84.32	12.12
Ag(4 wt%)/Cu_2O	4.99	82.39	12.62
Ag(5 wt%)/Cu_2O	6.84	80.67	12.49

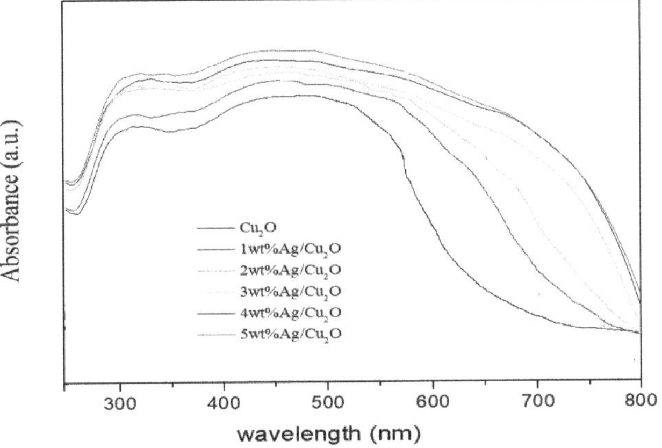

Figure 7. UV–Vis diffuse reflectance spectra of Cu_2O, Ag(1 wt%)/Cu_2O, Ag(2 wt%)/Cu_2O, Ag(3 wt%)/Cu_2O, Ag(4 wt%)/Cu_2O and Ag(5 wt%)/Cu_2O semiconductors.

Table 2. The band gap energy of semiconductors by UV–Vis DRS analyses.

Semiconductor	Band Gap Energy (eV)
Cu_2O	2.17
Ag(1 wt%)/Cu_2O	2.06
Ag(2 wt%)/Cu_2O	1.92
Ag(3 wt%)/Cu_2O	1.75
Ag(4 wt%)/Cu_2O	1.55
Ag(5 wt%)/Cu_2O	1.43

3.3. *Effect of Scavenger Dosages on Photocatalytic Oxidation by Ag/Cu_2O Assisted with Persulfate*

Equivalent concentrations of benzene, 1-propanol and methanol were individually blended with nitrobenzene in wastewater to disclose reactive radicals under photocatalysis by Ag/Cu_2O with assistance of persulfate. As demonstrated in Figure 9, nitrobenzene removal efficiency was sharply faded upon the addition of benzene, due to a high reaction rate constant for benzene and sulfate radicals (3×10^9 M^{-1} s^{-1}) [37]. Alternatively, 1-propanol and methanol slightly suppressed the nitrobenzene removal rate, on account of moderate rate constants for 1-propanol and methanol, with sulfate radicals being 6.0×10^7 M^{-1} s^{-1} and 3.2×10^6 M^{-1} s^{-1}, respectively [57]. It deserves noting that the extent of nitrobenzene removal percentages faded corresponds to the reactive activity for various scavengers and sulfate radicals. It was revealed that sulfate radicals were principal oxidants for nitrobenzene degradation in wastewater.

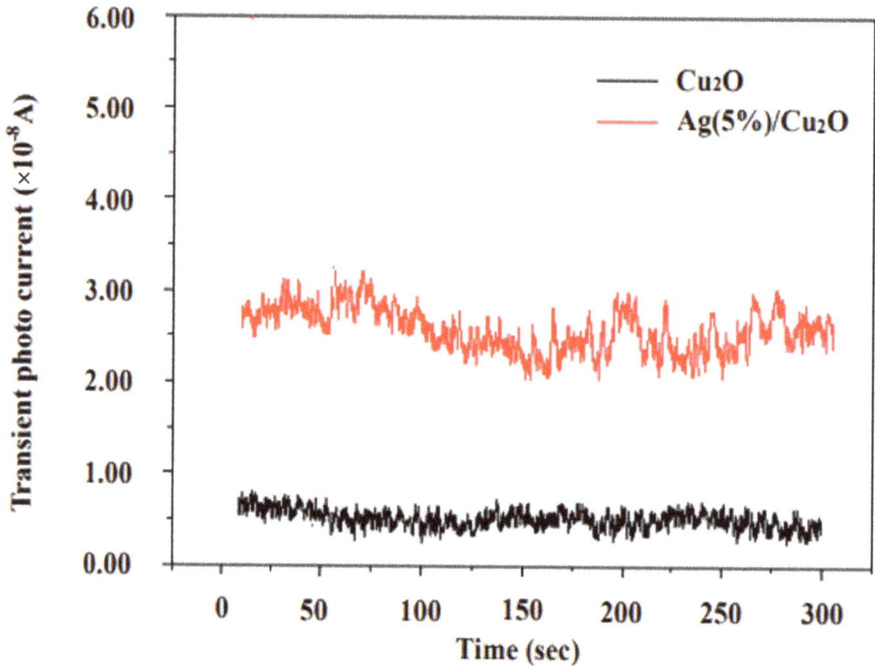

Figure 8. The transient photocurrent analyses of Cu_2O and Ag(5%)/Cu_2O under visible light irradiation.

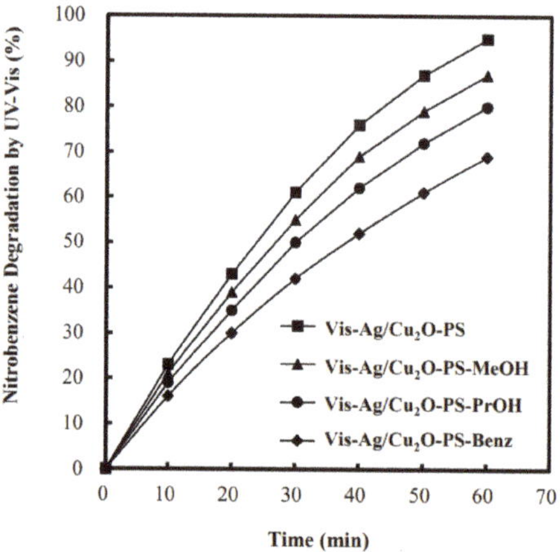

Figure 9. Effect of coexistence of benzene, 1-propanol and methanol, respectively, on the nitrobenzene degradation efficiency.

3.4. Effect of Persulfate Concentrations on Photocatalytic Oxidation by Ag/Cu_2O Assisted with Persulfate

The optimal persulfate concentration for nitrobenzene elimination should be determined in consideration of commercialization. As presented in Figure 10a, the time flow patterns of TOC removal efficiency were dependent on persulfate concentrations. Undoubtedly, increasing persulfate concentrations enhanced nitrobenzene removal rates, wherein high sulfate radical yields could be sensibly expected. Nonetheless, the nitrobenzene removal efficiency faded under an excess persulfate concentration (70 mM). This phenomenon may be interpreted with probable side reactions for the overdosage of persulfate anions and sulfate radicals [38,58]. Further, the photocatalytic oxidation of nitrobenzene was performed in the existence of benzene scavengers to discriminate sulfate radicals yields under various persulfate concentrations (refer to Figure 10b). Definitely, the scavenging effect dramatically displays an analogy between sulfate radical yields and TOC removal efficiency. Accordingly, sulfate radicals were likely to be responsible for nitrobenzene oxidation.

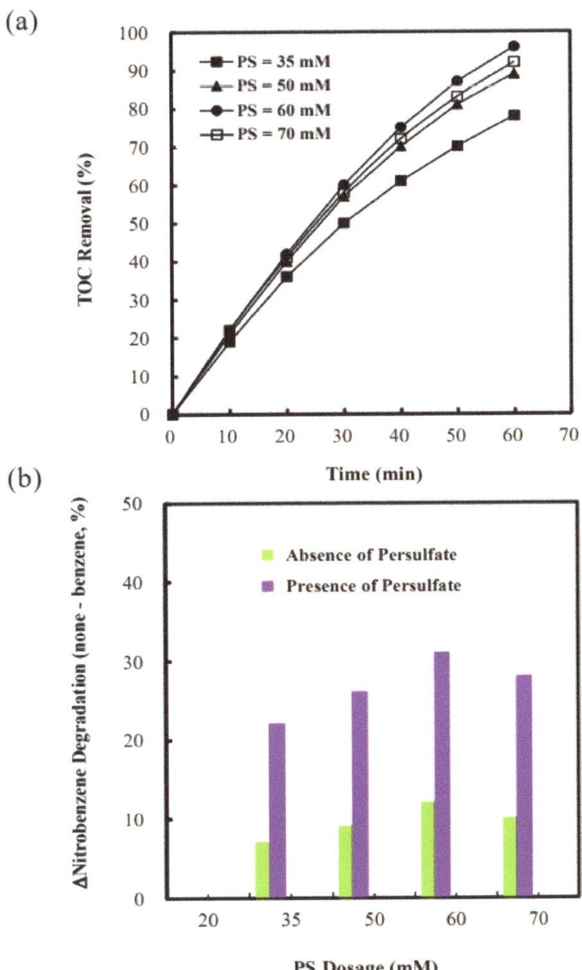

Figure 10. (**a**) Effect of persulfate concentrations on the TOC removal efficiency under the conditions of visible light power = 103.2 W, Ag(5 wt%)/Cu_2O dosage = 1.20 g L^{-1} and T = 318 K. (**b**) The difference of nitrobenzene degradation efficiency between the absence of benzene and presence of benzene monitored by UV–Vis and served as scavenging effect under the conditions of visible light power = 103.2 W, Ag(5 wt%)/Cu_2O dosage = 1.20 g L^{-1} and T = 318 K.

3.5. *Effect of Ag/Cu_2O Dosage on Photocatalytic Oxidation by Ag/Cu_2O Assisted with Persulfate*

An optimal dosage of Ag/Cu_2O semiconductor needs to be essentially established from the process design viewpoint. Figure 11a illustrates the time flow patterns of TOC removal efficiency as functions of Ag/Cu_2O dosages. Evidently, the nitrobenzene degradation rate raised with an increment of Ag/Cu_2O dosages, whereas it reduced under overdosages of Ag/Cu_2O ($\geq$1.50 g L^{-1}). The improvement in the nitrobenzene removal efficiency could be attributed to high sulfate radical yields, caused by the massive activation of persulfate anions via the photocatalysis of Ag/Cu_2O. Conversely, lesser semiconductors received optical energy because of scatter of visible light irradiation, resulting from excessive dosages of Ag/Cu_2O powder [59]. Nitrobenzene decomposition efficiency likewise displayed a similar trend as benzene scavenging behaviors (refer to Figure 11b). The outcomes convince us that sulfate radicals were chief oxidants toward nitrobenzene destruction. Especially, the optimal conditions for complete mineralization of nitrobenzene were determined as follows: visible light power = 103.2 W, persulfate concentration = 60 mM, Ag/Cu_2O dosage = 1.35 g L^{-1} and T = 318 K. In this work, the photocatalytic stability of Ag/Cu_2O was proved via repetitions of five tests (shown in Figure 12). Evidently, nitrobenzene removal efficiency reached almost 98% during the overall experiment. That convinces us of the feasibility for the potential application of Ag/Cu_2O to industrial wastewater treatment.

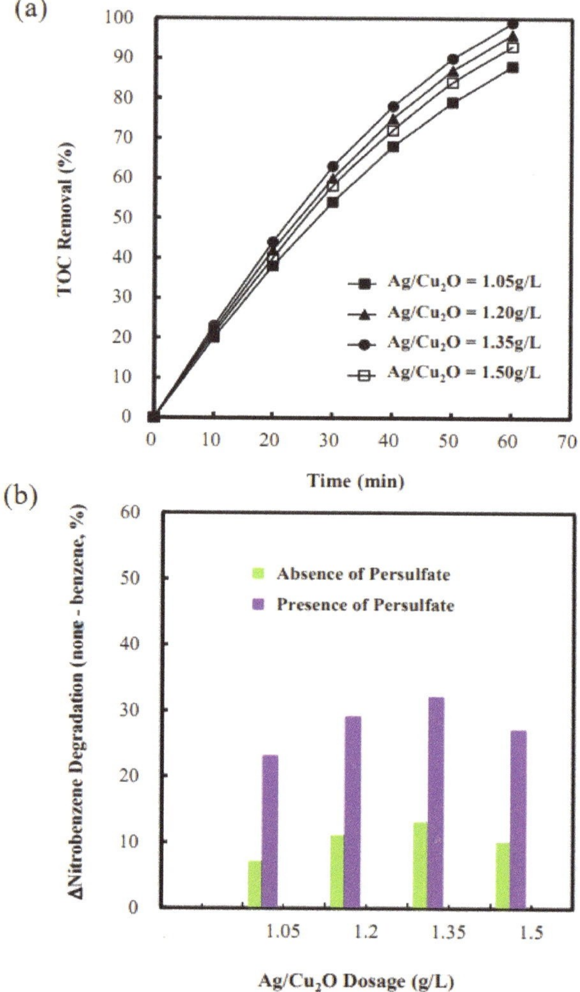

Figure 11. (**a**) Effect of Ag(5 wt%)/Cu_2O dosages on the TOC removal efficiency under the conditions of visible light power = 103.2 W, persulfate concentration = 60 mM and T = 318 K. (**b**) The difference

of nitrobenzene degradation efficiency between the absence of benzene and presence of benzene monitored by UV–Vis and served as scavenging effect under the conditions of visible light power = 103.2 W, persulfate concentration = 60 mM and T = 318 K.

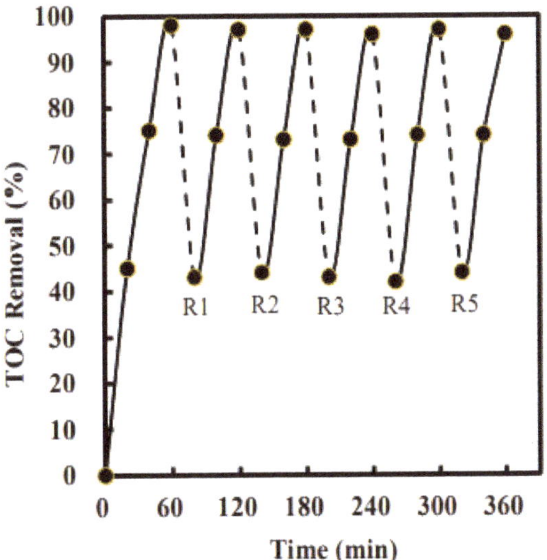

Figure 12. The photocatalytic stability of Ag(5 wt%)/Cu_2O examined by means of repetitions of five tests.

3.6. Reaction Pathways of Photocatalytic Oxidation of Nitrobenzene by Ag/Cu_2O Assisted with Persulfate

All reaction intermediates extracted from photocatalytic oxidation of nitrobenzene using Ag/Cu_2O with assistance of persulfate were examined by a GC-MS spectrometer. Table 3 summarizes the ingredients procured, including nitrobenzene used for raw materials, phenol, 2-nitrophenol, 3-nitrophenol, 4-nitrophenol, hydroquinone and *p*-benzoquinone. In consideration of the source of 2-nitrophenol, 3-nitrophenol and 4-nitrophenol, it was believed that nitrobenzene underwent O_2 addition, followed with $HO_2\bullet$ elimination for the generation of hydroxycyclohexadienyl radicals and sequential hydroxylated compounds [57,60]. Phenol was clearly monitored as an intermediate because of the probable occurrence of nitrophenol denitration [61]. Afterward, phenol ordinarily executed oxidation reaction to hydroquinone, accompanied with successive hydrogen abstraction to *p*-benzoquinone. Ultimately, nitrobenzene would be mineralized into nitrate ions (analyzed by UV–Vis 313 nm), water and carbon dioxide. Based on most degradation intermediates cautiously identified, the hypothesized pathways for photocatalytic oxidation of nitrobenzene by Ag/Cu_2O aided with persulfate is demonstrated in Figure 13.

Table 3. Compositions of nitrobenzene and reaction intermediates identified by GC-MS.

Component	m/z (Relative Abundance, %)
Feedstock	
Nitrobenzene	50 (15.7), 51 (37.7), 65 (13.6), 74 (9.0), 77 (100), 78 (7.5), 93 (16.9), 123(70.2)
Reaction intermediate	
Phenol	38 (5.4), 39 (12.5), 40 (6.9), 55(6.4), 63 (6.5), 65 (21.0), 66 (27.4), 94 (100), 95 (7.7)
2-Nitrophenol	39 (15.7), 53 (9.8), 63 (20.2), 64 (13.9), 65 (25.5), 81 (19.6), 109 (18.2), 139 (100)
3-Nitrophenol	39 (35.9), 53 (10.7), 63 (14.7), 64 (7.9), 65 (63.7), 81 (15.8), 93 (51.4), 139 (100)
4-Nitrophenol	39 (44.2), 53 (23.3), 63 (28.1), 65 (79.9), 81 (33.0), 93 (26.9), 109 (67.1), 139 (100)
Hydroquinone	39 (6.9), 53 (14.4), 54 (12.9), 55 (10.5), 81 (25.3), 82 (12.3), 110 (100), 143 (9.6)
p-Benzoquinone	26 (18.1), 52 (17.9), 53 (17.1), 54 (63.3), 80 (28.2), 82 (36.3), 108 (100), 110 (12.1)

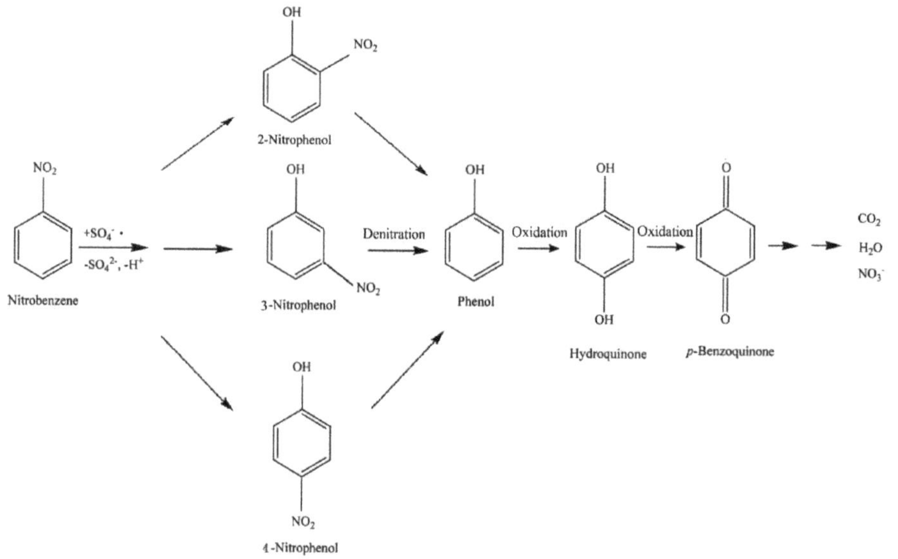

2-Nitrophenol
Nitrobenzene
$+SO_4^{-}\cdot$
$-SO_4^{2-}, -H^+$
3-Nitrophenol
Denitration
Phenol
Oxidation
Oxidation
Hydroquinone
p-Benzoquinone
CO_2
H_2O
NO_3^-
4-Nitrophenol

Figure 13. Overall reaction pathways of nitrobenzene in wastewater by photocatalysis of Ag/Cu_2O with assistance of persulfate.

4. Conclusions

In light of the above discussions, nitrobenzene contaminants were principally mineralized via reactive sulfate radicals, induced from persulfate anions activated effectively by the photocatalysis of Ag/Cu_2O semiconductors. It was intensely supported by benzene scavenger, wherein sulfate radical yields display an analogy with nitrobenzene removal efficiency. As far as GC-MS analyses are concerned, the overall reaction pathways on nitrobenzene oxidation could be proposed as follows. Firstly, nitrobenzene was transformed into hydroxycyclohexadienyl radicals, followed with oxidation step into 2-nitrophenol, 3-nitrophenol and 4-nitrophenol separately. Sequentially, nitrophenol-related compounds executed the denitration procedure to phenol, which was oxidized further for the synthesis of hydroquinone and *p*-benzoquinone. As expected, nitrobenzene would be nearly mineralized into carbon dioxide, nitrate ions and water. The striking results persuade us that the photocatalysis of Ag/Cu_2O coupled with persulfate is an effective and synergistic manner for disposal of industrial effluents.

Author Contributions: Conceptualization, W.-S.C. and J.-Y.C.; methodology, J.-Y.C.; software, J.-Y.C.; validation, W.-S.C., J.-Y.C.; formal analysis, J.-Y.C.; investigation, W.-S.C.; resources, W.-S.C.; data curation, J.-Y.C.; writing—original draft preparation, W.-S.C.; writing—review and editing, W.-S.C.; visualization, J.-Y.C.; supervision, W.-S.C.; project administration, W.-S.C.; funding acquisition, W.-S.C. All authors have read and agreed to the published version of the manuscript.

Funding: This research received no external funding.

Institutional Review Board Statement: Not applicable.

Informed Consent Statement: Not applicable.

Data Availability Statement: The study did not contain any data reported.

Conflicts of Interest: The authors declare no conflict of interest.

References

1. Weissermel, K.; Arpe, H.-J. *Ullmann's Encyclopedia of Industrial Chemistry*, 5th ed.; VCH: Weinheim, Germany, 1991; Volume A17.
2. Holder, J.W. Nitrobenzene carcinogenicity in animals and human hazard evaluation. *Toxicol. Ind. Health* **1999**, *15*, 445–457.
3. Wang, G.X.; Zhang, X.Y.; Yao, C.Z.; Tian, M.Z. Acute toxicity and mutagenesis of three metabolites mixture of nitrobenzene in mice. *Toxicol. Ind. Health* **2011**, *27*, 167–171. [CrossRef]
4. Zhu, L.; Ma, B.; Zhang, L. The study of distribution and fate of nitrobenzene in a water/sediment microcosm. *Chemosphere* **2007**, *69*, 1579–1585. [CrossRef]
5. Carlos, L.; Nichela, D.; Triszcz, J.M.; Felice, J.I.; Einschlag, F.S.G. Nitration of nitrobenzene in Fenton's processes. *Chemosphere* **2010**, *80*, 340–345. [CrossRef] [PubMed]
6. Jiang, B.C.; Lu, Z.Y.; Liu, F.Q.; Li, A.M.; Dai, J.J.; Xu, L.; Chu, L.M. Inhibiting 1,3-dinitrobenzene formation in Fenton oxidation of nitrobenzene through a controllable reductive pretreatment with zero-valent iron. *Chem. Eng. J.* **2011**, *174*, 258–265. [CrossRef]
7. Zhang, Y.; Zhang, K.; Dai, C.; Zhou, X.; Si, H. An enhanced Fenton reaction catalyzed by natural heterogeneous pyrite for nitrobenzene degradation in an aqueous solution. *Chem. Eng. J.* **2014**, *244*, 438–445. [CrossRef]
8. Nichela, D.A.; Berkovic, A.M.; Costante, M.R.; Juliarena, M.P.; Einschlag, F.S.G. Nitrobenzene degradation in Fenton-like systems using Cu(II) as catalyst. Comparison between Cu(II)- and Fe(III)-based systems. *Chem. Eng. J.* **2013**, *228*, 1148–1157. [CrossRef]
9. Duan, H.; Liu, Y.; Yin, X.; Bai, J.; Qi, J. Degradation of nitrobenzene by Fenton-like reaction in a H_2O_2/schwertmannite system. *Chem. Eng. J.* **2016**, *283*, 873–879. [CrossRef]
10. Sun, Y.; Yang, Z.; Tian, P.; Sheng, Y.; Xu, J.; Han, Y.F. Oxidative degradation of nitrobenzene by a Fenton-like reaction with Fe-Cu bimetallic catalysts. *Appl. Catal. B Environ.* **2019**, *244*, 1–10. [CrossRef]
11. Elshafei, G.M.S.; Yehia, F.Z.; Dimitry, O.I.H.; Badawi, A.M.; Eshaq, G. Ultrasonic assisted-Fenton-like degradation of nitrobenzene at neutral pH using nanosized oxide of Fe and Cu. *Ultrason. Sonochem.* **2014**, *21*, 1358–1365. [CrossRef]
12. Anotai, J.; Sakulkittimasak, P.; Boonrattanakij, N.; Lu, M.C. Kinetics of nitrobenzene oxidation and iron crystallization in fluidized-bed Fenton process. *J. Hazard. Mater.* **2009**, *165*, 874–880. [CrossRef] [PubMed]
13. Ratanatamskul, C.; Chintitanun, S.; Masomboon, N.; Lu, M.C. Inhibitory effect of inorganic ions on nitrobenzene oxidation by fluidized-bed Fenton process. *J. Mol. Catal. A Chem.* **2010**, *331*, 101–105. [CrossRef]
14. Nitoi, I.; Oancea, P.; Raileanu, M.; Crisan, M.; Constantin, L.; Cristea, I. UV-VIS photocatalytic degradation of nitrobenzene from water using heavy metal doped titania. *J. Ind. Eng. Chem.* **2015**, *21*, 677–682. [CrossRef]
15. Shen, X.Z.; Liu, Z.C.; Xie, S.M.; Guo, J. Degradation of nitrobenzene using titania photocatalysts co-doped with nitrogen and cerium under visible light illumination. *J. Hazard. Mater.* **2009**, *162*, 1193–1198. [CrossRef]
16. Tayade, R.J.; Bajaj, H.C.; Jasra, R.V. Photocatalytic removal of organic contaminants from water exploiting tuned band gap photocatalysts. *Desalination* **2011**, *275*, 160–165. [CrossRef]
17. Weavers, L.K.; Liang, F.H.; Hoffmann, M.R. Aromatic compound degradation in water using a combination of sonolysis and ozonolysis. *Environ. Sci. Technol.* **1998**, *32*, 2727–2733. [CrossRef]
18. Zhao, L.; Ma, W.; Ma, J.; Wen, G.; Liu, Q. Relationship between acceleration of hydroxyl radical initiation and increase of multiple-ultrasonic field amount in the process of ultrasound catalytic ozonation for degradation of nitrobenzene in aqueous solution. *Ultrason. Sonochem.* **2015**, *22*, 198–204. [CrossRef]
19. Zhao, L.; Ma, J.; Sun, Z.Z.; Zhai, X.D. Catalytic ozonation for the degradation of nitrobenzene in aqueous solution by ceramic honeycomb-supported manganese. *Appl. Catal. B Environ.* **2008**, *83*, 256–264. [CrossRef]
20. Zhao, L.; Ma, J.; Sun, Z.Z.; Liu, H. Influencing mechanism of temperature on the degradation of nitrobenzene in aqueous solution by ceramic honeycomb catalytic ozonation. *J. Hazard. Mater.* **2009**, *167*, 1119–1125. [CrossRef]
21. Chen, Y.; Li, H.; Liu, W.; Tu, Y.; Zhang, Y.; Han, W.; Wang, L. Electrochemical degradation of nitrobenzene by anodic oxidation on the constructed TiO_2-NTs/SnO_2-Sb/PbO_2 electrode. *Chemosphere* **2014**, *113*, 48–55. [CrossRef]
22. Xia, K.; Xie, F.; Ma, Y. Degradation of nitrobenzene in aqueous solution by dual-pulse ultrasound enhanced electrochemical process. *Ultrason. Sonochem.* **2014**, *21*, 549–553. [CrossRef]
23. Ji, Y.; Shi, Y.; Wang, L.; Lu, J. Denitration and renitration processes in sulfate radical-mediated degradation of nitrobenzene. *Chem. Eng. J.* **2017**, *315*, 591–597. [CrossRef]
24. Guo, J.; Zhu, L.; Sun, N.; Lan, Y. Degradation of nitrobenzene by sodium persulfate activated with zero-valent zinc in the presence of low frequency ultrasound. *J. Taiwan Inst. Chem. Eng.* **2017**, *78*, 137–143. [CrossRef]
25. Pan, Y.; Zhou, M.; Li, X.; Xu, L.; Tang, Z.; Sheng, X.; Li, B. Highly efficient persulfate oxidation process activated with pre-magnetization Fe^0. *Chem. Eng. J.* **2017**, *318*, 50–56. [CrossRef]
26. Zhang, Y.; Xu, X.; Pan, Y.; Xu, L.; Zhou, M. Pre-magnetized Fe^0 activated persulphate for the degradation of nitrobenzene in groundwater. *Sep. Purif. Technol.* **2019**, *212*, 555–562. [CrossRef]
27. De Luca, A.; He, X.; Dionysiou, D.D.; Dantas, R.F.; Esplugas, S. Effects of bromide on the degradation of organic contaminants with UV and Fe^{2+} activated persulfate. *Chem. Eng. J.* **2017**, *318*, 206–213. [CrossRef]
28. Chen, W.S.; Shih, Y.C. Mineralization of aniline in aqueous solution by sono-activated peroxydisulfate enhanced with PbO semiconductor. *Chemosphere* **2020**, *239*, 124686. [CrossRef]
29. Zheng, Z.; Huang, B.; Wang, Z.; Guo, M.; Qin, X.; Zhang, X.; Wang, P.; Dai, Y. Crystal faces of Cu_2O and their stabilities in photocatalytic reactions. *J. Phys. Chem. C* **2009**, *113*, 14448–14453. [CrossRef]

30. Su, Y.; Li, H.; Ma, H.; Robertson, J.; Nathan, A. Controlling surface termination and facet orientation in Cu_2O nanoparticles for high photocatalytic activity: A combined experimental and DFT study. *ACS Appl. Mater. Interfaces* **2017**, *9*, 8100–8106. [CrossRef]
31. Tang, H.; Liu, X.; Xiao, M.; Huang, Z.; Tan, X. Effect of particle size and morphology on surface thermodynamics and photocatalytic thermodynamics of nano-Cu_2O. *J. Environ. Chem. Eng.* **2017**, *5*, 4447–4453. [CrossRef]
32. Chu, C.Y.; Huang, M.H. Facet-dependent photocatalytic properties of Cu_2O crystals probed by electron, hole and radical scavengers. *J. Mater. Chem. A* **2017**, *5*, 15116–15123. [CrossRef]
33. Chen, W.S.; Huang, S.L. Photocatalytic degradation of bisphenol-A in aqueous solution by calcined PbO semiconductor irradiated with visible light. *Desalin. Water Treat.* **2020**, *190*, 147–155. [CrossRef]
34. He, S.L.; Wang, L.P.; Zhang, J.; Hou, M.F. Fenton pre-treatment of wastewater containing nitrobenzene using ORP for indicating the endpoint of reaction. *Procedia Earth Planet. Sci.* **2009**, *1*, 1268–1274. [CrossRef]
35. Satdeve, N.S.; Ugwekar, R.P.; Bhanvase, B.A. Ultrasound assisted preparation and characterization of Ag supported on ZnO nanoparticles for visible light degradation of methylene blue dye. *J. Mol. Liq.* **2019**, *291*, 111313. [CrossRef]
36. Chen, W.S.; Huang, C.P. Mineralization of aniline in aqueous solution by electro-activated persulfate oxidation enhanced with ultrasound. *Chem. Eng. J.* **2015**, *266*, 279–288. [CrossRef]
37. Liang, C.J.; Su, H.W. Identification of sulfate and hydroxyl radicals in thermally activated persulfate. *Ind. Eng. Chem. Res.* **2009**, *48*, 5558–5562. [CrossRef]
38. Lin, H.; Wu, J.; Zhang, H. Degradation of bisphenol A in aqueous solution by a novel electro/ Fe^{3+}/ peroxydisulfate process. *Sep. Purif. Technol.* **2013**, *117*, 18–23. [CrossRef]
39. Wang, B.; Fu, T.; An, B.; Liu, Y. UV light-assisted persulfate activation by Cu^0-Cu_2O for the degradation of sulfamerazine. *Sep. Purif. Technol.* **2020**, *251*, 117321. [CrossRef]
40. Xi, Q.; Gao, G.; Jin, M.; Zhang, Y.; Zhou, H.; Wu, C.; Zhao, Y.; Wang, L.; Guo, P.; Xu, J. Design of graphitic carbon nitride supported Ag-Cu_2O composites with hierarchical structures for enhanced photocatalytic properties. *Appl. Surf. Sci.* **2019**, *471*, 714–725. [CrossRef]
41. Sakar, M.; Balakumar, S. Reverse Ostwald ripening process induced dispersion of Cu_2O nanoparticles in silver-matrix and their interfacial mechanism mediated sunlight driven photocatalytic properties. *J. Photochem. Photobiol. A Chem.* **2018**, *356*, 150–158. [CrossRef]
42. Zhang, M.; Wang, J.; Wang, Y.; Zhang, J.; Han, X.; Chen, Y.; Wang, Y.; Karim, Z.; Hu, W.; Deng, Y. Promoting the charge separation and photoelectrocatalytic water reduction kinetics of Cu_2O nanowires via decorating dual-cocatalysts. *J. Mater. Sci. Technol.* **2021**, *62*, 119–127. [CrossRef]
43. Chen, L.; Guo, S.; Dong, L.; Zhang, F.; Gao, R.; Liu, Y.; Wang, Y.; Zhang, Y. SERS effect on the presence and absence of rGO for Ag@Cu_2O core-shell. *Mater. Sci. Semicond. Process.* **2019**, *91*, 290–295. [CrossRef]
44. Sharma, K.; Maiti, K.; Kim, N.H.; Hui, D.; Lee, J.H. Green synthesis of glucose-reduced graphene oxide supported Ag-Cu_2O nanocomposites for the enhanced visible-light photocatalytic activity. *Compos. Part. B* **2018**, *138*, 35–44. [CrossRef]
45. Biesinger, M.C.; Lau, L.W.M.; Gerson, A.R.; Smart, R.S.C. Resolving surface chemical states in XPS analysis of first row transition metals, oxides and hydroxides: Sc, Ti, V, Cu and Zn. *Appl. Surf. Sci.* **2010**, *257*, 887–898. [CrossRef]
46. Poulston, S.; Parlett, P.M.; Stone, P.; Bowker, M. Surface oxidation and reduction of CuO and Cu_2O studied using XPS and XAES. *Surf. Interface Anal.* **1996**, *24*, 811–820. [CrossRef]
47. Liu, S.; Zhao, X.; Zeng, H.; Wang, Y.; Qiao, M.; Guan, W. Enhancement of photoelectrocatalytic degradation of diclofenac with persulfate activated by Cu cathode. *Chem. Eng. J.* **2017**, *320*, 168–177. [CrossRef]
48. Liu, X.; Chen, J.; Liu, P.; Zhang, H.; Li, G.; An, T.; Zhao, H. Controlled growth of CuO/Cu_2O hollow microsphere composites as efficient visible-light-active photocatalysts. *Appl. Catal. A Gen.* **2016**, *521*, 34–41. [CrossRef]
49. Kumar, S.; Parlett, C.M.A.; Isaacs, M.A.; Jowett, D.V.; Douthwaite, R.E.; Cockett, M.C.R.; Lee, A.F. Facile synthesis of hierarchical Cu_2O nanocubes as visible light photocatalysts. *Appl. Catal. B Environ.* **2016**, *189*, 226–232. [CrossRef]
50. Anbu, P.; Gopinath, S.C.B.; Yun, H.S.; Lee, C.-G. Temperature-dependent green biosynthesis and characterization of silver nanoparticles using balloon flower plants and their antibacterial potential. *J. Mol. Struct.* **2019**, *1177*, 302–309. [CrossRef]
51. Han, Z.; Ren, L.; Cui, Z.; Chen, C.; Pan, H.; Chen, J. Ag/ZnO flower heterostructures as a visible-light driven photocatalyst via surface plasmon resonance. *Appl. Catal. B Environ.* **2012**, *126*, 298–305. [CrossRef]
52. Zhang, X.; Wang, Y.; Hou, F.; Li, H.; Yang, Y.; Zhang, X.; Yang, Y.; Wang, Y. Effects of Ag loading on structural and photocatalytic properties of flower-like ZnO microspheres. *Appl. Surf. Sci.* **2017**, *391*, 476–483. [CrossRef]
53. Maji, S.K.; Mukherjee, N.; Dutta, A.K.; Srivastava, D.N.; Paul, P.; Karmakar, B.; Mondal, A.; Adhikary, B. Deposition of nanocrystalline CuS thin film from a single precursor: Structural, optical and electrical properties. *Mater. Chem. Phys.* **2011**, *130*, 392–397. [CrossRef]
54. Štengl, V.; Grygar, T.M. The simplest way to Iodine-doped anatase for photocatalysts activated by visible light. *Int. J. Photoenergy* **2011**, *2011*, 685935. [CrossRef]
55. Kamaraj, E.; Somasundaram, S.; Balasubramani, K.; Eswaran, M.P.; Muthuramalingam, R.; Park, S. Facile fabrication of CuO-Pb_2O_3 nanophotocatalysts for efficient degradation of Rose Bengal dye under visible light irradiation. *Appl. Surf. Sci.* **2018**, *433*, 206–212. [CrossRef]

56. Muthukumaran, M.; Niranjani, S.; Samuel Barnabas, K.; Narayanan, V.; Raju, T.; Venkatachalam, K. Green route synthesis and characterization of cuprous oxide (Cu_2O): Visible light irradiation photocatalytic activity of MB dye. *Mater. Today Proc.* **2019**, *14*, 563–568. [CrossRef]
57. Neta, P.; Huie, R.E.; Ross, A.B. Rate constants for reactions of inorganic radicals in aqueous solution. *J. Phys. Chem. Ref. Data* **1988**, *17*, 1027–1284. [CrossRef]
58. Hou, L.W.; Zhang, H.; Xue, X.F. Ultrasound enhanced heterogeneous activation of peroxydisulfate by magnetite catalyst for the degradation of tetracycline in water. *Sep. Purif. Technol.* **2012**, *84*, 147–152. [CrossRef]
59. Jyothi, K.P.; Yesodharan, S.; Yesodharan, E.P. Ultrasound (US), ultraviolet light (UV) and combination (US + UV) assisted semiconductor catalysed degradation of organic pollutants in water: Oscillation in the concentration of hydrogen peroxide formed in situ. *Ultrason. Sonochem.* **2014**, *21*, 1787–1796. [CrossRef] [PubMed]
60. Anipsitakis, G.P.; Dionysiou, D.D.; Gonzalez, M.A. Cobalt-mediated activation of peroxymonosulfate and sulfate radical attack on phenolic compounds Implications of chlorine ions. *Environ. Sci. Technol.* **2006**, *40*, 1000–1007. [CrossRef] [PubMed]
61. Zhou, J.; Xiao, J.; Xiao, D.; Guo, Y.; Fang, C.; Lou, X.; Wang, Z.; Liu, J. Transformations of chloro and nitro groups during the peroxymonosulfate-based oxidation of 4-chloro-2-nitrophenol. *Chem. Eng. J.* **2015**, *134*, 446–451. [CrossRef]

 *photochem*

Review

Green Synthesis of Heterogeneous Visible-Light-Active Photocatalysts: Recent Advances

Alessio Zuliani * and Camilla Maria Cova

Center for Colloid and Surface Science (CSGI) & Department of Chemistry, University of Florence, Via della Lastruccia 3, Sesto Fiorentino, 50019 Florence, Italy; cova@csgi.unifi.it
* Correspondence: zuliani@csgi.unifi.it

Abstract: The exploitation of visible-light active photocatalytic materials can potentially change the supply of energy and deeply transform our world, giving access to a carbon neutral society. Currently, most photocatalysts are produced through low-ecofriendly, energy dispersive, and fossil-based synthesis. Over the last few years, research has focused on the development of innovative heterogeneous photocatalysts by the design of sustainable and green synthetic approaches. These strategies range from the use of plant extracts, to the valorization and recycling of metals inside industrial sludges or from the use of solventless techniques to the elaboration of mild-reaction condition synthetic tools. This mini-review highlights progresses in the development of visible-light-active heterogeneous photocatalysts based on two different approaches: the design of sustainable synthetic methodologies and the use of biomass and waste as sources of chemicals embedded in the final photoactive materials.

Keywords: photocatalysis; visible light; biomass; waste; green chemistry; nanocatalysis

Citation: Zuliani, A.; Cova, C.M. Green Synthesis of Heterogeneous Visible-Light-Active Photocatalysts: Recent Advances. *Photochem* **2021**, *1*, 147–166. https://doi.org/photochem1020009

Received: 18 June 2021
Accepted: 20 July 2021
Published: 27 July 2021

1. Introduction

Photochemistry plays a crucial role in the so-defined "Earth green transition", i.e., the transition towards a climate-neutral economy [1,2]. Indeed, thanks to photocatalytically active materials, it is possible to exploit sunlight as an energetical source for unlimited functions ranging from the evolution of hydrogen through water splitting to the degradation of pollutants or to photopolymerization [3–8]. Considering that the sun continuously irradiates the planet with 120,000 terawatts, amounting to almost 6000 times the Earth's energy consumption, the potentialities of photochemistry are really unlimited [9]. However, the direct use of sunlight is still poorly exploited, while the use of the "stocked" form of solar energy, i.e., fossil fuels, represents the main power source of humanity. Indeed, the sun has been giving energy (in the form of light and heat) to the planet for over 4.5 billion years, and this irradiation has been captured and chemically transformed into fossil fuels over millions of years. Despite the sun being expected to keep shining on the Earth for at least another four billion years before becoming a full-blown giant red star whose brightness will burn the planet, the rate of production of fossil fuels is enormously slower that the rate of consumption. As a result, fossil fuels are classified as non-renewable resources (at least on a decade-scale and not on an eon-scale) [10]. Furthermore, and more importantly, the use of fossil fuels implies the release, in just a few years, of carbon accumulated over millions of years [11]. This carbon emission, mainly in the form of carbon dioxide, is faster that the ability of the planet to "capture" and fix it again, with the consequence of an atmosphere filled with carbon dioxide. A truly novel situation for mankind, generating unexpected consequences [12–15]. As a result, sunlight should be exploited in a faster and more direct way. One option could be the proper use of biomass, which is one of the greenest ways to produce energy—in the (direct) form of electricity (e.g., by burning biomass) or as hydrogen, chemical products, and biodiesels, among others [16,17]. It should be noted that, similar to the use of fossil fuels, the use of biomass consists of exploiting solar energy already converted into chemical bonds by the action of chlorophyll, i.e., nature's photocatalyst, but in the form of renewable and carbon neutral resources. However, the use of biomass entails some important limits, such as the possible detriment of food production or, as in the case of the use of biodiesel, the release of particulates into the environment [18,19]. On the other hand, photochemistry offers, at least on a theoretical level, the possibility of directly using sunlight, avoiding all the drawbacks rising from the utilization of biomass [20].

As a result, over the last few years, national and international policies, such as those by the UN or EU, but more importantly researchers' innovative spirit, have driven the investigation into visible light active photoactive materials for numerous achievements. Special focus should be put on photocatalysts developed by green and environmentally friendly synthesis. In fact, this approach follows the principles of sustainable development and fully fills the idea of green transition: there is no real benefit in developing a material with high photocatalytic activity, but with a negative impact on the environment due to its low-sustainable production. Many research groups have, therefore, developed different green and environmentally friendly photocatalysts, as summarized in recent reviews (Table 1).

Table 1. Recent reviews describing green photocatalytic materials.

Title	Ref.
Bio-inspired and biomaterials-based hybrid photocatalysts for environmental detoxification: A review	[21]
Green synthesis: Photocatalytic degradation of textile dyes using metal and metal oxide nanoparticles-latest trends and advancements	[22]
Recent Development of Photocatalysts Containing Carbon Species: A Review	[23]
Graphene-Based Materials as Efficient Photocatalysts for Water Splitting	[24]
Lignin-Based Composite Materials for Photocatalysis and Photovoltaics	[25]

The purpose of this mini-review is to provide a summary of the most recent insights into the development of innovative photocatalysts by focusing on visible light active materials prepared through the valorization of biomass and waste: as far as we know, a similar review is not present in the literature. In detail, this work describes, in two separate sections, photocatalysts produced via sustainable processes (i.e., focusing on the synthetic steps) and photocatalysts produced by englobing in the final product biomass or waste materials (i.e., using them as sources of chemicals).

2. A Green Synthetic Approach

A primary approach to producing sustainable photocatalysts is through environmentally friendly and green synthesis. This approach follows, as much as possible, the 12 principles of green chemistry [26,27], and suggests a rapid solution for the sustainable preparation of photocatalysts.

According to the literature, a captivating method for the green synthesis of photocatalysts is the reduction as well as stabilization/capping of metal ions using different plant extracts with high content of polyphenols, or more in general of phytochemicals. This approach is clearly not limited to the production of photocatalysts, but is also exploited for the preparation of many other catalysts and materials, such as nanocatalysts, electrodes, etc. [28]. It has to be remarked that when vegetable extracts are used for the production of photocatalysts, sunlight can be considered the central core of the methodology, being the source of energy used for the production of the phytochemicals (though photosynthesis) and activation of the produced photocatalyst, as illustrated in Figure 1.

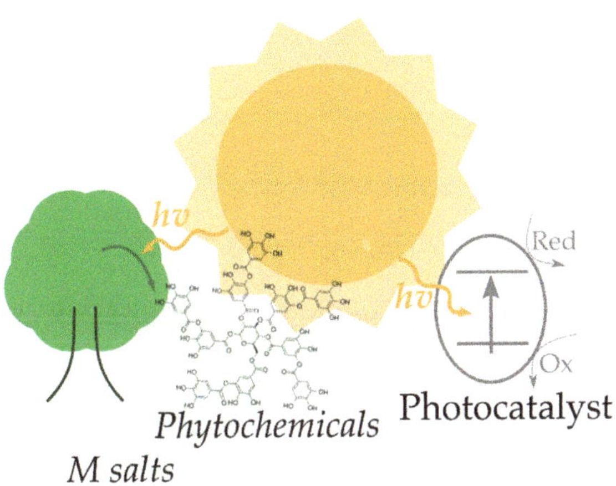

Figure 1. Conceptual illustration of the role of the sun in the synthesis of phytochemicals and the activation of photocatalysts produced using phytochemicals as stabilizing/reducing agents.

Other approaches include solventless techniques, such as using milling tools, the use of environmentally friendly reactants, the use of mild reaction conditions, etc. Between most studied photocatalysts, a special focus should be put on those based on silver, iron, zinc, and titanium dioxide, whether supported/in combination with other compounds or in their pure form.

2.1. Silver-Based Photocatalsyts

Recently, Bi et al. described a solid-state milling approach for the preparation of silver iodide/bismuth oxide acetate (AgI/BiOAc) [29]. In details, the novel photocatalyst was prepared via a facile, green, and environmentally friendly one-pot milling method. The photocatalyst showed enhanced visible-light photocatalytic performances for the degradation of different common dyes such as methyl violet, methyl orange, malachite green, and colorless bisphenol A, in comparison with the same compound prepared via a traditional method. The iodine content was demonstrated to be capable of optimizing the energy band structure while the in situ preparation of AgI/BiOAc resulted in the catalyst possess closely contacted interfaces, a beneficial effect for the transfer and recombination of electrons and holes.

More recently, Nehru and coworkers studied the synthesis of silver nanoparticles supported over titanium dioxide (Ag@TiO_2) using aloe vera (*Aloe perfoliata*) gel as a capping and reducing agent [30]. The material was tested in the photodegradation of picric acid under visible light irradiation. Additionally, anticancer activity against lung cancer cell lines was determined, and it was proved that the adsorption of visible light enhanced the anticancer sensitivity by killing and inhibiting cancer cell reproduction. Similarly, Ag nanoparticles, (40 nm average crystalline size according to XRD) showing photocatalytic and antimicrobial activity, were prepared using jasmine (*Jasminum officinale*) extract [31]. In more detail, the particles were synthesized through a simple, green, eco-friendly, nontoxic, and cost-effective method using the extract of jasmine flower as a capping and stabilizing agent. The so-synthesized particles showed good performances in the degradation of methyl blue, a standard dye for degradation test, under visible light irradiation. Additionally, the particles were proved to possess antimicrobial characteristics for both grams positive and negative bacteria. Other recently employed plant extracts for the synthesis of silver-based photocatalyst include Nīm (*Azadirachta indica*) tree fruit extract [32], solanum surattense (*Solanum virginianum*) leaves extract [33], chili (*Capsicum annuum*) extract [34], apple (from *Malus domestica* tree) and grape (from *Vitis vinifera* tree) extract [35], garlic (*Allium sativum*) extract [36], devil pepper (*Rauvolfia tetraphylla*) leaves extract [37], and small-flowered black hawthorn (*Crataegus pentagyna*) fruit extract [38]. Some plant extracts were also used for the preparation of different bimetallic photocatalysts, such Ag/CuO [39]. In details, *Cyperus pangorei* leaves extract was used to synthetize Ag/CuO nanoparticles by a co-precipitation method starting from Cu^{2+} and Ag^{2+} ions. The photocatalytic activity of Ag/CuO was studied in the Rhodamine B dye degradation under visible light irradiation, proving that Ag-doped with CuO improved the catalytic performances in comparison with pure CuO. The material was also tested against Gram-positive (*Staphylococcus aureus-S. aureus*) and Gram-negative (*Escherichia coli-E. coli*) bacteria.

Another green synthetic approach for the preparation of silver-based photocatalyst involved an ion exchange approach [40]. Specifically, a novel plasmonic $Ag@AgVO_3/BiVO_4$ heterostructure was prepared via an in situ topotactic sustainable ion exchange-reduction method, as illustrated in Figure 2.

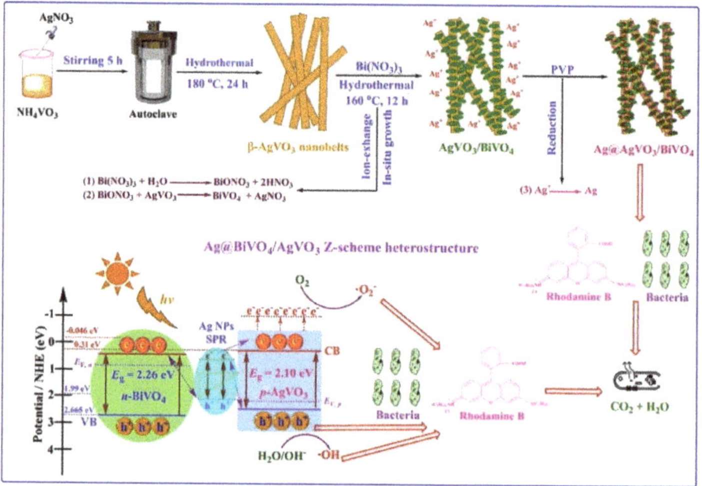

Figure 2. Schematic representation of preparation and utilization of $Ag@AgVO_3/BiVO_4$ heterostructure. Taken from [40] with the permission of Elsevier.

The novel ternary heterostructure was composed of $AgVO_3$ nanobelts loaded with $BiVO_4$ nanosheets and Ag nanoparticles, which grew in situ on the surface of $AgVO_3$ through a topotactic transformation with the assistance of polyvinylpyrrolidone (PVP). PVP not only made the recycling of Ag possible, but also induced a surface plasmon resonance effect in the composite, via a novel green approach. The photocatalytic tests for the degradation of rhodamine B indicated that the novel $Ag@AgVO_3/BiVO_4$ composite material showed a significantly enhanced photocatalytic activity under visible light irradiation compared to $AgVO_3$, $Ag@BiVO_4$ and $Ag@AgVO_3/BiVO_4$. This behavior was ascribed to the SPR effect of Ag NPs and to the large specific surface area combined with the stable heterostructure. The photocatalyst was also demonstrated to possess antimicrobial properties.

Liu and coworkers proposed another green method for the preparation of silver nanoparticles (50 nm) supported on AgCl using microbial culture broths and visible light irradiation [41]. In detail, the nanoparticles were prepared by first forming a colloidal dispersion of AgCl in a microbe-free aqueous lysogeny broth (Miller) solution. Then, an in situ photoreduction driven by sunlight formed small plasmonic Ag@AgCl. The biomolecular ligands of the microbial solution broth worked as a promoter of the reduction process, making the formation of novel Ag@AgCl structures with improved photochemical activity possible. Remarkably, these structures were not obtainable via a standard chemical reduction process, as shown in Figure 3.

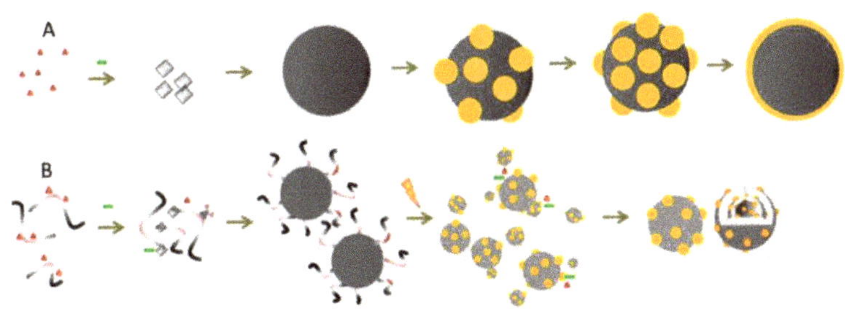

Figure 3. Schematic representation of the synthesis of Ag@AgCl: (**A**) Growth of Ag@AgCl in a typical chemical reduction process; (**B**) biogenic synthesis using microbial culture broths and photoreduction. Taken from [41] and reproduced by permission of The Royal Society of Chemistry.

Table 2 summarizes last works for the preparation of silver-based photocatalysts via sustainable synthetic approaches.

Table 2. Silver-based photocatalysts prepared through green synthetic approaches.

Photocatalyst	Green Synthetic Approach	Application	Ref.
AgI/BiOAc	One-pot solventless milling approach	Dye degradation	[29]
Ag/TiO_2	Use of plant extract gel (*Aloe perfoliata*) as a capping and reducing agent	Picric acid degradation; anticancer activity	[30]
Ag nanoparticles	Use of leaves extract (*Jasminum officinale*) as a capping and stabilizing agent	Dye degradation; antimicrobial activity	[31]
Ag/AgCl nanocomposites	Use of fruit extract (*Azadirachta indica*) as a capping and stabilizing agent	Dye degradation; antimicrobial activity	[32]
Ag nanoparticles	Use of leaves extract (*Solanum virginianum*) as a stabilizing and reducing agent	Dye degradation; antibacterial activity	[33]
Ag/Ag_2O nanoparticles	Use of fruit extract (*Capsicum annuum*) as a stabilizing and reducing agent	Dye degradation	[34]
Ag/CuO	Use of leaves extract (*Cyperus pangorei*) as a stabilizing and reducing agent	Dye degradation; antibacterial activity	[39]
Ag nanoparticles	Use of fruit extracts (*Malus domestica* and *Vitis vinifera*) as stabilizing and reducing agents	Dye degradation; antibacterial activity	[35]
Ag/CeO_2	Use of plant extract (*Allium sativum*) as a stabilizing agent	Dye degradation; antibacterial activity	[36]
Ag nanoparticles	Use of leaves extract (*Rauvolfia tetraphylla*) as a stabilizing agent	Dye degradation; antibacterial activity, LED preparation	[37]
Fe_3O_4/SiO_2/Cu_2O-Ag nanocomposites	Use of fruit extract (*Crataegus pentagyna*) as a capping and reducing agent	Magnetically recoverable photocatalyst for dye degradation	[38]
Ag@$AgVO_3$/$BiVO_4$ heterostructure	Recycling of Ag during the synthetic step and introduction of the surface plasmon resonance effect	Dye degradation; antibacterial activity	[40]
Ag@AgCl	Use of microbial culture broths (tryptic soy broth (TSB) and Lysogeny broth (LB)) for enhancing the photoreduction of silver precursor	Dye degradation	[41]

2.2. Iron-Based Photocatalsyts

The use of plant extracts is also largely employed for the preparation of iron-based photocatalysts. For example, photocatalytically active hematite nanorods, Fe_2O_3, were prepared using the resin of the stems of banana (*Musa paradisiaca* Linn) flowers [42]. During the one-pot synthetic phase, the natural resin acted as both an oxygen source as well as structure stabilizing agent. The so-obtained material was characterized with different techniques, and it was demonstrated that the rods have lengths between 528 and 72 nm, as illustrated in SEM images in Figure 4. The photocatalytic efficiency of the catalyst was investigated by solar induced reduction of dichromate and degradation of malachite green as well as via photoelectrochemical water splitting experiments. In addition, the material was proved to possess potential to work against lung cancer.

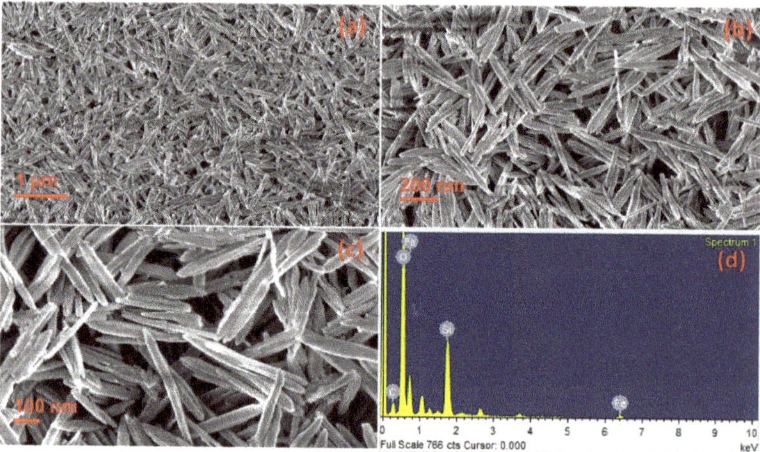

Figure 4. (**a**–**c**) SEM images of hematite nanorods and (**d**) EDS spectrum of the nanorods. Taken from [42] with the permission of Elsevier.

Aravind's research group developed a novel solution for the preparation of iron oxide nanoparticles using a waste generated from the avocado industry [43]. Specifically, phytochemicals found in avocado fruit (*Persea americana*) rind extract were used as reducing and stabilizing agent for the formation of iron nanoparticles starting from iron nitrate (III), in an eco-friendly and cost competitive method. The nanoparticles were employed as a photocatalyst for the decolorization of a sequence of different commonly employed dyes, including malachite green, Congo red, crystal violet, safranin, and methyl orange. In addition, the iron oxide nanoparticles showed antibacterial response against some Gram-negative and Gram-positive bacteria (such as *Escherichia coli*, *Streptococcus* sp., *Staphylococcus* sp., *Salmonella* sp., *Bacillus* sp., *Pseudomonas* sp. and *Proteus* sp.). Similarly, Rawat et al. prepared multi-structured Fe_2O_3 nanoparticles using *Rhaphiolepis indicia* leaves extracts. In detail, the eco-friendly green synthesis was carried out by using the hydroxyl groups of the polyphenols as a reducing agent, with the carboxyl and carbonyl groups as capping agents. Thanks to different characterization techniques, including XRD, FTIR, TEM and SEM, it was proved that multi-structured Fe_2O_3 was formed by nano-scaled hexagonal, rectangular slabs, pentagonal plates, and rods. The so-produced Fe_2O_3 nanostructure showed visible-light photocatalytic behavior for the degradation of Reactive Yellow-86 dye, a pollutant found in textile wastewater. The removal efficiencies were found to also be stable after five recycles, proving the good reusability of the photocatalyst [44].

Other natural extracts for the production of iron nanoparticles include, for example, billygoat-weed (*Ageratum conyzoides*) plant extract [45] or roselle (*Hibiscus sabdariffa*) flower extract [46]. Table 3 summarizes iron-based visible-light-active photocatalysts prepared using plant extracts.

Table 3. Iron-based photocatalysts prepared through green synthetic approaches.

Photocatalyst	Green Synthetic Approach	Application	Ref.
Fe_2O_3 nanorods	Use of a natural resin (*Musa Paradisiaca Linn*) as an oxygen source and stabilizing agent	Dye degradation; photoelectrochemical water splitting; antibacterial activity	[42]
Iron oxide nanoparticles	Use of food waste (*Persea americana*) as a stabilizing and reducing agent	Dye degradation; antibacterial activity	[43]
Fe_3O_4 multistructured nanoparticles	Use of leaves extract (*Rhaphiolepis indicia*) as a capping and reducing agent	Dye degradation	[44]
Iron oxide nanoparticles	Use of plant extract (*Ageratum conyzoides*) as a stabilizing and reducing agent	Dye degradation, antibacterial activity	[45]
Iron oxide nanoparticles	Use of leaves extract (*Hibiscus sabdariffa*) as a stabilizing and reducing agent	Dye degradation	[46]

2.3. Zinc-Based Photocatalsyts

The preparation of ZnO photocatalysts via green methodologies has received many attentions in the last decade [47]. More in details, the use of different natural extracts has gained interest due to the easiness of the preparation of ZnO nanoparticles employing them. It must be remarked that pure ZnO catalysts are active only in the UV region, due to their large band gap. However, recently proposed synthetic green solutions might be modified to prepare visible-light active doped/modified ZnO nanoparticles and deserve to be mentioned. For example, ZnO nanoparticles (with an average diameter of 18–25 nm) were prepared by using the leaf extract of the plant *Ruta chalepensis*, as illustrated in Figure 5.

Figure 5. Schematic representation of the preparation and utilization of ZnO nanoparticles. Taken from [47] with the permission of Elsevier.

The material showed a band gap of 2.86 eV (remarkably lower than the band gap of chemically synthetized ZnO of 3.35 eV) due to lower charge transfer resistance.

Additionally, Yang and coworkers prepared ZnO nanoparticles with peculiar quaker ladies flower type structure using a mixture of extracts from different plants of the family *Araliaceae* [48]. In detail, the extract was derived from four of the roots of panax plants including *Panax ginseng, Acanthopanax senticosus, Kalopanax septemlobus and Dendropanax morbifera*. Similarly, Zheng et al. developed a synthesis of ZnO nanoparticles using roselle flower (*Hibiscus sabdariffa*) and oil palm (*Elaeis guineensis*) leaf extracts as stabilizing and reducing agents in combination with NaOH [49]. The particles were found to be in the size of 10–15 nm and in the form of agglomerated spherical shape. Other plant extracts used for the preparation of ZnO nanoparticles include, among others, *Becium grandiflorum* leaves extract, [50] sea buckthorn (*Hippophae*) fruit extract [51], sugar leaf (*Stevia rebaudiana)* extract [52], or Green tea [53].

In order to directly prepare visible-light active Zn-base photocatalysts, the researchers' efforts focused on the design of green methodologies for the simultaneous combination of different metals to produce composites/supported systems active under visible light irradiation.

For example, p-Co_3O_4/n-ZnO composite catalysts were produced by a co-precipitation method using hyacinth (*Eichhornia Crassipes*) plant extract [54]. The use of plant extract during the synthesis was proved to enhance the catalytic performances for the degradation of methylene blue under visible light irradiation. In addition, the formation of p-n junction simplifies the photogenerated electron–hole separation and further improved the catalytic efficiency. With a similar approach, Shaheen and coworkers prepared CuO/ZnO nanoparticles photocatalytically active for the degradation of methylene blue [55]. In details, they used a penicillin extract (*Penicillium corylophilum As-1*) to synthesize the photocatalyst in a fast, green, and easy way, as shown in Figure 6.

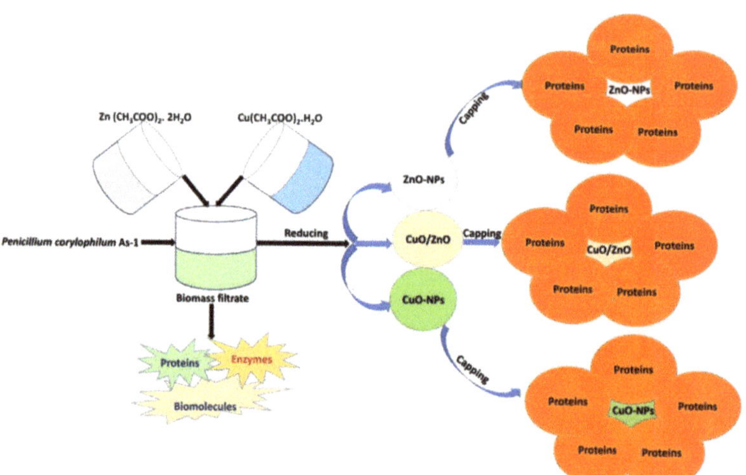

Figure 6. Schematic representation of the preparation and utilization of CuO/ZnO nanoparticles. Taken from [55] with the permission of Elsevier.

Ag-ZnO nanocomposites (15–25 nm) were likewise prepared using potato peel extract (*Solanum tuberosum*) and used for dye degradation under visible light irradiation [56]. Other recently described green methodologies for the preparation of ZnO based photoactive composites include, as reported in Table 4, the preparation of Ag-ZnO nanoparticles using *Excoecaria agallocha* leaf extract under a controlled ultrasound cavitation technique [57], the synthesis of ZnO/$GdCoO_3$ nanocomposites in a two-phase system utilizing *Myristica fragrans* leaf extracts via a high-speed stirring method [58], the preparation of Au-ZnO hetero-nanostructures using employing pecan nuts (*Carya illinoinensis*) leaves extract [59], or the synthesis of Co-doped ZnO through the accumulation of cobalt ion onto *Eichhornia crassipes* plant tissue for different days and combined with zinc [60].

Table 4. Zinc-based photocatalysts prepared through green synthetic approaches.

Photocatalyst	Green Synthetic Approach	Application	Ref.
ZnO nanoparticles	Use of leaves extract (*Ruta chalepensis*) as a stabilizing agent	Dye degradation (UV irradiation)	[47]
ZnO nano-flowers	Use of roots extract (*Araliaceae* family) as a stabilizing agent	Dye degradation (UV irradiation)	[48]
ZnO nanoparticles	Use of leaves extract (*Hibiscus sabdariffa* and *Elaeis guineensis*) as stabilizing agents	Dye degradation (UV irradiation)	[49]
ZnO nanoparticles	Use of leaves extract (*Becium grandiflorum*) as a stabilizing agent	Dye degradation (UV irradiation); antimicrobial activity	[50]
ZnO nanoparticles	Use of fruit extract (*Hippophae*) as a stabilizing agent	Dye degradation (UV irradiation)	[51]
ZnO nanoparticles	Use of leaves extract (*Stevia rebaudiana*) as a stabilizing agent	Dye degradation (UV irradiation)	[52]
Zn-TiO_2	Use of plant extract (Green Tea) as a reducing agent	Dye degradation (UV irradiation)	[53]
p-Co_3O_4/n-ZnO	Use of plant extract (*Eichhornia Crassipes*) as a stabilizing and oxygen source agent	Dye degradation	[54]
CuO/ZnO nanoparticles	Use of fungi extract (*Penicillium corylophilum As-1*) as a stabilizing and reducing agent	Dye degradation	[55]
Ag-ZnO nanocomposites	Use of plant extract (*Solanum tuberosum*) as a stabilizing and reducing agent	Dye degradation	[56]
Ag-ZnO nanoparticles	Use of leaves extract (*Excoecaria agallocha*) as a stabilizing agent	Dye degradation; antibacterial activity	[57]
ZnO/$GdCoO_3$ nanocomposites	Use of leaves extract (*Myristica fragrans*) as a stabilizing agent	Dye degradation	[58]
Au-ZnO hetero-nanostructures	Use of leaves extract (*Carya illinoinensis*) as a stabilizing agent	Dye degradation	[59]
Co/ZnO	Use of *Eichhornia crassipes* plant tissue for the accumulation of Co and sequential combination with ZnO	Dye degradation	[60]

2.4. Titanium Dioxide-Based Photocatalsyts

The synthesis of titanium dioxide, perhaps the most studied photocatalytic compound [4,5,61–66], has greatly evolved in the past years, and numerous green and sustainable synthetic approaches have been proposed. For example, recently, Tayebi et al. prepared TiO_2 based on hexagonal mesoporous silicate (HMS) and loaded by different concentrations of natural polyphenol oak gall tannin [67]. The use of tannin prevented the aggregation of TiO_2 nanoparticles, resulting in an enhanced photocatalytic performance for the degradation of an anionic dye (i.e., Direct yellow 86, used in paper and textile industries). Many other plant extracts have been proposed as capping, stabilizing, and reducing agents, for the production of TiO_2 nanoparticles [68–72], while other approaches include, for example, solventless mechanochemical techniques [73,74]. Wang and coworkers prepared colored TiO_2 (green, gray, orange, and yellow) through a novel mechanochemical method consisting of milling titanium dioxide with or without melamine [75]. As shown in Figure 7, the so-prepared materials showed some or all the phases of titanium dioxide.

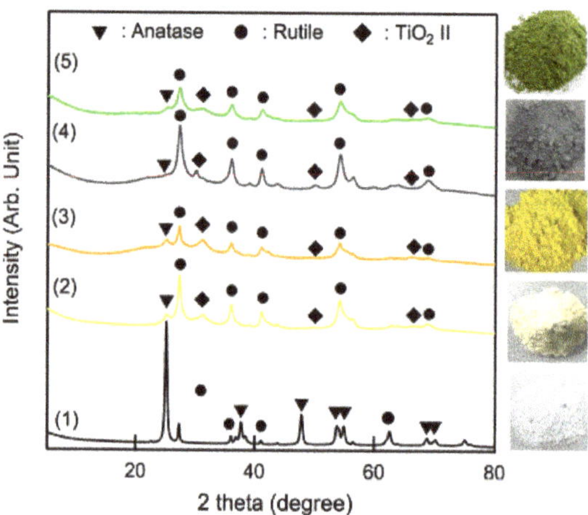

Figure 7. XRD patterns (plus photographs) of TiO_2 produced via mechanochemical approach. (1) Commercially available TiO_2 P25 before ball milling, (2) P25 milled in air, (3) P25 milled with melamine in air, (4) P25 milled in Ar, and (5) P25 milled with melamine in Ar; taken from [75] Copyright (2020), American Chemical Society.

The different colors derived from the reduced Eg (~2.3 eV) and the presence of chemical and physical dopants and tailored the photocatalytic activity. TiO_2 photocatalyst were tested in the degradation of methyl red under visible light irradiation, and no loadings, such as noble metals, metals, and metal oxides were needed. Notably, the materials had a photocatalytic activity up to five-time higher than commercially available P25 TiO_2.

2.5. Miscellaneous

Other photocatalysts prepared via green and sustainable synthesis include composites made of different oxides, such as Cerium Oxide [76], Copper Oxide [77–79], Tin Oxide [80], or Bismuth Vanadate [81]. For example, Raki et al. [76] prepared pure CeO_2 Mn/CeO_2 nanocatalsyts (9–11 nm) using seed extract of Senna Alexandrina (*Cassia angustifolia*) as reducing and capping agents.

3. A Focus on Materials: Waste and Bioderived Materials

A second important approach for the preparation of green photocatalysts involves the use of biomass or waste as sources of chemicals during the synthetic steps. Differently from the first approach, described as the design of sustainable and environmentally friendly synthesis, this latter methodology focuses primary on the sources of materials for the preparation of photocatalysts. This approach strongly improves the sustainability of a process, as it could produce a photocatalyst (with all the benefits reported in the introduction section) by exploiting largely available (i.e., biomass) or scarcely/non-valorized (i.e., waste) materials. The photocatalytically active systems prepared using this tactic may be divided into materials derived from biomass and waste used as sources of carbon, and materials derived from waste and biomass used as sources of other chemicals, as schematically illustrated in Figure 8.

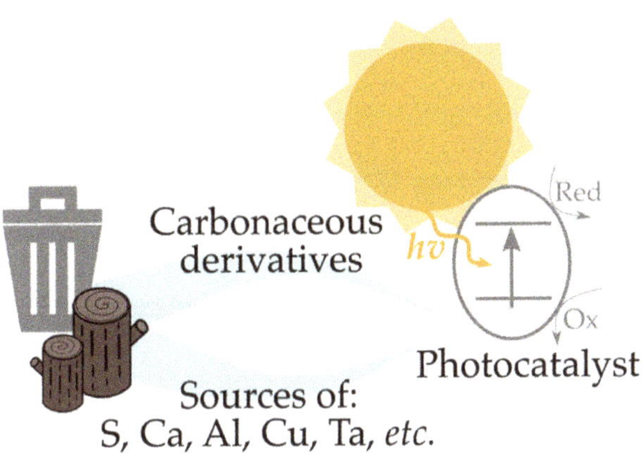

Figure 8. Conceptual valorization of waste and biomass to photocatalyst by exploitation of carbon or other chemicals content.

3.1. Biomass and Waste as Sources of Carbon

The most diffuse method for the preparation of biomass/waste-based photocatalysts is centered on the utilization of carbonaceous materials such as activated carbon, carbon dots, carbon nanotube/nanofiber, graphene, fullerene, g-C_3N_4, and carbon sponges/aerogels, derived from a large variety of biomass or waste [82–91]. For example, in a very recent work, a C-ZnO/MoS_2/mesoporous carbon nanocomposite was successfully prepared by a two-step solution-processed synthetic protocol using sucrose as a carbon source for the preparation of a mesoporous carbonaceous supporting material [92]. The novel ternary composite exhibited a well-interconnected 3D mesoporous microstructure assembled by carbon nanosheets, loaded with quasi zero-dimensional ZnO nanoparticles and two-dimensional MoS_2 nanosheets. The material showed enhanced visible-light-driven photocatalytic performance with remarkably high photo-corrosion resistance, as demonstrated by photodegradation tests of methyl orange (Figure 9).

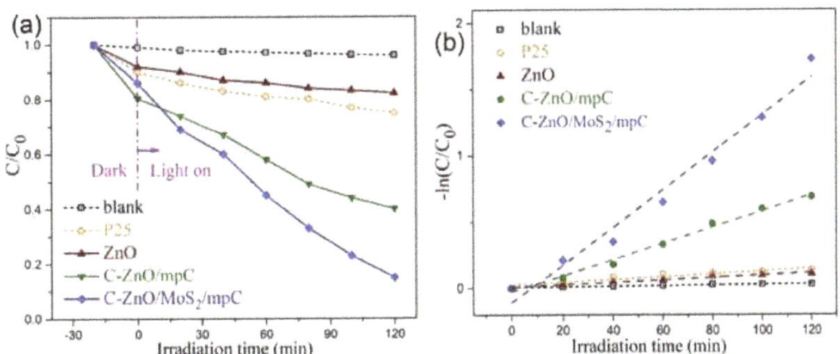

Figure 9. Photodegradation tests for methyl orange degradation using C-ZnO/MoS_2/mesoporous carbon nanocomposite. (**a**) Relative concentrations (C/C0) of methyl orange as a function of the irradiation time under visible-light irradiation; (**b**) kinetic plots. Taken from [92] with the permission of Elsevier.

Importantly, the incorporation of carbon in the composite radically promoted the photoactivity and photostability of the photocatalyst, thanks to few positive synergistic effects, such as increased surface area and active reaction sites, boosted surface charge utilization efficiency, and lowered bandgap.

ZnO/carbon xerogel composite were prepared using tannin as carbon source [93]. The diffuse reflectance test demonstrated that light absorption was significantly enhanced for the composite, while the solar light-driven photodegradation tests revealed that the synthesized composite achieved almost complete degradation of Rhodamine B.

Also, ternary Z-scheme C-doped graphitic carbon nitride/tungsten oxide (C-doped g-C_3N_4/WO_3) was successfully fabricated via an hydrothermal impregnation, using cellulose nanocrystal as carbon source [94]. The material exhibited narrower bandgap, enhanced visible-light absorption and separation of charge carrier, faster interfacial charge transfer, good oxidation/reduction capacities, and consequentially improved global photocatalytic activity performance.

Zhang's research group developed a novel graphitized carbon nitride photocatalysts using the desulfurized waste liquid extracting salt from coking plants, such as ammonium thiocyanate, ammonium thiosulfate, and ammonium sulfate. The addition of the salts provoked a large release of sulfur-containing gas during the pyrolysis, significantly increasing the specific surface area and the pore volume of the photocatalysts. In addition, pyrolysis with sulfur-containing salts resulted in the incorporation of sulfur in the carbon, widening the band gap of the photocatalysts to 2.94 eV (i.e., enhancing its visible light activity). The photocatalysts showed improved NOx removal efficiency under visible light irradiation [95].

Despite the literature reporting many other types of visible light active carbonaceous photocatalysts derived from biomass or waste, such as a recently published lignin-based photocatalysts [96], a photocatalyst derived from wood flour waste (illustrated in Figure 10) [97], or a $MnFe_2O_4$-based photocatalyst supported over coal derived from industrially produced fly ash [98], or many others [99–101], specific insights about the preparation and utilization of these materials have already been fully described elsewhere and are not repeated here [23–25,102–105].

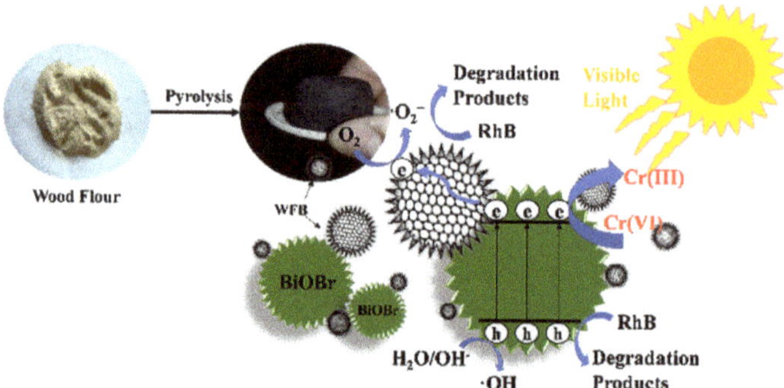

Figure 10. Synthesis and use of a carbonaceous photocatalyst derived from wood flour waste. Taken from [97] with the permission of Elsevier.

3.2. Biomass and Waste as Sources of (Carbon) and Chemicals

Beside the use of biomass or waste as mere sources of carbon, a more elaborated method consists in producing visible-light active photocatalysts using waste and biomass also as sources of other chemicals.

For example, in the past years, Luque's research group has developed different novel compounds using waste pig bristles as sulfur (present in the bristles in the form of cysteine, methionine, and cysteic acid) and carbon source, including photocatalytically active Cu_2S [106–108]. The photocatalysts was specifically prepared through a fast and easy microwave-assisted synthesis. Cu_2S showed a narrow bulk band gap of 1.2 eV and was therefore tested for the degradation of methyl red under visible light irradiation, as shown in Figure 11.

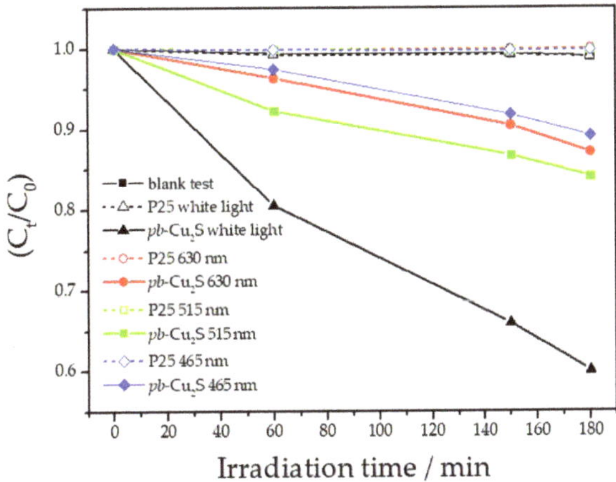

Figure 11. Photodegradation of methyl red catalyzed by Cu_2S produced from waste pig bristles as a function of time at different wavelength irradiation, taken from [106]. Reproduced by permission of The Royal Society of Chemistry.

The results demonstrated the possibility to prepare sulfide materials, with proved photocatalytic activity under visible light irradiation [109], with a largely produced waste (approximatively 225 k ton of pig bristles are produced every year in the EU), which is still poorly valorized. Alternatively, other sulfides, such as iron sulfide (pyrite, FeS_2)-based photocatalysts can be produced using marcasite waste, a waste produced in the jewelry industry. According to Wasanapiarnpong et al., an iron sulfide-based photocatalysts can be used for the photocatalytic degradation of lignin [110].

In the last years, the literature has reported the use of waste eggshells as a source of calcium carbonate for the preparation of electrodes or many types of catalysts, including photocatalysts [111–115]. Indeed, eggshells are a waste largely produced in the dairy industry and its reuse and valorization could substitute the standardly employed low sustainable processes for their end-life treatment [116]. For example, calcium oxide nanoparticles were prepared using waste eggshells by a simple calcination process [117]. The so-produced nanoparticles were examined for the photocatalytic dye degradation of methylene blue and Toluidine blue in aqueous solutions. Remarkably, the catalyst could be recycled up to seven cycles without significant losses of activity.

Recently, Gil and coworkers developed a novel strategy for the recovery and reutilization of the aluminum present in the saline slags generated during Al processing [118]. With this approach the researchers prepared different valuable materials, proposing a sustainable alternative to the disposal of this waste sludge. For instance, they employed saline slags in a co-precipitation method to synthesize a sequence of samples containing zinc and various proportions of aluminum/titanium [119]. The materials were used as photocatalysts for the removal of diclofenac and salicylic acid (probably the most consumed non-steroidal anti-inflammatory drugs) from wastewater.

With a similar aim of valorizing an industrial waste, Ferretti et al., prepared nanometric TiO_2-based magnetic catalyst using fly ash produced from the treatment of iron and steel [120]. The photocatalysts was prepared via a first one-step hydrothermal activation, during which the precursor was transformed into a zeolite with good magnetic properties, and sequentially TiO_2 nanoparticles were supported on the zeolites through ultrasounds treatment. The photocatalytic activity of the novel catalysts was verified through the abatement of methylene blue.

Additionally, waste printed circuit boards (PCBs) can be used to produce photocatalysts. This type of recycling process can prevent the end life treatment of a largely-produced waste, since 20–50 million tons of waste electric and electronic equipment are produced worldwide each year [121]. For example, Qian's research group used PCBs as copper source to prepare a Cu_2O-basedphotocatalyst, and sequentially employed it for the photocatalytic reduction of Cr(VI) under visible light irradiation [122]. Similarly, the metals contained in waste capacitors can be reused to prepare Ta_2O_5 based or $BaTiO_3$-based visible-light active photocatalysts [123,124].

Table 5 reports last achievements in the development of visible-light-active photocatalysts using biomass and waste as chemical sources.

Table 5. Photocatalysts prepared through waste and biomass as sources of chemicals.

Photocatalyst	Green Synthetic Approach	Application	Ref.
Cu_2S	Microwave assisted valorization of waste pig bristles as a sulfur and carbon source	Dye degradation	[106]
FeS_2/titanium dioxide	Marcasite waste from jewelry industry as a source of FeS_2	Lignin degradation	[110]
CaO	Solvent free valorization of waste eggshells as a calcium source	Dye degradation	[117]
ZnTiFeAl-hydrotalcites	Use of saline slags, generated during Al processing, as an Al source	Diclofenac and salicylic acid degradation	[119]
Cu_2O	Waste printed circuit boards (PCBs) as sources of copper	Reduction of Cr(VI)	[122]
Ta_2O_5-based photocatalyst	Waste capacitors as sources of Ta	Hydrogen production	[123]
Nb-Pb co-doped and Ag-Pd-Sn-Ni loaded BaTiO3	Milling of waste capacitors (as sources of $BaTiO_3$, Ag, Pd, Sn, Ni, Nb, and Pb)	Hydrogen production	[124]

4. Summary and Outlook

National and international policies as well as the public pressure for the shifting to a sustainable society are sensibly changing the research and development of materials. With no doubts, this change is also positively altering the design of new photocatalytic compounds. This mini-review has summarized some of the most recent and innovative approaches for the green production of visible-light-active photocatalysts, dividing the most recent works in function of the mere synthetic approach, i.e., with a focus on the sustainable characteristics of synthesis (e.g., the use of naturally derived nanoparticles stabilizing and reducing agents instead of classically used fossil-derived counterparts); or in function of the use of waste or biomass materials as sources of chemicals (e.g., the exploitation of waste slugs as sources of metals). All these methods have a common view of a transversal thinking aimed at avoiding low-eco-friendly procedures to produce materials (photocatalysts) designed for a green scope (i.e., Sun-driven photocatalysis). To date, researchers have proposed many innovative and out-of-the-box solutions for the green preparation of visible-light active photocatalysts, which further development may definitively change the way we know photochemistry. In addition, some investigation paths are still barely explored and may be open to the design of an infinitive number of new materials, such as in the case of the direct use of a waste as a photocatalyst (e.g., waste copper slag used as photocatalysts for the degradation of toxic alcohols) [125].

Nevertheless, major restrictions still limit the large-scale application of these novel materials. Indeed, the industrial manufacturing of green photocatalysts is still economically inconvenient and has no investor's interest. Clearly, fossil-based low-sustainable manufacturing procedures are too cheap to be substituted by procedures which scale-up have uncertain outcomes. Additionally, most of the world's chemical plants for the production of materials (in general) were designed to operate with high (or at least medium)-grade-purity raw materials, having intrinsic characteristics that are difficult to fund when operating with biomass or waste (considering also all troubles related to the different physical and chemical properties such as different viscosity, for example when operating with a solution of stabilizing agents such as EDTA in comparison with phenolic enriched solutions). It is quite obvious that the purification of biomass and waste, for example, for the selective extraction of chemicals (such as Sulphur in the case of [106]) will be even more complicated and increase the production costs of the so-derived photocatalysts (although they would probably be more active, due to the removal of impurities).

Finally, it should be mention that the practice of considering and evaluating the end-life treatment of the novel photocatalysts (or green synthetic approaches), as well as the determination of objective data of the overall sustainability of the processes, is still poorly diffused. Indeed, if on a hand all the discussed methods are claimed to decrease the environmental impact of the production of photocatalysts, on the other hand, any end-life considerations or at least any green metrics of the processes are rarely reported and discussed. Evidently, the use of biomass or waste during the processes could have positive effect on the environment, but these hypotheses should be supported by critical and objective factors. For example, green metrics such as E-Factor [126] can give a rapid and objective view of the sustainability of the process. Additionally, more detailed analysis, such as Life Cycle Assessments (LCA) [127], can deeply describe the potential impact of the novel photocatalysts on the environment. Parallelly, biodegradability or composability tests of the waste eventually produced during the preparation of the photocatalysts and may highlight critical aspects that must be modified. Additionally, some considerations of the production costs of the novel green photocatalysts in comparison with the standardly employed procedures should be described.

Author Contributions: Conceptualization, A.Z. and C.M.C.; writing—original draft preparation, A.Z. and C.M.C.; writing—review and editing, A.Z. and C.M.C.; supervision, A.Z. All authors have read and agreed to the published version of the manuscript.

Funding: This research received no external funding.

Institutional Review Board Statement: Not applicable.

Informed Consent Statement: Not applicable.

Conflicts of Interest: The authors declare no conflict of interest.

References

1. Skjaerseth, J.B. Towards a European Green Deal: The evolution of EU climate and energy policy mixes. *Int. Environ. Agreem. Politics Law Econ.* **2021**, *21*, 25–41. [CrossRef]
2. Ohnesorge, L.; Rogge, E. Europe's Green Policy: Towards a Climate Neutral Economy by Way of Investors' Choice. *Eur. Co. Law* **2021**, *18*, 34–39.
3. Kudo, A.; Miseki, Y. Heterogeneous photocatalyst materials for water splitting. *Chem. Soc. Rev.* **2009**, *38*, 253–278. [CrossRef] [PubMed]
4. Chiarello, G.L.; Zuliani, A.; Ceresoli, D.; Martinazzo, R.; Selli, E. Exploiting the Photonic Crystal Properties of TiO_2 Nanotube Arrays To Enhance Photocatalytic Hydrogen Production. *Acs Catal.* **2016**, *6*, 1345–1353. [CrossRef]
5. Dozzi, M.V.; Chiarello, G.L.; Pedroni, M.; Livraghi, S.; Giamello, E.; Selli, E. High photocatalytic hydrogen production on Cu(II) pre-grafted Pt/TiO_2. *Appl. Catal. B Environ.* **2017**, *209*, 417–428. [CrossRef]
6. Dutta, S.; Biswas, S.; Maji, R.C.; Saha, R. Environmentally Sustainable Fabrication of Cu1.94S-rGO Composite for Dual Environmental Application: Visible-Light-Active Photocatalyst and Room-Temperature Phenol Sensor. *ACS Sustain. Chem. Eng.* **2018**, *6*, 835–845. [CrossRef]
7. Lu, N.; Zhang, Z.; Wang, Y.; Liu, B.; Guo, L.; Wang, L.; Huang, J.; Liu, K.; Dong, B. Direct evidence of IR-driven hot electron transfer in metal-free plasmonic $W_{18}O_{49}$/Carbon heterostructures for enhanced catalytic H_2 production. *Appl. Catal. B Environ.* **2018**, *233*, 19–25. [CrossRef]
8. Lucena, R.; Conesa, J.C. V-Substituted $ZnIn_2S_4$: A (Visible+NIR) Light-Active Photocatalyst. *Photochem* **2021**, *1*, 1–9. [CrossRef]
9. Gratzel, M. Recent Advances in Sensitized Mesoscopic Solar Cells. *Acc. Chem. Res.* **2009**, *42*, 1788–1798. [CrossRef]
10. Zuliani, A.; Ivars, F.; Luque, R. Advances in Nanocatalyst Design for Biofuel Production. *Chemcatchem* **2018**, *10*, 1968–1981. [CrossRef]
11. Zhang, X.C.; Caldeira, K. Time scales and ratios of climate forcing due to thermal versus carbon dioxide emissions from fossil fuels. *Geophys. Res. Lett.* **2015**, *42*, 4548–4555. [CrossRef]
12. Fargione, J.; Hill, J.; Tilman, D.; Polasky, S.; Hawthorne, P. Land clearing and the biofuel carbon debt. *Science* **2008**, *319*, 1235–1238. [CrossRef] [PubMed]

13. Held, I.M.; Soden, B.J. Robust responses of the hydrological cycle to global warming. *J. Clim.* **2006**, *19*, 5686–5699. [CrossRef]
14. Harley, C.D.G.; Hughes, A.R.; Hultgren, K.M.; Miner, B.G.; Sorte, C.J.B.; Thornber, C.S.; Rodriguez, L.F.; Tomanek, L.; Williams, S.L. The impacts of climate change in coastal marine systems. *Ecol. Lett.* **2006**, *9*, 228–241. [CrossRef]
15. Al-Ghussain, L. Global warming: Review on driving forces and mitigation. *Environ. Prog. Sustain. Energy* **2019**, *38*, 13–21. [CrossRef]
16. Okkerse, C.; van Bekkum, H. From fossil to green. *Green Chem.* **1999**, *1*, 107–114. [CrossRef]
17. Gao, M.M.; Zhu, L.L.; Peh, C.K.; Ho, G.W. Solar absorber material and system designs for photothermal water vaporization towards clean water and energy production. *Energy Environ. Sci.* **2019**, *12*, 841–864. [CrossRef]
18. Zhu, L.; Cheung, C.S.; Zhang, W.G.; Huang, Z. Effect of charge dilution on gaseous and particulate emissions from a diesel engine fueled with biodiesel and biodiesel blended with methanol and ethanol. *Appl. Therm. Eng.* **2011**, *31*, 2271–2278. [CrossRef]
19. Shukla, P.C.; Gupta, T.; Labhsetwar, N.K.; Agarwal, A.K. Trace metals and ions in particulates emitted by biodiesel fuelled engine. *Fuel* **2017**, *188*, 603–609. [CrossRef]
20. Balzani, V.; Credi, A.; Venturi, M. Photochemical conversion of solar energy. *Chemsuschem* **2008**, *1*, 26–58. [CrossRef]
21. Kumar, A.; Sharma, G.; Naushad, M.; Al-Muhtaseb, A.H.; Garcia-Penas, A.; Mola, G.T.; Si, C.L.; Stadler, F.J. Bio-inspired and biomaterials-based hybrid photocatalysts for environmental detoxification: A review. *Chem. Eng. J.* **2020**, *382*. [CrossRef]
22. Nagajyothi, P.C.; Vattikuti, S.V.P.; Devarayapalli, K.C.; Yoo, K.; Shim, J.; Sreekanth, T.V.M. Green synthesis: Photocatalytic degradation of textile dyes using metal and metal oxide nanoparticles-latest trends and advancements. *Crit. Rev. Environ. Sci. Technol.* **2020**, *50*, 2617–2723. [CrossRef]
23. Shi, Z.J.; Ma, M.G.; Zhu, J.F. Recent Development of Photocatalysts Containing Carbon Species: A Review. *Catalysts* **2019**, *9*, 20. [CrossRef]
24. Albero, J.; Mateo, D.; Garcia, H. Graphene-Based Materials as Efficient Photocatalysts for Water Splitting. *Molecules* **2019**, *24*, 906. [CrossRef] [PubMed]
25. Khan, A.; Nair, V.; Colmenares, J.C.; Glaser, R. Lignin-Based Composite Materials for Photocatalysis and Photovoltaics. *Top. Curr. Chem.* **2018**, *376*, 1–31. [CrossRef] [PubMed]
26. Anastas, P.T.; Bartlett, L.B.; Kirchhoff, M.M.; Williamson, T.C. The role of catalysis in the design, development, and implementation of green chemistry. *Catal. Today* **2000**, *55*, 11–22. [CrossRef]
27. Anastas, P.; Eghbali, N. Green Chemistry: Principles and Practice. *Chem. Soc. Rev.* **2010**, *39*, 301–312. [CrossRef]
28. Zuliani, A.; Cano, M.; Calsolaro, F.; Puente Santiago, A.R.; Giner-Casares, J.J.; Rodríguez-Castellón, E.; Berlier, G.; Cravotto, G.; Martina, K.; Luque, R. Improving the electrocatalytic performance of sustainable Co/carbon materials for the oxygen evolution reaction by ultrasound and microwave assisted synthesis. *Sustain. Energy Fuels* **2021**, *5*, 720–731. [CrossRef]
29. Jia, X.M.; Han, Q.F.; Zheng, M.Y.; Bi, H.P. One pot milling route to fabricate step-scheme AgI/I-BiOAc photocatalyst: Energy band structure optimized by the formation of solid solution. *Appl. Surf. Sci.* **2019**, *489*, 409–419. [CrossRef]
30. Hariharan, D.; Thangamuniyandi, P.; Christy, A.J.; Vasantharaja, R.; Selvakumar, P.; Sagadevan, S.; Pugazhendhi, A.; Nehru, L.C. Enhanced photocatalysis and anticancer activity of green hydrothermal synthesized Ag@TiO_2 nanoparticles. *J. Photochem. Photobiol. B Biol.* **2020**, *202*, 111636. [CrossRef] [PubMed]
31. Aravind, M.; Ahmad, A.; Ahmad, I.; Amalanathan, M.; Naseem, K.; Mary, S.M.M.; Parvathiraja, C.; Hussain, S.; Algarni, T.S.; Pervaiz, M.; et al. Critical green routing synthesis of silver NPs using jasmine flower extract for biological activities and photocatalytical degradation of methylene blue. *J. Environ. Chem. Eng.* **2021**, *9*, 104877. [CrossRef]
32. Panchal, P.; Meena, P.; Nehra, S.P. A rapid green synthesis of Ag/AgCl-NC photocatalyst for environmental applications. *Environ. Sci. Pollut. Res.* **2021**, *28*, 3972–3982. [CrossRef]
33. Mani, M.; Chang, J.H.; Gandhi, A.D.; Vizhi, D.K.; Pavithra, S.; Mohanraj, K.; Mohanbabu, B.; Babu, B.; Balachandran, S.; Kumaresan, S. Environmental and biomedical applications of AgNPs synthesized using the aqueous extract of Solanum surattense leaf. *Inorg. Chem. Commun.* **2020**, *121*, 108228. [CrossRef]
34. El-Shabasy, R.; Yosri, N.; El-Seedi, H.; Shoueir, K.; El-Kemary, M. A green synthetic approach using chili plant supported Ag/Ag_2O@P_{25} heterostructure with enhanced photocatalytic properties under solar irradiation. *Optik* **2019**, *192*, 162943. [CrossRef]
35. Parthibavarman, M.; Bhuvaneshwari, S.; Jayashree, M.; BoopathiRaja, R. Green Synthesis of Silver (Ag) Nanoparticles Using Extract of Apple and Grape and with Enhanced Visible Light Photocatalytic Activity. *Bionanoscience* **2019**, *9*, 423–432. [CrossRef]
36. Ayodhya, D.; Veerabhadram, G. Green synthesis of garlic extract stabilized Ag@CeO_2 composites for photocatalytic and sonocatalytic degradation of mixed dyes and antimicrobial studies. *J. Mol. Struct.* **2020**, *1205*, 127611. [CrossRef]
37. Vinay, S.P.; Udayabhanu; Nagarju, G.; Chandrappa, C.P.; Chandrasekhar, N. Enhanced photocatalysis, photoluminescence, and anti-bacterial activities of nanosize Ag: Green synthesized via Rauvolfia tetraphylla (devil pepper). *Sn Appl. Sci.* **2019**, *1*, 1–14. [CrossRef]
38. Ebrahimzadeh, M.A.; Mortazavi-Derazkola, S.; Zazouli, M.A. Eco-friendly green synthesis and characterization of novel Fe_3O_4/SiO_2/Cu_2O-Ag nanocomposites using Crataegus pentagyna fruit extract for photocatalytic degradation of organic contaminants. *J. Mater. Sci. Mater. Electron.* **2019**, *30*, 10994–11004. [CrossRef]
39. Parvathiraja, C.; Shailajha, S. Bioproduction of CuO and Ag/CuO heterogeneous photocatalysis-photocatalytic dye degradation and biological activities. *Appl. Nanosci.* **2021**, *11*, 1411–1425. [CrossRef]

40. Ju, P.; Wang, Y.; Sun, Y.; Zhang, D. In-situ green topotactic synthesis of a novel Z-scheme $Ag@AgVO_3/BiVO_4$ heterostructure with highly enhanced visible-light photocatalytic activity. *J. Colloid Interface Sci.* **2020**, *579*, 431–447. [CrossRef]
41. Shen, Z.; Liu, B.; Pareek, V.; Wang, S.; Li, X.; Liu, L.; Liu, S. Sustainable synthesis of highly efficient sunlight-driven Ag embedded AgCl photocatalysts. *Rsc Adv.* **2015**, *5*, 80488–80495. [CrossRef]
42. Ramar, K.; Ahamed, A.J.; Muralidharan, K. Robust green synthetic approach for the production of iron oxide nanorods and its potential environmental and cytotoxicity applications. *Adv. Powder Technol.* **2019**, *30*, 2636–2648. [CrossRef]
43. Kamaraj, M.; Kidane, T.; Muluken, K.U.; Aravind, J. Biofabrication of iron oxide nanoparticles as a potential photocatalyst for dye degradation with antimicrobial activity. *Int. J. Environ. Sci. Technol.* **2019**, *16*, 8305–8314. [CrossRef]
44. Kaur, N.; Singh, J.; Kumar, S.; Singh, P.; Al-Rashed, S.; Kaur, H.; Rawat, M. An efficient and viable photodegradation of a textile Reactive yellow-86 dye under direct sunlight by multi-structured Fe_2O_3 encapsulated with phytochemicals of R. Indica. *J. Mater. Sci. Mater. Electron.* **2020**, *31*, 21233–21247. [CrossRef]
45. Madivoli, E.S.; Kareru, P.G.; Maina, E.G.; Nyabola, A.O.; Wanakai, S.I.; Nyang'au, J.O. Biosynthesis of iron nanoparticles using Ageratum conyzoides extracts, their antimicrobial and photocatalytic activity. *Sn Appl. Sci.* **2019**, *1*, 1–11. [CrossRef]
46. Alshehri, A.; Malik, M.A.; Khan, Z.; Al-Thabaiti, S.A.; Hasan, N. Biofabrication of Fe nanoparticles in aqueous extract of Hibiscus sabdariffa with enhanced photocatalytic activities. *RSC Adv.* **2017**, *7*, 25149–25159. [CrossRef]
47. Kumar, M.R.A.; Ravikumar, C.R.; Nagaswarupa, H.P.; Purshotam, B.; Gonfa, B.; Murthy, H.C.A.; Sabir, F.K.; Tadesse, S. Evaluation of bi-functional applications of ZnO nanoparticles prepared by green and chemical methods. *J. Environ. Chem. Eng.* **2019**, *7*. [CrossRef]
48. Kaliraj, L.; Ahn, J.C.; Rupa, E.J.; Abid, S.; Lu, J.; Yang, D.C. Synthesis of panos extract mediated ZnO nano-flowers as photocatalyst for industrial dye degradation by UV illumination. *J. Photochem. Photobiol. B Biol.* **2019**, *199*, 111588. [CrossRef]
49. Lau, G.E.; Abdullah, C.A.C.; Ahmad, W.; Assaw, S.; Zheng, A.L.T. Eco-Friendly Photocatalysts for Degradation of Dyes. *Catalysts* **2020**, *10*, 1129. [CrossRef]
50. Kahsay, M.H. Synthesis and characterization of ZnO nanoparticles using aqueous extract of Becium grandiflorum for antimicrobial activity and adsorption of methylene blue. *Appl. Water Sci.* **2021**, *11*, 1–12. [CrossRef]
51. Rupa, E.J.; Kaliraj, L.; Abid, S.; Yang, D.C.; Jung, S.K. Synthesis of a Zinc Oxide Nanoflower Photocatalyst from Sea Buckthorn Fruit for Degradation of Industrial Dyes in Wastewater Treatment. *Nanomaterials* **2019**, *9*, 1692. [CrossRef] [PubMed]
52. Seo, K.H.; Markus, J.; Soshnikova, V.; Oh, K.H.; Anandapadmanaban, G.; Perez, Z.E.J.; Mathiyalagan, R.; Kim, Y.J.; Yang, D.C. Facile and green synthesis of zinc oxide particles by Stevia Rebaudiana and its in vitro photocatalytic activity. *Inorg. Nano-Met. Chem.* **2019**, *49*, 1–6. [CrossRef]
53. Tavakoli, F.; Badiei, A. Facile Synthesis of Zn-TiO_2 Nanostructure, Using Green Tea as an Eco-Friendly Reducing Agent for Photodegradation of Organic Pollutants in Water. *Pollution* **2018**, *4*, 687–696. [CrossRef]
54. Aragaw, S.G.; Sabir, F.K.; Andoshe, D.M.; Zelekew, O.A. Green synthesis of p-Co_3O_4/n-ZnO composite catalyst withEichhornia Crassipesplant extract mediated for methylene blue degradation under visible light irradiation. *Mater. Res. Express* **2020**, *7*, 095508. [CrossRef]
55. Fouda, A.; Salem, S.S.; Wassel, A.R.; Hamza, M.F.; Shaheen, T.I. Optimization of green biosynthesized visible light active CuO/ZnO nano-photocatalysts for the degradation of organic methylene blue dye. *Heliyon* **2020**, *6*, e04896. [CrossRef]
56. Alharthi, F.A.; Alghamdi, A.A.; Al-Zaqri, N.; Alanazi, H.S.; Alsyahi, A.A.; El Marghany, A.; Ahmad, N. Facile one-pot green synthesis of Ag-ZnO Nanocomposites using potato peeland their Ag concentration dependent photocatalytic properties. *Sci. Rep.* **2020**, *10*, 1–14. [CrossRef]
57. Khan, M.S.; Dhavan, P.P.; Jadhav, B.L.; Shimpi, N.G. Ultrasound-Assisted Green Synthesis of Ag-Decorated ZnO Nanoparticles UsingExcoecaria agallochaLeaf Extract and Evaluation of Their Photocatalytic and Biological Activity. *Chemistryselect* **2020**, *5*, 12660–12671. [CrossRef]
58. Yulizar, Y.; Eprasatya, A.; Apriandanu, D.O.B.; Yunarti, R.T. Facile synthesis of ZnO/$GdCoO_3$ nanocomposites, characterization and their photocatalytic activity under visible light illumination. *Vacuum* **2021**, *183*, 109821. [CrossRef]
59. Ahmad, M.; Rehman, W.; Khan, M.M.; Qureshi, M.T.; Gul, A.; Haq, S.; Ullah, R.; Rab, A.; Menaa, F. Phytogenic fabrication of ZnO and gold decorated ZnO nanoparticles for photocatalytic degradation of Rhodamine B. *J. Environ. Chem. Eng.* **2021**, *9*, 104725. [CrossRef]
60. Zelekew, O.A.; Aragaw, S.G.; Sabir, F.K.; Andoshe, D.M.; Duma, A.D.; Kuo, D.H.; Chen, X.Y.; Desissa, T.D.; Tesfamariam, B.B.; Feyisa, G.B.; et al. Green synthesis of Co-doped ZnO via the accumulation of cobalt ion onto Eichhornia crassipes plant tissue and the photocatalytic degradation efficiency under visible light. *Mater. Res. Express* **2021**, *8*, 025010. [CrossRef]
61. Konstantinou, I.K.; Albanis, T.A. TiO_2-assisted photocatalytic degradation of azo dyes in aqueous solution: Kinetic and mechanistic investigations—A review. *Appl. Catal. B Environ.* **2004**, *49*, 1–14. [CrossRef]
62. Mahshid, S.; Askari, M.; Ghamsari, M.S. Synthesis of TiO_2 nanoparticles by hydrolysis and peptization of titanium isopropoxide solution. *J. Mater. Process. Technol.* **2007**, *189*, 296–300. [CrossRef]
63. Fujishima, A.; Zhang, X.T.; Tryk, D.A. TiO_2 photocatalysis and related surface phenomena. *Surf. Sci. Rep.* **2008**, *63*, 515–582. [CrossRef]
64. Wen, Z.H.; Ci, S.Q.; Mao, S.; Cui, S.M.; Lu, G.H.; Yu, K.H.; Luo, S.L.; He, Z.; Chen, J.H. TiO_2 nanoparticles-decorated carbon nanotubes for significantly improved bioelectricity generation in microbial fuel cells. *J. Power Sources* **2013**, *234*, 100–106. [CrossRef]

65. Schneider, J.; Matsuoka, M.; Takeuchi, M.; Zhang, J.L.; Horiuchi, Y.; Anpo, M.; Bahnemann, D.W. Understanding TiO_2 Photocatalysis: Mechanisms and Materials. *Chem. Rev.* **2014**, *114*, 9919–9986. [CrossRef]
66. Chiarello, G.L.; Dozzi, M.V.; Selli, E. TiO_2-based materials for photocatalytic hydrogen production. *J. Energy Chem.* **2017**, *26*, 250–258. [CrossRef]
67. Binaeian, E.; Seghatoleslami, N.; Chaichi, M.J.; Tayebi, H.A. Preparation of titanium dioxide nanoparticles supported on hexagonal mesoporous silicate (HMS) modified by oak gall tannin and its photocatalytic performance in degradation of azo dye. *Adv. Powder Technol.* **2016**, *27*, 1047–1055. [CrossRef]
68. Nabi, G.; Raza, W.; Tahir, M.B. Green Synthesis of TiO_2 Nanoparticle Using Cinnamon Powder Extract and the Study of Optical Properties. *J. Inorg. Organomet. Polym. Mater.* **2020**, *30*, 1425–1429. [CrossRef]
69. Ahmad, W.; Jaiswal, K.K.; Soni, S. Green synthesis of titanium dioxide (TiO_2) nanoparticles by using Mentha arvensis leaves extract and its antimicrobial properties. *Inorg. Nano Met. Chem.* **2020**, *50*, 1032–1038. [CrossRef]
70. Sonker, R.K.; Hitkari, G.; Sabhajeet, S.R.; Sikarwar, S.; Rahul; Singh, S. Green synthesis of TiO_2 nanosheet by chemical method for the removal of Rhodamin B from industrial waste. *Mater. Sci. Eng. B Adv. Funct. Solid-State Mater.* **2020**, *258*, 114577. [CrossRef]
71. Aslam, M.; Abdullah, A.Z.; Rafatullah, M. Recent development in the green synthesis of titanium dioxide nanoparticles using plant-based biomolecules for environmental and antimicrobial applications. *J. Ind. Eng. Chem.* **2021**, *98*, 1–16. [CrossRef]
72. Goncalves, R.A.; Toledo, R.P.; Joshi, N.; Berengue, O.M. Green Synthesis and Applications of ZnO and TiO_2 Nanostructures. *Molecules* **2021**, *26*, 2236. [CrossRef] [PubMed]
73. Zhu, Y.Y.; Ling, Q.; Liu, Y.F.; Wang, H.; Zhu, Y.F. Photocatalytic H-2 evolution on MoS2-TiO_2 catalysts synthesized via mechanochemistry. *Phys. Chem. Chem. Phys.* **2015**, *17*, 933–940. [CrossRef]
74. Sun, S.J.; Ding, H.; Luo, Q.; Chen, S.J. The preparation of silica-TiO_2 composite by mechanochemistry method and its properties as a pigment. *Mater. Res. Innov.* **2015**, *19*, S269–S272. [CrossRef]
75. Wang, Y.F.; Saitow, K. Mechanochemical Synthesis of Red-Light-Active Green TiO_2 Photocatalysts with Disorder: Defect-Rich, with Polymorphs, and No Metal Loading. *Chem. Mater.* **2020**, *32*, 9190–9200. [CrossRef]
76. Antony, D.; Yadav, R. Facile fabrication of green nano pure CeO_2 and Mn-decorated CeO_2 with Cassia angustifoliaseed extract in water refinement by optimal photodegradation kinetics of malachite green. *Environ. Sci. Pollut. Res.* **2021**, *28*, 18589–18603. [CrossRef]
77. Rabiee, N.; Bagherzadeh, M.; Kiani, M.; Ghadiri, A.M.; Etessamifar, F.; Jaberizadeh, A.H.; Shakeri, A. Biosynthesis of Copper Oxide Nanoparticles with Potential Biomedical Applications. *Int. J. Nanomed.* **2020**, *15*, 3983–3999. [CrossRef] [PubMed]
78. Alhebshi, N.; Huang, H.; Ghandour, R.; Alghamdi, N.K.; Alharbi, O.; Aljurban, S.; He, J.H.; Al-Jawhari, H. Green synthesized CuxO@Cu nanocomposites on a Cu mesh with dual catalytic functions for dye degradation and hydrogen evaluation. *J. Alloy. Compd.* **2020**, *848*, 156284. [CrossRef]
79. Navada, K.M.; Nagaraja, G.K.; D'Souza, J.N.; Kouser, S.; Ranjitha, R.; Manasa, D.J. Phyto assisted synthesis and characterization of Scoparia dulsisL.leaf extract mediated porous nano CuO photocatalysts and its anticancer behavior. *Appl. Nanosci.* **2020**, *10*, 4221–4240. [CrossRef]
80. Fatimah, I.; Sahroni, I.; Muraza, O.; Doong, R.A. One-pot biosynthesis of SnO_2 quantum dots mediated by Clitoria ternatea flower extract for photocatalytic degradation of rhodamine B. *J. Environ. Chem. Eng.* **2020**, *8*, 103879. [CrossRef]
81. Mohamed, H.E.A.; Sone, B.T.; Khamlich, S.; Coetsee-Hugo, E.; Swart, H.C.; Thema, T.; Sbiaa, R.; Dhlamini, M.S. Biosynthesis of $BiVO_4$ nanorods using Callistemon viminalis extracts: Photocatalytic degradation of methylene blue. *Mater. Today Proc.* **2021**, *36*, 328–335. [CrossRef]
82. Munoz-Batista, M.J.; Kubacka, A.; Fernandez-Garcia, M. Effect of g-C_3N_4 loading on TiO_2-based photocatalysts: UV and visible degradation of toluene. *Catal. Sci. Technol.* **2014**, *4*, 2006–2015. [CrossRef]
83. Shi, M.J.; Wei, W.; Jiang, Z.F.; Han, H.K.; Gao, J.R.; Xie, J.M. Biomass-derived multifunctional TiO_2/carbonaceous aerogel composite as a highly efficient photocatalyst. *RSC Adv.* **2016**, *6*, 25255–25266. [CrossRef]
84. Wei, W.; Hu, H.H.; Huang, Z.Y.; Jiang, Z.F.; Lv, X.M.; Xie, J.M.; Kong, L.R. $BiPO_4$ nanorods anchored in biomass-based carbonaceous aerogel skeleton: A 2D–3D heterojunction composite as an energy-efficient photocatalyst. *J. Supercrit. Fluids* **2019**, *147*, 33–41. [CrossRef]
85. Djellabi, R.; Zhao, X.; Bianchi, C.L.; Su, P.D.; Ali, J.; Yang, B. Visible light responsive photoactive polymer supported on carbonaceous biomass for photocatalytic water remediation. *J. Clean. Prod.* **2020**, *269*, 122286. [CrossRef]
86. Yu, X.N.; Liu, Z.X.; Wang, Y.M.; Luo, H.Y.; Tang, X. Fabrication of corncob-derived biomass charcoal decorated g-C_3N_4 photocatalysts for removing 2-mercaptobenzothiazole. *New J. Chem.* **2020**, *44*, 15908–15918. [CrossRef]
87. Zhu, Z.; Ma, C.C.; Yu, K.S.; Lu, Z.Y.; Liu, Z.; Huo, P.W.; Tang, X.; Yan, Y.S. Synthesis Ce-doped biomass carbon-based g-C_3N_4 via plant growing guide and temperature-programmed technique for degrading 2-Mercaptobenzothiazole. *Appl. Catal. B Environ.* **2020**, *268*, 118432. [CrossRef]
88. Liang, W.; Pan, J.H.; Duan, X.J.; Tang, H.; Xu, J.; Tang, G.G. Biomass carbon modified flower-like Bi_2WO_6 hierarchical architecture with improved photocatalytic performance. *Ceram. Int.* **2020**, *46*, 3623–3630. [CrossRef]
89. Wang, T.; Liu, X.Q.; Han, D.L.; Ma, C.C.; Wei, M.B.; Huo, P.W.; Yan, Y.S. Biomass derived the V-doped carbon/Bi_2O_3 composite for efficient photocatalysts. *Environ. Res.* **2020**, *182*, 108998. [CrossRef] [PubMed]

90. Sun, J.; Lin, X.M.; Xie, J.; Zhang, Y.Z.; Wang, Q.; Ying, Z.R. Facile synthesis of novel ternary g-C_3N_4/ferrite/biochar hybrid photocatalyst for efficient degradation of methylene blue under visible-light irradiation. *Colloids Surf. A Physicochem. Eng. Asp.* **2020**, *606*, 125556. [CrossRef]
91. Mondol, B.; Sarker, A.; Shareque, A.M.; Dey, S.C.; Islam, M.T.; Das, A.K.; Shamsuddin, S.M.; Molla, M.A.; Sarker, M. Preparation of Activated Carbon/TiO_2 Nanohybrids for Photodegradation of Reactive Red-35 Dye Using Sunlight. *Photochem* **2021**, *1*, 54–66. [CrossRef]
92. Islam, S.E.; Hang, D.R.; Chen, C.H.; Chou, M.M.C.; Liang, C.T.; Sharma, K.H. Rational design of hetero-dimensional C-ZnO/MoS_2 nanocomposite anchored on 3D mesoporous carbon framework towards synergistically enhanced stability and efficient visible-light-driven photocatalytic activity. *Chemosphere* **2021**, *266*, 129148. [CrossRef]
93. dos Santos, G.S.; Goes, C.M.; de Sousa, J.G.M.; de Moraes, N.P.; Chaguri, L.; Rodrigues, L.A. Evaluation of ZnO/Carbon Xerogel Composite as a Photocatalyst for Solar and Visible Light Degradation of the Rhodamine B Dye. *J. Nanosci. Nanotechnol.* **2021**, *21*, 2292–2301. [CrossRef]
94. Zhao, C.; Ran, F.L.; Dai, L.; Li, C.Y.; Zheng, C.Y.; Si, C.L. Cellulose-assisted construction of high surface area Z-scheme C-doped g-C_3N_4/WO_3 for improved tetracycline degradation. *Carbohydr. Polym.* **2021**, *255*, 117343. [CrossRef]
95. Duan, L.Y.; Li, G.Q.; Zhang, S.T.; Wang, H.Y.; Zhao, Y.L.; Zhang, Y.F. Preparation of S-doped g-C_3N_4 with C vacancies using the desulfurized waste liquid extracting salt and its application for NOx removal. *Chem. Eng. J.* **2021**, *411*, 128551. [CrossRef]
96. Donar, Y.O.; Bilge, S.; Sinag, A. Utilisation of lignin as a model biomass component for preparing a highly active photocatalyst under UV and visible light. *Mater. Sci. Semicond. Process.* **2020**, *118*, 105151. [CrossRef]
97. Geng, A.B.; Xu, L.J.; Gan, L.; Mei, C.T.; Wang, L.J.; Fang, X.Y.; Li, M.R.; Pan, M.Z.; Han, S.G.; Cui, J.Q. Using wood flour waste to produce biochar as the support to enhance the visible-light photocatalytic performance of BiOBr for organic and inorganic contaminants removal. *Chemosphere* **2020**, *250*, 126291. [CrossRef]
98. Mushtaq, F.; Zahid, M.; Mansha, A.; Bhatti, I.A.; Mustafa, G.; Nasir, S.; Yaseen, M. $MnFe_2O_4$/coal fly ash nanocomposite: A novel sunlight-active magnetic photocatalyst for dye degradation. *Int. J. Environ. Sci. Technol.* **2020**, *17*, 4233–4248. [CrossRef]
99. Shen, X.L.; Zhu, Z.L.; Zhang, H.; Di, G.L.; Chen, T.; Qiu, Y.L.; Yin, D.Q. Carbonaceous composite materials from calcination of azo dye-adsorbed layered double hydroxide with enhanced photocatalytic efficiency for removal of Ibuprofen in water. *Environ. Sci. Eur.* **2020**, *32*, 1–15. [CrossRef]
100. Kumar, A.; Choudhary, P.; Kumar, K.; Krishnan, V. Plasmon induced hot electron generation in two dimensional carbonaceous nanosheets decorated with Au nanostars: Enhanced photocatalytic activity under visible light. *Mater. Chem. Front.* **2021**, *5*, 1448–1467. [CrossRef]
101. Daulbayev, C.; Sultanov, F.; Korobeinyk, A.V.; Yeleuov, M.; Azat, S.; Bakbolat, B.; Umirzakov, A.; Mansurov, Z. Bio-waste-derived few-layered graphene/$SrTiO_3$/PAN as efficient photocatalytic system for water splitting. *Appl. Surf. Sci.* **2021**, *549*, 149176. [CrossRef]
102. Leary, R.; Westwood, A. Carbonaceous nanomaterials for the enhancement of TiO_2 photocatalysis. *Carbon* **2011**, *49*, 741–772. [CrossRef]
103. Ong, W.J.; Tan, L.L.; Ng, Y.H.; Yong, S.T.; Chai, S.P. Graphitic Carbon Nitride (g-C_3N_4)-Based Photocatalysts for Artificial Photosynthesis and Environmental Remediation: Are We a Step Closer To Achieving Sustainability? *Chem. Rev.* **2016**, *116*, 7159–7329. [CrossRef] [PubMed]
104. Nalid, N.R.; Majid, A.; Tahir, M.B.; Niaz, N.A.; Khalid, S. Carbonaceous-TiO_2 nanomaterials for photocatalytic degradation of pollutants: A review. *Ceram. Int.* **2017**, *43*, 14552–14571. [CrossRef]
105. Ge, J.; Zhang, Y.; Park, S.J. Recent Advances in Carbonaceous Photocatalysts with Enhanced Photocatalytic Performances: A Mini Review. *Materials* **2019**, *12*, 1916. [CrossRef]
106. Zuliani, A.; Munoz-Batista, M.J.; Luque, R. Microwave-assisted valorization of pig bristles: Towards visible light photocatalytic chalcocite composites. *Green Chem.* **2018**, *20*, 3001–3007. [CrossRef]
107. Cova, C.M.; Zuliani, A.; Santiago, A.R.P.; Caballero, A.; Munoz-Batista, M.J.; Luque, R. Microwave-assisted preparation of Ag/Ag_2S carbon hybrid structures from pig bristles as efficient HER catalysts. *J. Mater. Chem. A* **2018**, *6*, 21516–21523. [CrossRef]
108. Cova, C.M.; Zuliani, A.; Munoz-Batista, M.J.; Luque, R. A Sustainable Approach for the Synthesis of Catalytically Active Peroxidase-Mimic ZnS Catalysts. *ACS Sustain. Chem. Eng.* **2019**, *7*, 1300–1307. [CrossRef]
109. Lee, S.L.; Chang, C.J. Recent Progress on Metal Sulfide Composite Nanomaterials for Photocatalytic Hydrogen Production. *Catalysts* **2019**, *9*, 457. [CrossRef]
110. Nawaukkaratharnant, N.; Sujaridworakun, P.; Mongkolkachit, C.; Wasanapiarnpong, T. Possible use of waste from marcasite jewelry industry as iron pyrite source incorporated with titanium dioxide for photodegradation of lignin under a halogen tungsten lamp. *Mater. Lett.* **2020**, *271*, 127778. [CrossRef]
111. Mosaddegh, E. Ultrasonic-assisted preparation of nano eggshell powder: A novel catalyst in green and high efficient synthesis of 2-aminochromenes. *Ultrason. Sonochemistry* **2013**, *20*, 1436–1441. [CrossRef]
112. Rahman, M.M.; Netravali, A.N.; Tiimob, B.J.; Rangari, V.K. Bioderived "Green" Composite from Soy Protein and Eggshell Nanopowder. *ACS Sustain. Chem. Eng.* **2014**, *2*, 2329–2337. [CrossRef]
113. Ferro, A.C.; Guedes, M. Mechanochemical synthesis of hydroxyapatite using cuttlefish bone and chicken eggshell as calcium precursors. *Mater. Sci. Eng. C Mater. Biol. Appl.* **2019**, *97*, 124–140. [CrossRef]

114. Zhang, X.Y.; He, X.S.; Kang, Z.W.; Cui, M.L.; Yang, D.P.; Luque, R. Waste Eggshell-Derived Dual-Functional CuO/ZnO/Eggshell Nanocomposites: (Photo)catalytic Reduction and Bacterial Inactivation. *Acs Sustain. Chem. Eng.* **2019**, *7*, 15762–15771. [CrossRef]
115. Li, Z.H.; Yang, D.P.; Chen, Y.S.; Du, Z.Y.; Guo, Y.L.; Huang, J.L.; Li, Q.B. Waste eggshells to valuable $Co_3O_4/CaCO_3$ materials as efficient catalysts for VOCs oxidation. *Mol. Catal.* **2020**, *483*, 110766. [CrossRef]
116. Luo, W.Q.; Ji, Y.P.; Qu, L.; Dang, Z.; Xie, Y.Y.; Yang, C.F.; Tao, X.Q.; Zhou, J.M.; Lu, G.N. Effects of eggshell addition on calcium-deficient acid soils contaminated with heavy metals. *Front. Environ. Sci. Eng.* **2018**, *12*, 4. [CrossRef]
117. Sree, G.V.; Nagaraaj, P.; Kalanidhi, K.; Aswathy, C.A.; Rajasekaran, P. Calcium oxide a sustainable photocatalyst derived from eggshell for efficient photo-degradation of organic pollutants. *J. Clean. Prod.* **2020**, *270*, 122294. [CrossRef]
118. Gil, A.; Korili, S.A. Management and valorization of aluminum saline slags: Current status and future trends. *Chem. Eng. J.* **2016**, *289*, 74–84. [CrossRef]
119. Santamaria, L.; Vicente, M.A.; Korili, S.A.; Gil, A. Saline slag waste as an aluminum source for the synthesis of Zn-Al-Fe-Ti layered double-hydroxides as catalysts for the photodegradation of emerging contaminants. *J. Alloys Compd.* **2020**, *843*, 156007. [CrossRef]
120. Alberti, S.; Caratto, V.; Peddis, D.; Belviso, C.; Ferretti, M. Synthesis and characterization of a new photocatalyst based on TiO_2 nanoparticles supported on a magnetic zeolite obtained from iron and steel industrial waste. *J. Alloys Compd.* **2019**, *797*, 820–825. [CrossRef]
121. Huang, K.; Guo, J.; Xu, Z.M. Recycling of waste printed circuit boards: A review of current technologies and treatment status in China. *J. Hazard. Mater.* **2009**, *164*, 399–408. [CrossRef] [PubMed]
122. Zhu, P.; Xia, B.; Li, H.H.; Ma, Y.; Qian, G.R. Visible-Light-Driven Photoreduction of Cr(VI) by Waste-Based Cu_2O Photocatalyst from Waste Printed Circuit Boards. *Environ. Eng. Sci.* **2021**, *38*, 565–574. [CrossRef]
123. Niu, B.; Chen, Z.Y.; Xu, Z.M. Recycling waste tantalum capacitors to synthesize high value-added Ta_2O_5 and polyaniline-decorated Ta_2O_5 photocatalyst by an integrated chlorination-sintering-chemisorption process. *J. Clean. Prod.* **2020**, *252*, 117206. [CrossRef]
124. Cao, Y.L.; Niu, B.; Sun, H.H.; Xu, Z.M. Synthesizing a high value-added composite photocatalyst using waste capacitors combined with PANI by a mechanical chemistry method. *Sustain. Energy Fuels* **2021**, *5*, 2916–2926. [CrossRef]
125. Montoya-Bautista, C.V.; Acevedo-Pena, P.; Zanella, R.; Ramirez-Zamora, R.M. Characterization and Evaluation of Copper Slag as a Bifunctional Photocatalyst for Alcohols Degradation and Hydrogen Production. *Top. Catal.* **2021**, *64*, 131–141. [CrossRef]
126. Sheldon, R.A. The E factor 25 years on: The rise of green chemistry and sustainability. *Green Chem.* **2017**, *19*, 18–43. [CrossRef]
127. Kleinekorte, J.; Fleitmann, L.; Bachmann, M.; Katelhon, A.; Barbosa-Povoa, A.; von der Assen, N.; Bardow, A. Life Cycle Assessment for the Design of Chemical Processes, Products, and Supply Chains. *Annu. Rev. Chem. Biomol. Eng.* **2020**, *11*, 203–233. [CrossRef] [PubMed]

Article

Preparation of Activated Carbon/TiO_2 Nanohybrids for Photodegradation of Reactive Red-35 Dye Using Sunlight

Bappy Mondol [1], Anupam Sarker [1], A. M. Shareque [1], Shaikat Chandra Dey [1], Mohammad Tariqul Islam [2], Ajoy Kumar Das [1], Sayed Md. Shamsuddin [1], Md. Ashraful Islam Molla [1,*] and Mithun Sarker [1,*]

1 Department of Applied Chemistry and Chemical Engineering, Faculty of Engineering and Technology, University of Dhaka, Dhaka 1000, Bangladesh; bappy279@gmail.com (B.M.); suptaaccesarker@gmail.com (A.S.); sharequetawsif@gmail.com (A.M.S.); shaikat@du.ac.bd (S.C.D.); ajoyappliedchemistry@gmail.com (A.K.D.); sdin@du.ac.bd (S.M.S.)

2 Department of Chemistry, American International University-Bangladesh, Dhaka 1229, Bangladesh; tariquldu@aiub.edu

* Correspondence: ashraful.acce@du.ac.bd (M.A.I.M.); mithun@du.ac.bd (M.S.); Tel.: +88-015-5235-9706 (M.A.I.M.); +88-017-9330-1288 (M.S.)

Abstract: Activated carbon/titanium dioxide (AC/TiO_2) nanohybrids were synthesized by a hydrothermal technique using various weight percent of commercial AC and were characterized by X-ray diffraction (XRD), field emission scanning electron microscopy (FESEM), Fourier transform infrared (FTIR) and thermogravimetric analysis (TGA). The synthesized nanohybrids were applied to photodegradation of Reactive Red-35 (RR-35) dye in aqueous solution using sunlight. Due to the synergistic effect of adsorption and photodegradation activity, AC/TiO_2 nanohybrids were more efficient in treating the aqueous dye solution than that of AC and TiO_2 The maximum (95%) RR-35 dye removal from the water was obtained with 20 wt% AC/TiO_2 within 30 min at natural pH of 5.6. The possible photodegradation mechanism of RR-35 dye with AC/TiO_2 was discussed from the scavenger test. Moreover, AC/TiO_2 was found to be suitable for long-term repeated applications through recyclability experiments. Therefore, AC/TiO_2 nanohybrid is a promising photocatalyst for treating azo dyes especially RR-35 from water.

Keywords: TiO_2; activated carbon; nanohybrid; photodegradation; azo dyes; sunlight

Citation: Mondol, B.; Sarker, A.; Shareque, A.M.; Dey, S.C.; Islam, M.T.; Das, A.K.; Shamsuddin, S.M..; Molla, M..A.I.; Sarker, M. Preparation of Activated Carbon/TiO_2 Nanohybrids for Photodegradation of Reactive Red-35 Dye Using Sunlight. *Photochem* **2021**, *1*, 54–66. https://doi.org/10.3390/photochem1010006

Academic Editor: Vincenzo Vaiano

Received: 20 April 2021
Accepted: 14 May 2021
Published: 18 May 2021

1. Introduction

Adverse effects are being observed in the environment, especially, in water quality due to the rapid urbanization and industrialization in the world. Hazardous wastes along with untreated toxic chemicals are usually flushed into rivers and ponds or nearby water reservoirs [1]. The dyes are considered hazardous water pollutants due to their toxic effects on the environment and complexity in removal [2]. Textile dyes, especially azo dyes not only affect the aesthetic merit of surface water but also reduces light penetration that hampers aquatic lives [2]. Moreover, azo dyes are usually mutagenic and carcinogenic [2,3]. Reactive Red-35 (RR-35) is a typical toxic azo dye and has widely been used in textile industries because of its vibrant colors, water-fastness and energy-efficient application techniques [4].

Several techniques have been applied for wastewater treatment, especially in azo dye removal from water [2,5]. However, the traditional physical and biological water treatment processes have some drawbacks such as prolonged tendency, high cost and lower efficiency [2]. Moreover, some non-traditional processes such as ozonation, radiation and membrane processes have some drawbacks; for example, ozonation and radiation processes generate toxic byproducts, which could be difficult to remove [6,7]. Therefore, cost-effective and environment-friendly methods are desired for the removal of toxic dyes from the effluent streams.

Photocatalytic degradation, an advanced oxidation water purification method, has attracted much interest to researchers since it can cost-effectively degrade the dyes without causing any further harm to nature [8]. To date, several metal oxides [9,10] including titanium dioxide (TiO_2) [11] have been selected as photocatalysts due to their less toxicity, chemical stability and distinct electronic structure (unoccupied conduction band and occupied valence band) [12]. Among them, TiO_2 is extensively used for a wide range of applications compared to other photocatalysts, due to its low cost, excellent aquatic stability and high redox potential [13,14]. However, some drawbacks such as easy agglomeration, lower adsorption capacity, rapid phase transformation (from highly active anatase to less active rutile phase) and separation of nano-sized TiO_2 from water limit the industrial applications of TiO_2 in wastewater treatment [15,16]. Therefore, the introduction of support material is needed to overcome these problems and to increase the efficiency of TiO_2 [16]. Several materials including activated carbon (AC) [17], clay [18], metal-organic frameworks [19], silica [20], etc., were very commonly used as a support for TiO_2. Among them, AC is a very effective support material especially in water remediation due to its excellent hydrophobicity and high surface area with different functionalities [18]. Thus, AC would not only work as effective support material but also continuously adsorb water pollutants such as dye molecules around the surface of TiO_2 to halt recombination and smooth degradation. Moreover, AC/TiO_2 nanocomposites are extensively used in several purposes such as the antimicrobial agent [21], photocatalytic degradation of pharmaceutical products [22] and dye removal [17]. Therefore, the possible applications of AC/TiO_2 in photocatalytic degradation of dye in water are very interesting considering the synergistic effect of the adsorption process mediated by AC and the photocatalytic activity of TiO_2.

Herein, AC/TiO_2 nanohybrids were synthesized via hydrothermal technique followed by calcination. The fabricated nanohybrids were thoroughly characterized and applied as a photocatalyst for the degradation of RR-35 in aqueous solution. Moreover, the plausible mechanisms for photodegradation of RR-35 were recommended based on the radical scavenger experiments.

2. Materials and Methods

2.1. Materials

Activated carbon (AC, Darco, 100 mesh, surface area 600 m^2/g) was purchased from Merck, Mumbai, India. Acetone (99%), $TiCl_3$ (15%) and NaOH flakes (97%) were purchased from Loba Chemie Pvt. Ltd, Mumbai, India. NH_4OH and HCl (37%) were purchased from Merck KGaA (64271 Darmstadt, Germany). H_2O_2 (35%) was collected from TCI, Tamil Nadu, India. Reactive Red-35 (RR-35) was purchased from Sisco Research Laboratories Pvt. Ltd. (SRL), Mumbai, India. All the chemicals are reagent grade and utilized without further purification.

2.2. Preparation of AC/TiO_2 Nanohybrids

In this study, the various proportions of AC/TiO_2 nanohybrids were successfully synthesized, according to the previous report with slight modification [23,24]. A 10 mL solution of $TiCl_3$ in 15% HCl was hydrolyzed using 20 mL deionized water and then a dropwise addition of aqueous NH_4OH (20 mL) produced dark blue precipitate, which turned yellowish due to the addition of H_2O_2 through constant stirring. After 30 min of vigorous stirring, the calculated amount of AC was added to the mixture. The obtained suspensions were left for 4 days for aging and then the precipitates were filtered, washed with distilled water followed by acetone and dried at 60 °C. Finally, calcination was done at 400 °C under air for 6 h to obtain desired AC/TiO_2 nanohybrids. For comparison, bare TiO_2 was also synthesized by a similar procedure without using AC.

2.3. Characterization of Studied Nanohybrids

An X-ray diffractometer (Ultima IV, Rigaku Corporation, Akishima, Japan) fitted with Cu Kα radiation (λ = 0.154 nm) was used to analyze the XRD patterns of the studied samples at room temperature (25 °C). The morphological features of the samples were examined by an analytical field emission scanning electron microscope (FESEM) (JEOL JSM-6490LA, Tokyo, Japan) operating at an accelerating voltage of 20 kV. The interaction between TiO_2 and AC was observed by recording transmission spectra on a Fourier transform infrared spectrometer (FTIR, Shimadzu IR Prestidge-21, Japan) in the wavenumber range of 4000–400 cm^{-1} with resolution 4 cm^{-1} and S/N ratio of over 40,000:1. The thermal stability of the samples was inspected by recording thermograms on a thermogravimetric analyzer (TG-00260, SHIMADZU Corp, Japan), which operated within 25–800 °C with 10 °C min^{-1} heating rate and 5 min hold time.

2.4. RR-35 Dye Removal Experiments

The dye removal performance of commercial AC, bare TiO_2 and AC/TiO_2 nanohybrids were carried out at various operating conditions such as pH, time and dye concentration using a standard solution of RR-35 (at 536 λ_{max} nm [25]) under solar irradiation in open-air. The remaining concentration of RR-35 after degradation at various conditions was measured by recording absorbance on a UV-1700 Spectrophotometer (Shimadzu, Japan). Batch experiments were performed under similar conditions on sunny days between 11:00 and 14:00 BST (Bangladesh standard time) for the investigation of the % removal of RR-35 using the studied photocatalysts (location coordinate: 23.7275 °N, 90.4019 °E; intensity of sunlight: 3.0 mW/cm^2; time of year: June to August). The intensity of light was measured by a UV radiometer with a sensor of 320–420 nm wavelength (UVR-400, Iuchi Co., Osaka, Japan). For adjusting the pH of the solutions before degradation experiments, dilute HCl (0.1 M) and NaOH (0.1 M) aqueous solutions were used. It should be noted that the natural solution pH of RR-35 was 5.6. Moreover, a reusability experiment was carried out to evaluate the potential of AC/TiO_2 for industrial application.

3. Results and Discussion

3.1. XRD Analysis

The XRD patterns of AC, TiO_2 and AC/TiO_2 nanohybrids are shown in Figure 1. As can be seen in Figure 1a, the XRD of AC shows two weak broad peaks at 25° and 44° corresponding to reflection in the (002) plane and the (100) plane, which indicates the existence of quartz phase at low concentration and two-dimensional reflections of disordered stacking of micrographites, respectively [26,27]. On the other hand, the XRD pattern of TiO_2 and AC/TiO_2 nanohybrids shows diffraction peaks at about 25.34°, 37.76°, 48.00°, 53.81° and 55.11° (2θ values), which could be indexed as (101), (004), (200), (105) and (211) planes, respectively, of an anatase TiO_2 [28,29]. It suggests that the anatase TiO_2 is predominantly formed in the nanohybrids when calcined at 400 °C [24]. The sharp peaks of the XRD pattern of synthesized nanohybrids acknowledge that there were no impurities present in the nanohybrids. Therefore, the XRD analysis strongly supports the successful fabrication of the nanohybrids. Interestingly, it was observed that the diffraction peak intensity was decreased and slight peak broadening was occurred in nanohybrids with increasing AC content (Figure 1b), similar to the previous report [22,23]. Therefore, the crystallite size of nanohybrids was decreased with increasing AC content. Based on the XRD results, the crystal size of the nanohybrids is determined using the Scherrer formula for the peak at $2\theta = 25.15°$. The crystal size of bare TiO_2, 10 wt% AC/TiO_2, 20 wt% AC/TiO_2 and 30 wt% AC/TiO_2 could be estimated as 26, 20, 16 and 15 nm, respectively. Notably, there was no characteristic peak that is observed in nanohybrids due to AC. The reason could be attributed to the fact that the main peak of AC might be shielded by the peak of anatase TiO_2 [30].

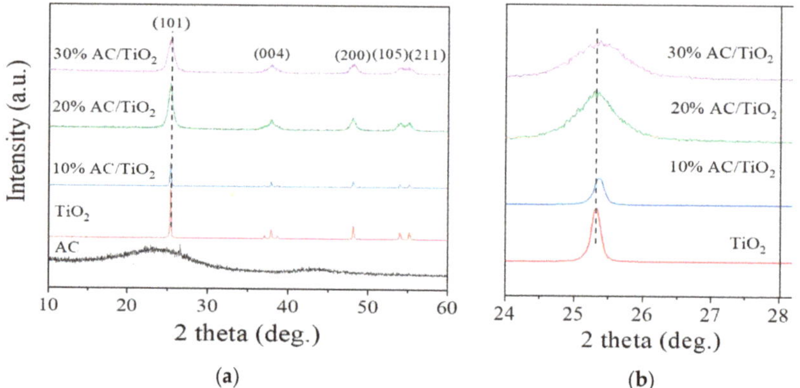

Figure 1. XRD pattern for (**a**) AC, TiO_2 and AC/TiO_2 nanohybrids and (**b**) elongated peak of TiO_2 and AC/TiO_2 nanohybrids.

3.2. FESEM Analysis

The morphologies of the AC, TiO_2 and AC/TiO_2 nanohybrids are investigated by FESEM micrographs and shown in Figure 2. The external surface of the AC clusters is flat, rough and porous whereas the aggregation of TiO_2 crystallites is significant in TiO_2, with a size of 26 nm (based on the Scherrer equation). From FESEM images of AC/TiO_2 nanohybrids, it is observed that the TiO_2 particle with the nanosized dimension is uniformly deposited on the AC surface and in the bulk of the AC and the aggregate size of TiO_2 particles strongly depended on the AC content. The FESEM images, shown in Figure 2, reveal that particle size of nanohybrids decreased with increasing AC content, which is also agreeable with the XRD results. Therefore, it is easily assumed that the nanohybrid had a relatively higher percentage of smaller particles than TiO_2, which might be beneficial for photocatalytic activity [31].

Figure 2. FESEM images of (**a**) AC, (**b**) TiO_2, (**c**) 10 wt% AC/TiO_2, (**d**) 20 wt% AC/TiO_2 and (**e**) 30 wt% AC/TiO_2.

3.3. FTIR Analysis

FTIR spectra of TiO_2 and AC/TiO_2 nanohybrids are shown together with AC in Figure 3. The main absorption peaks were located at 3430, 1058, 710 and 530 cm^{-1}. The bands observed in AC and synthesized nanohybrids, located at 3430 cm^{-1} were due to –OH stretching. Bare TiO_2 and nanohybrids revealed an intense peak contributing to the stretching vibrations of Ti–O bonds at 710 and 530 cm^{-1}, which is the characteristic peak of TiO_2 [28]. Interestingly, with the addition of AC, a new peak located at 1058 cm^{-1} appeared and the peak at 530 cm^{-1} shifted to lower wavenumbers, similar to the previous report [23]. Besides, with increasing AC content, the width and intensity of the peak at 1058 cm^{-1} was increased. Thus, it can be concluded that the appearance of this peak is related to AC in the AC/TiO_2 composites [23].

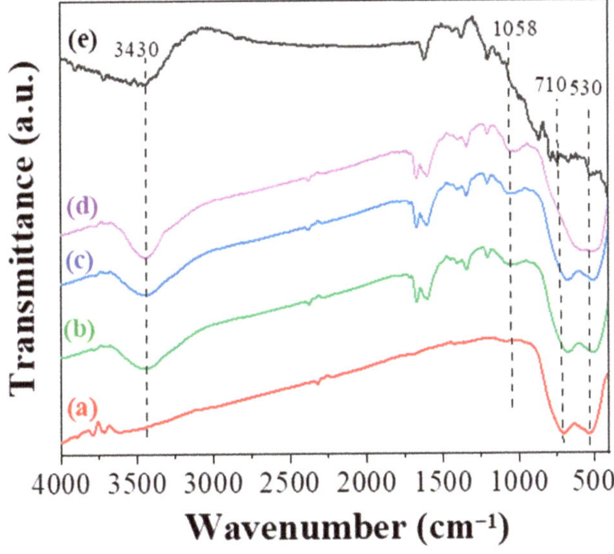

Figure 3. FTIR spectra of (a) TiO_2, (b) 10% AC/TiO_2, (c) 20% AC/TiO_2, (d) 30% AC/TiO_2 and (e) AC.

3.4. TG Analysis

To investigate the amount of TiO_2 in the AC/TiO_2 composites synthesized at different ratios, thermogravimetric (TG) tests were carried out in a range of 25–800 °C under N_2 flow. As shown in the TG curves (Figure 4), the gentle weight loss of commercial AC, TiO_2 and AC/TiO_2 nanohybrids at temperatures ranging from 20 to 200 °C is observed due to the desorption of physisorbed water. According to the TG curves, the weight loss of the nanohybrids increased with increasing AC content. The final weight loss was 15.6% for pure AC, which is agreeable with earlier reported results [32] and 6.6%, 4.3%, 2.4% and 0.60% for 30 wt% AC/TiO_2, 20 wt% AC/TiO_2, 10 wt% AC/TiO_2 and TiO_2.

3.5. Effect of AC Content

The effect of AC content in nanohybrids for the removal of RR-35 from water under dark and sunlight irradiation is depicted in Figure 5. To quantify the relative contributions of both adsorption and photodegradation towards the overall removal of RR-35, control experiments in dark were carried out for 30 min. As shown in Figure 5, the adsorption of RR-35 by AC, TiO_2 and 20% AC/TiO_2 under dark were calculated to be 42%, 9% and 34% respectively. The respective values for commercial AC and bare TiO_2 were 47% and 26% under sunlight irradiation within 30 min. The removal efficiency of AC in dark and under sunlight is nearly similar which indicates that the removal of RR-35 using AC was due to adsorption rather than photocatalysis. Additionally, the TiO_2 was not much effective for the degradation of RR-35 due to a large band gap, similar to the earlier report for the degradation of methyl orange dye [33]. Commercial AC showed better performance than TiO_2, which might be due to the adsorption of RR-35 on AC. Additionally, it was noticed that the photocatalytic efficiency of nanohybrids was increased compared with commercial AC and TiO_2 because of the synergistic effect of both adsorption and photocatalysis due to a combination of AC and TiO_2 [34]. Among studied nanohybrids, 20 wt% AC/TiO_2 showed the best performance in RR-35 removal (95%) due to the formation of nanoparticles with good crystallinity [33]. Probably, all of the TiO_2 nanoparticles might take part in interaction with AC in the case of 20wt% AC/TiO_2, which was important for maximum removal of RR-35 [33,35]. Therefore, further investigation was carried out using 20 wt% AC/TiO_2 as a photocatalyst for degradation of RR-35.

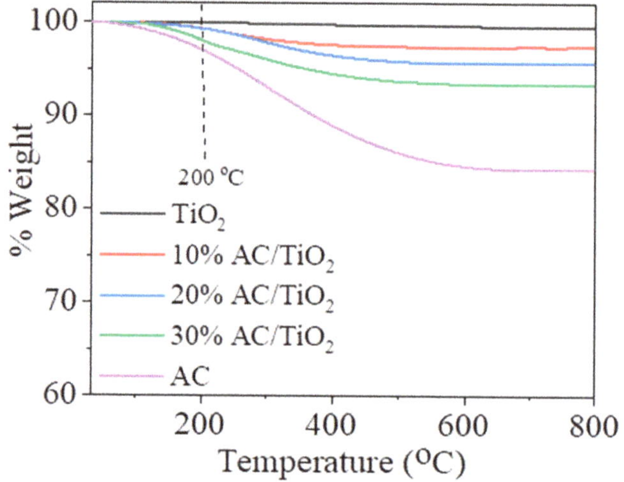

Figure 4. TGA of AC, TiO_2 and AC/TiO_2 nanohybrids.

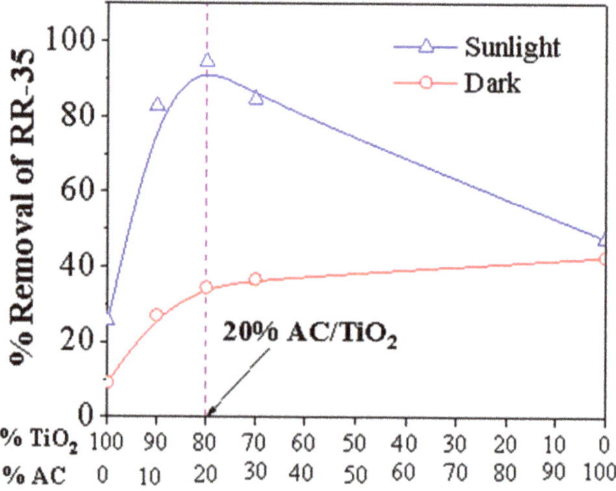

Figure 5. Effect of AC content in nanohybrids for the removal of RR-35 from water (initial concentration of RR-35: 20 mg L^{-1}; pH: 5.6) under dark and sunlight irradiation (time: 30 min).

3.6. Effect of pH

Solution pH strongly controls the surface charge of a nanohybrid and thus the photodegradation of dye is also greatly affected. The degradation of RR-35 by 20 wt% AC/TiO_2 was recorded within a pH range of 2–10 and the results are displayed in Figure 6. The results revealed that the degradation of dye increased with solution pH up to a certain level (at pH 6.0) and then decreased with solution pH. The maximum removal of RR-35 at pH 6.0 was 99%, after 30 min of the sunlight irradiation. At natural pH 5.6, the RR-35 removal was observed 95%, which was almost similar to the percent of removal obtained at pH 6.0. Therefore, further investigation such as the effect of time, dye concentration, scavenger test and reusability experiment was carried at natural solution pH (5.6) of RR-35. The point of zero charge value (pzc) of TiO_2 and AC/TiO_2 is reported as 6.8 and 7.6 respectively [25,36], which means that the surface of AC/TiO_2 becomes positive below this pH value (pH < 7.6) and negative above this pH value (pH > 7.6). RR-35 is an anionic dye in aqueous solution; therefore, at pH 5.6–7.0, the positively charged 20 wt% AC/TiO_2 surface could adsorb more dye molecules through electrostatic interaction. Due to the enhanced adsorption, more dye molecules are exposed to the active sites of the photocatalyst. The lower rate of photodegradation of RR-35 at alkaline media was found probably due to the presence of superfluous negative charge on the catalyst surface, which could repel the anionic dye molecules. Moreover, at a very low pH (from 2.0 to 4.0) the photodegradation of RR-35 decreased probably due to the protonation of dye molecules, which consequently reduced the adsorption of dye molecules on the AC/TiO_2 surface [36].

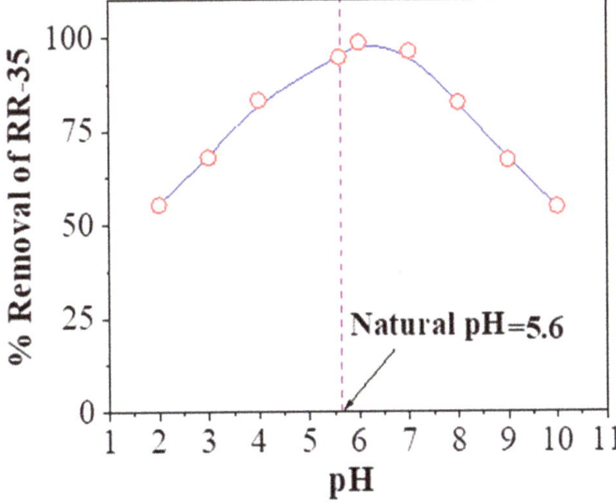

Figure 6. Effect of pH on the removal of RR-35 with 20 wt% AC/TiO_2 (initial concentration of RR-35: 20 mg/L^{-1}; pH: 2–10; irradiation time: 30 min).

3.7. *Effect of Time*

The effect of irradiation time on photodegradation of RR-35 using 20 wt% AC/TiO_2 was investigated by evaluating the percentage removal of RR-35 at different periods as shown in Figure 7. The result from Figure 7 revealed that 95% dye was removed by 5 mg of 20 wt% AC/TiO_2 in 30 min at pH 5.6 under sunlight irradiation. It was evident that the rate of degradation was higher during the initial stage (up to 10 min). Afterward, the degradation rate started decreasing gradually and the degradation finally reached equilibrium after 30 min. It might be due to the gradual decline in the concentration of RR-35 and simultaneous formation of degradation products with irradiation time. As a result, maximum degradation (95%) of RR-35 was found after 30 min of sunlight irradiation.

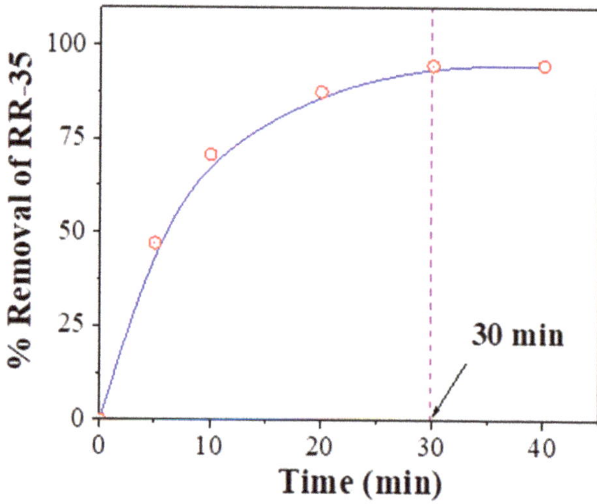

Figure 7. Effect of time on the removal of RR-35 with 20 wt% AC/TiO_2 (concentration of RR-35: 20 mg L^{-1}; pH: 5.6; irradiation time: 0–30 min).

3.8. *Effect of Initial Concentration of RR-35*

It is necessary to study the dependence of the dye removal efficiency on the initial dye concentration given its practical application. The dye removal rate of different initial RR-35 concentrations with 20 wt% AC/TiO_2 is presented in Figure 8. From Figure 8 it was observed that the dye removal rate rises with the increase in initial dye concentration from 10 to 20 mg L^{-1} and a further increase in dye concentration leads to the decrease in removal rate [37]. Additionally, with increasing the initial dye concentration up to 40 mg L^{-1}, the degradation amounts of RR-35 increased gradually, and exceeding 40 mg L^{-1}, the degradation amounts of RR-35 were almost the same. The high concentration gradient provides an increase of driving force to move RR-35 from the bulk to the nanohybrid surface. This mass transfer process continued until the nanohybrid surface was saturated with RR-35 molecules [38]. It could be stated that saturation probably occurred at 40 mg L^{-1} RR-35 concentration. Besides, at a concentration above 40 mg L^{-1}, a significant amount of sunlight may be absorbed by the dye molecules itself, which could reduce the photodegradation efficiency of AC/TiO_2 [39]. In this study, 20 mg L^{-1} of RR-35 solution was selected to evaluate the photodegradation of dye for the following experiment owing to the concentration of real wastewater.

3.9. *Role of Radical Scavenger*

Radical scavenger experiments were carried out to determine the reactive species responsible for the photodegradation of RR-35 dye. In this study, iso-propanol (IP), ascorbic acid (AA) and oxalic acid (OA) were introduced to be the scavengers of hydroxyl radicals ($^{\bullet}OH$), superoxide radical ($^{\bullet}O_2^-$) and holes (h^+), respectively [18,40–42]. The three scavenging experiments were similar to the RR-35 dye removal process. The effect of the radical scavengers on the RR-35 photodegradation over 20 wt% AC/TiO_2 using sunlight is displayed in Figure 9. From the figure, 95% removal of RR-35 was found with no scavenger (NS), while 56% removal was found with IP, 78% with AA and 85% with OA, respectively. The results strongly indicated that both $^{\bullet}OH$ and $^{\bullet}O_2^-$ radicals played a significant role in the photodegradation of RR-35, whereas, h^+ showed a minor effect in the degradation process under sunlight.

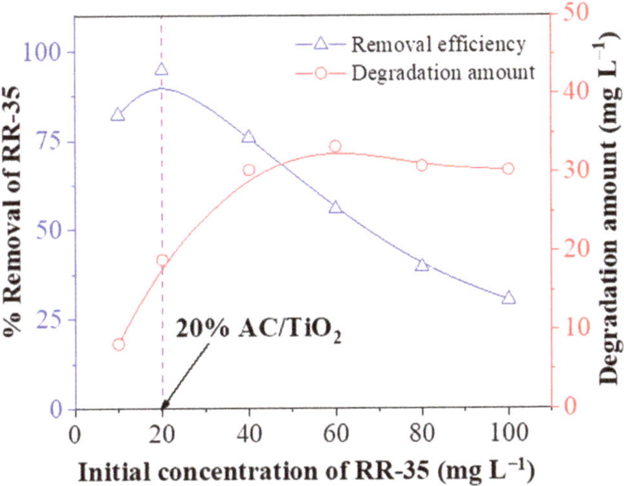

Figure 8. Effect of initial dye concentration on the removal of RR-35 with 20 wt% AC/TiO_2 (concentration of RR-35: 10–100 mg L^{-1}; pH: 5.6; irradiation time: 30 min).

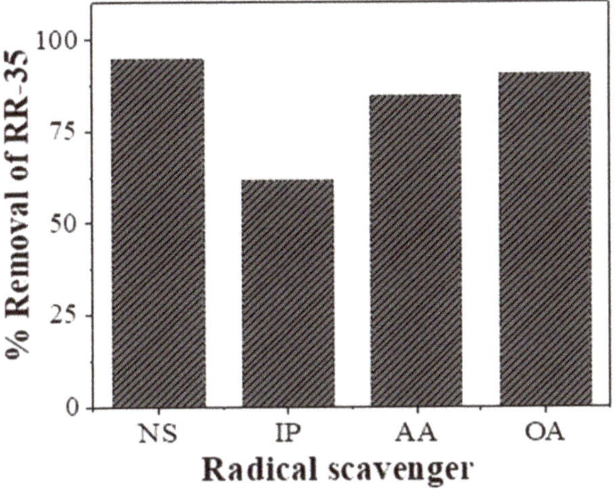

Figure 9. Role of radical scavengers on the RR-35 photodegradation with 20 wt% AC/TiO_2 (concentration of RR-35: 20 mg L^{-1}; concentration of all scavengers: 1 mmol L^{-1}, pH: 5.6; irradiation time: 30 min).

3.10. Photodegradation Mechanism

The fundamental mechanism of the AC/TiO_2 nanohybrids exhibit dual functions such as absorption and photodegradation to remove the RR-35 dye from an aqueous solution [34]. The proposed schematic illustration of the photodegradation mechanism of RR-35 dye by 20 wt% AC/TiO_2 using sunlight is presented in Figure 10. Under sunlight irradiation, the RR-35 removal process requires the dye to be adsorbed on the AC/TiO_2 surface before immediate photodegradation [43]. The efficiency of photodegradation can be affected by the adsorption performance of substrate AC and the photocatalytic activity of TiO_2 nanoparticles [34]. Moreover, the AC introduced to the composite structure acts as an electron sink, thereby decreasing the recombination of photogenerated charge carriers and finally enhancing the photocatalytic activity of the AC/TiO_2 [44]. During the sunlight irradiation, the AC/TiO_2 (considering the semiconductor of TiO_2, 3.2 eV) can absorb the UV light energy to create the conduction band electrons (e^-_{CB}) and valence band holes (h^+_{VB}) [30,45]. Then, the photogenerated holes (h^+_{VB}) are responsible for the oxidation of RR-35 dyes and water (H_2O) molecules adsorbed on the surface of the AC/TiO_2 and the formation of hydroxyl radicals ($^\bullet OH$) [44]. The photoexcited e^-_{CB} attack the dissolved oxygen (O_2) to generate superoxide radicals ($^\bullet O_2{}^-$), which can act as a precursor of many reactive species in photochemical reactions [46]. Moreover, $^\bullet O_2{}^-$ is an important source of $^\bullet OH$ radical and avoids the recombination of the electron/hole pair in the photocatalyst, thus improving the photodegradation process [47]. The photogenerated holes (h^+_{VB}), photoexcited electrons (e^-_{CB}), hydroxyl radicals ($^\bullet OH$) and superoxide radical ions ($^\bullet O_2{}^-$) take part in heterogeneous-photocatalytic reactions. These species can lead to the photodegradation of RR-35 molecules to intermediates, and then go further degradation to CO_2 and H_2O. Thus, the photodegradation of dye pollutants is one of the suitable processes because of its abundance advantages, such as breaking down pollutants and achieving complete mineralization to convert into non-toxic products [44]. Therefore, we expect the following photodegradation reactions of RR-35 dye by AC/TiO_2 under sunlight irradiation [45,47,48]:

$$AC/TiO_2 + \text{Sunlight}\ (hv) \rightarrow e^-_{CB} + h^+_{VB} \tag{1}$$

$$h^+_{VB} + OH^- \rightarrow {}^\bullet OH \tag{2}$$

$$O_2 + e^-_{CB} \rightarrow {}^\bullet O_2{}^- \tag{3}$$

$${}^\bullet O_2{}^- + e^-_{CB} + 2H^+ \rightarrow H_2O_2 \tag{4}$$

$$\text{RR-35 dye} + {}^\bullet OH/{}^\bullet O_2{}^- \rightarrow \text{Intermediates} \rightarrow CO_2 + H_2O \tag{5}$$

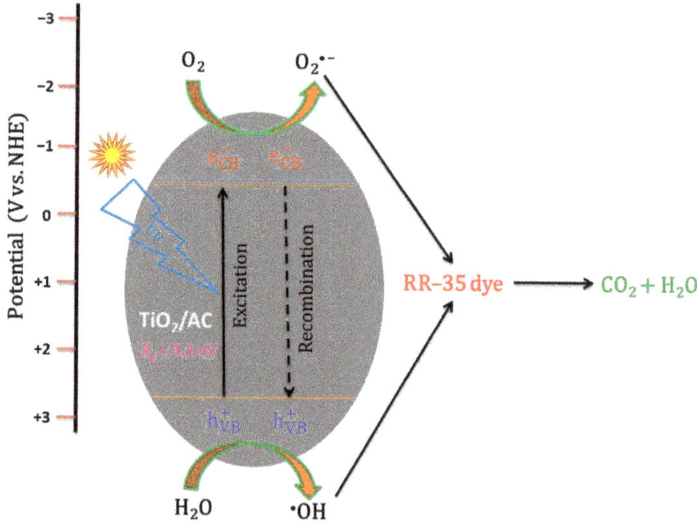

Figure 10. Schematic illustration of the photodegradation mechanism of RR-35 by 20 wt% AC/TiO_2 using sunlight.

3.11. *Reusability of AC/TiO$_2$*

Reusability is one of the most important characteristics of a material for removing pollutants from industrial effluents since it is related to the cost-effectiveness of the photocatalyst. The reusability of 20 wt% AC/TiO_2 was evaluated up to four cycles by simple alkali (0.01 M NaOH) washing [34]. The alkali-treated nanohybrid was washed with distilled water and dried for the next cycle. The result of the reusability experiment is illustrated in Figure 11. The graph clearly illustrated that the performance of the recycled 20 wt%

AC/TiO_2 for RR-35 removal remained almost consistent. The removal activity of the nanohybrid gradually decreased up to four cycles due to the poisoning effect of degraded products and the blocking of sunlight irradiation [49]. Therefore, it can be concluded that the AC/TiO_2 is suitable for long-term repeated use for the removal of RR-35 dye from the aqueous environment.

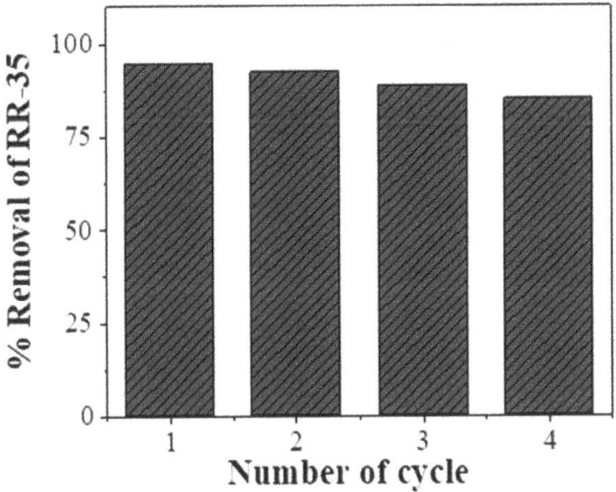

Figure 11. Reusability of 20 wt% AC/TiO_2 for RR-35 dye removal from water (concentration of RR-35: 20 mg L^{-1}; pH: 5.6; irradiation time: 30 min).

4. Conclusions

In this study, a cost-effective method has been reported for the preparation of AC/TiO_2 nanohybrids. The best removal efficiency for RR-35 dye was obtained by 20 wt% AC/TiO_2. The remarkable removal of RR-35 with AC/TiO_2 can be explained by both the adsorption and photodegradation activity of nanohybrids. RR-35 dye photodegradation with AC/TiO_2 using sunlight was mainly controlled by the $^{\bullet}OH$ and $^{\bullet}O_2^-$ radicals whereas, the h^+ played a minor effect in the process. Moreover, AC/TiO_2 nanohybrids can be regenerated by simple alkali washing. Therefore, AC/TiO_2 can be suggested as an efficient photocatalyst for RR-35 dye removal from water based on the facile synthesis, very fast with nearly complete removal, and excellent reusability.

Author Contributions: Conceptualization, M.A.I.M. and M.S.; methodology, B.M.; software, A.S. and A.M.S.; validation, B.M.; formal analysis, M.T.I., S.M.S., M.A.I.M. and M.S.; investigation, B.M.; resources, A.K.D.; data curation, B.M.; writing—original draft preparation, B.M., A.S., A.M.S. and S.C.D.; writing—review and editing, S.C.D., M.T.I., A.K.D., M.A.I.M. and M.S.; visualization, A.S., A.M.S. and S.M.S.; supervision, A.K.D.; project administration, A.K.D. All authors have read and agreed to the published version of the manuscript.

Funding: This research received no external funding.

Data Availability Statement: The data presented in this study are available in article.

Acknowledgments: The first author is grateful to the Ministry of Science and Technology, Bangladesh for providing the National Science and Technology Fellowship as a motivation for conducting this research. We are thankful to the Centre for Advanced Research in Sciences (CARS), Dhaka, Bangladesh for providing partial analytical support to carry out this research.

Conflicts of Interest: The authors declare no conflict of interest.

References

1. Wang, Q.; Yang, Z. Industrial water pollution, water environment treatment, and health risks in China. *Environ. Pollut.* **2016**, *218*, 358–365. [CrossRef] [PubMed]
2. Katheresan, V.; Kansedo, J.; Lau, S.Y. Efficiency of various recent wastewater dye removal methods: A review. *J. Environ. Chem. Eng.* **2018**, *6*, 4676–4697. [CrossRef]
3. Alves de Lima, R.O.; Bazo, A.P.; Salvadori, D.M.F.; Rech, C.M.; de Palma Oliveira, D.; de AragãoUmbuzeiro, G. Mutagenic and carcinogenic potential of a textile azo dye processing plant effluent that impacts a drinking water source. *Mutat. Res.* **2007**, *626*, 53–60. [CrossRef] [PubMed]
4. Srinivasan, S.V.; Murthy, D.V.S. Statistical optimization for decolorization of textile dyes using Trametes versicolor. *J. Hazard. Mater.* **2009**, *165*, 909–914. [CrossRef]
5. Mezohegyi, G.; van der Zee, F.P.; Font, J.A.; Fortuny, A.; Fabregat, A. Towards advanced aqueous dye removal processes: A short review on the versatile role of activated carbon. *J. Environ. Manag.* **2012**, *102*, 148–164. [CrossRef] [PubMed]
6. Ike, I.A.; Lee, Y.; Hur, J. Impacts of advanced oxidation processes on disinfection byproducts from dissolved organic matter upon post-chlor(am)ination: A critical review. *Chem. Eng. J.* **2019**, *375*, 121929. [CrossRef]
7. Sarker, M.; Bhadra, B.N.; Seo, P.W.; Jhung, S.H. Adsorption of benzotriazole and benzimidazole from water over a Co-based metal azolate framework MAF-5(Co). *J. Hazard. Mater.* **2017**, *324*, 131–138. [CrossRef] [PubMed]
8. Zangeneh, H.; Zinatizadeh, A.A.L.; Habibi, M.; Akia, M.; Isa, M.H. Photocatalytic oxidation of organic dyes and pollutants in wastewater using different modified titanium dioxides: A comparative review. *J. Ind. Eng. Chem.* **2015**, *26*, 1–36. [CrossRef]
9. Gusain, R.; Gupta, K.; Joshi, P.; Khatri, O.P. Adsorptive removal and photocatalytic degradation of organic pollutants using metal oxides and their composites: A comprehensive review. *Adv. Colloid Interface Sci.* **2019**, *272*, 102009. [CrossRef]
10. Gnanasekaran, L.; Hemamalini, R.; Saravanan, R.; Ravichandran, K.; Gracia, F.; Agarwal, S.; Gupta, V.K. Synthesis and characterization of metal oxides (CeO_2, CuO, NiO, Mn_3O_4, SnO_2 and ZnO) nanoparticles as photo catalysts for degradation of textile dyes. *J. Photochem. Photobiol. B* **2017**, *173*, 43–49. [CrossRef]
11. Al-Amin, M.; Dey, S.C.; Rashid, T.U.; Ashaduzzaman, M.; Shamsuddin, S.M. Solar assisted photocatalytic degradation of reactive azo dyes in presence of anatase titanium dioxide. *Int. J. Latest Res. Eng. Technol.* **2016**, *2*, 14–21.
12. Dong, S.; Feng, J.; Fan, M.; Pi, Y.; Hu, L.; Han, X.; Liu, M.; Sun, J.; Sun, J. Recent developments in heterogeneous photocatalytic water treatment using visible light-responsive photocatalysts: A review. *RSC Adv.* **2015**, *5*, 14610–14630. [CrossRef]
13. Lee, S.-Y.; Park, S.-J. TiO_2 photocatalyst for water treatment applications. *J. Ind. Eng. Chem.* **2013**, *19*, 1761–1769. [CrossRef]
14. Paul, S.C.; Dey, S.C.; Molla, M.A.I.; Islam, M.S.; Debnath, S.; Miah, M.Y.; Ashaduzzaman, M.; Sarker, M. Nanomaterials as electrocatalyst for hydrogen and oxygen evolution reaction: Exploitation of challenges and current progressions. *Polyhedron* **2021**, *193*, 114871. [CrossRef]
15. Dong, H.; Zeng, G.; Tang, L.; Fan, C.; Zhang, C.; He, X.; He, Y. An overview on limitations of TiO_2-based particles for photocatalytic degradation of organic pollutants and the corresponding countermeasures. *Water Res.* **2015**, *79*, 128–146. [CrossRef] [PubMed]
16. Chen, J.; Qiu, F.; Xu, W.; Cao, S.; Zhu, H. Recent progress in enhancing photocatalytic efficiency of TiO_2-based materials. *Appl. Catal. A Gen.* **2015**, *495*, 131–140. [CrossRef]
17. Asiltürk, M.; Şener, Ş. TiO_2-activated carbon photocatalysts: Preparation, characterization and photocatalytic activities. *Chem. Eng. J.* **2012**, *180*, 354–363. [CrossRef]
18. Hasan, A.K.M.M.; Dey, S.C.; Rahman, M.M.; Zakaria, A.M.; Sarker, M.; Ashaduzzaman, M.; Shamsuddin, S.M. A kaolinite/TiO_2/ZnO-based novel ternary composite for photocatalytic degradation of anionic azo dyes. *Bull. Mater. Sci.* **2020**, *43*, 1–9. [CrossRef]
19. Wang, C.-C.; Wang, X.; Liu, W. The synthesis strategies and photocatalytic performances of TiO_2/MOFs composites: A state-of-the-art review. *Chem. Eng. J.* **2020**, *391*, 123601. [CrossRef]
20. Sun, Z.; Bai, C.; Zheng, S.; Yang, X.; Frost, R.L. A comparative study of different porous amorphous silica minerals supported TiO_2 catalysts. *Appl. Catal. A Gen.* **2013**, *458*, 103–110. [CrossRef]
21. Suba, V.; Saravanabhavan, M.; Krishna, L.S.; Kaleemulla, S.; Kumar, E.R.; Rathika, G. Evaluation of curcumin assistance in the antimicrobial and photocatalytic activity of a carbon based TiO_2 nanocomposite. *New J. Chem.* **2020**, *44*, 15895–15907. [CrossRef]
22. Martins, A.C.; Cazetta, A.L.; Pezoti, O.; Souza, J.R.B.; Zhang, T.; Pilau, E.J.; Asefa, T.; Almeida, V.C. Sol-gel synthesis of new TiO_2/activated carbon photocatalyst and its application for degradation of tetracycline. *Ceram. Int.* **2017**, *43*, 4411–4418. [CrossRef]
23. Liu, S.X.; Chen, X.Y.; Chen, X. A TiO_2/AC composite photocatalyst with high activity and easy separation prepared by a hydrothermal method. *J. Hazard. Mat.* **2007**, *143*, 257–263. [CrossRef] [PubMed]
24. Ambrus, Z.; Bala´zs, N.; Alapi, T.; Wittmann, G.; Sipos, P.; Dombi, A.; Mogyoro´si, K. Synthesis, structure and photocatalytic properties of Fe(III)-doped TiO_2 prepared from $TiCl_3$. *Appl. Catal. B. Environ.* **2008**, *81*, 27–37. [CrossRef]
25. Bansal, P.; Sud, D. Photocatalytic degradation of commercial dye, CI Reactive Red 35 in aqueous suspension: Degradation pathway and identification of intermediates by LC/MS. *J. Mol. Catal. A Chem.* **2013**, *374–375*, 66–72. [CrossRef]
26. Sarker, M.; Ahmed, I.; Jhung, S.H. Adsorptive removal of herbicides from water over nitrogen-doped carbon obtained from ionic liquid@ZIF-8. *Chem. Eng. J.* **2017**, *323*, 203–211. [CrossRef]
27. Zhu, Z.; Li, A.; Zhong, S.; Liu, F.; Zhang, Q. Preparation and characterization of polymer-based spherical activated carbons with tailored pore structure. *J. Appl. Polym. Sci.* **2008**, *109*, 1692–1698. [CrossRef]

28. Liu, Y.; Yang, S.; Hong, J.; Sun, C. Low-temperature preparation and microwave photocatalytic activity study of TiO_2-mounted activated carbon. *J. Hazard. Mater.* **2007**, *142*, 208–215. [CrossRef]
29. Huang, D.; Miyamoto, Y.; Matsumoto, T.; Tojo, T.; Fan, T.; Ding, J.; Guo, Q.; Zhang, D. Preparation and characterization of high-surface-area TiO_2/activated carbon by low-temperature impregnation. *Sep. Purif. Technol.* **2011**, *78*, 9–15. [CrossRef]
30. Xing, B.; Shi, C.; Zhang, C.; Yi, G.; Chen, L.; Guo, H.; Huang, G.; Cao, J. Preparation of TiO_2/activated carbon composites for photocatalytic degradation of RhB under UV light irradiation. *J. Nanomater.* **2016**, *2016*, 8393648. [CrossRef]
31. Xu, Y.-J.; Zhuang, Y.; Fu, X.Z. New insight for enhanced photocatalytic activity of TiO_2 by doping carbon nanotubes: A case study on degradation of benzene and methyl orange. *J. Phys. Chem. C* **2010**, *114*, 2669–2676. [CrossRef]
32. Anyika, C.; Asri, N.A.M.; Majid, Z.A.; Yahya, A.; Jaafar, J. Synthesis and characterization of magnetic activated carbon developed from palm kernel shells. *Nanotechnol. Environ. Eng.* **2017**, *2*, 1–25. [CrossRef]
33. Saravanan, R.; Aviles, J.; Gracia, F.; Mosquera, E.; Gupta, V.K. Crystallinity and lowering band gap induced visible light photocatalytic activity of TiO_2/CS (Chitosan) nanocomposites. *Int. J. Biol. Macromol* **2018**, *109*, 1239–1245. [CrossRef]
34. Xue, G.; Liu, H.H.; Chen, Q.Y.; Hills, C.; Tyrer, M.; Innocent, F. Synergy between surface adsorption and photocatalysis during degradation of humic acid on TiO_2/activated carbon composites. *J. Hazard. Mater.* **2011**, *186*, 765–772. [CrossRef] [PubMed]
35. Dey, S.C.; Moztahida, M.; Sarker, M.; Ashaduzzaman, M.; Shamsuddin, S.M. pH-triggered interfacial interaction of kaolinite/chitosan nanocomposites with anionic azo dye. *J. Compos. Sci.* **2019**, *3*, 39. [CrossRef]
36. Atout, H.; Bouguettoucha, A.; Chebli, D.; Gatica, J.M.; Vidal, H.; Yeste, M.P.; Amrane, A. Integration of adsorption and photocatalytic degradation of methylene blue using TiO_2 supported on granular activated carbon. *Arab. J. Sci. Eng.* **2017**, *42*, 1475–1486. [CrossRef]
37. Sakib, A.A.M.; Masum, S.M.; Hoinkis, J.; Islam, R.; Molla, M.A.I. Synthesis of CuO/ZnO nanocomposites and their application in photodegradation of toxic textile dye. *J. Compos. Sci.* **2019**, *3*, 91. [CrossRef]
38. Lang, X.; Wang, T.; Sun, M.; Chen, X.; Liu, Y. Advances and applications of chitosan-based nanomaterials as oral delivery carriers: A review. *Int. J. Biol. Macromol.* **2020**, *154*, 433–445. [CrossRef]
39. Molla, M.A.I.; Ahsan, S.; Tateishi, I.; Furukawa, M.; Katsumata, H.; Suzuki, T.; Kaneco, S. Degradation, kinetics, and mineralization in the solar photocatalytic treatment of aqueous amitrole solution with titanium dioxide. *Environ. Eng. Sci.* **2018**, *35*, 401–407. [CrossRef]
40. Santhosh, C.; Malathi, A.; Daneshvar, E.; Kollu, P.; Bhatnagar, A. Photocatalytic degradation of toxic aquatic pollutants by novel magnetic 3D-TiO_2@HPGA nanocomposite. *Sci. Rep.* **2018**, *8*, 1–15. [CrossRef]
41. Molla, M.A.I.; Furukawa, M.; Tateishi, I.; Katsumata, H.; Kaneco, S. Fabrication of Ag-doped ZnO by mechanochemical combustion method and their application into photocatalytic Famotidine degradation. *J. Environ. Sci. Health. Part A* **2019**, *54*, 914–923. [CrossRef] [PubMed]
42. Su, J.; Zhu, L.; Geng, P.; Chen, G. Self-assembly graphitic carbon nitride quantum dots anchored on TiO_2 nanotube arrays: An efficient heterojunction for pollutants degradation under solar light. *J. Hazard. Mater.* **2016**, *316*, 159–168. [CrossRef] [PubMed]
43. Omri, A.; Benzina, M. Almond shell activated carbon: Adsorbent and catalytic support in the phenol degradation. *Environ. Monit. Assess.* **2014**, *186*, 3875–3890. [CrossRef] [PubMed]
44. Baca, M.; Wenelska, K.; Mijowska, E.; Kaleńczuk, R.J.; Zielińska, B. Physicochemical and photocatalytic characterization of mesoporous carbon/titanium dioxide spheres. *Diam. Relat. Mater.* **2020**, *101*, 107551. [CrossRef]
45. Molla, M.A.I.; Tateishi, I.; Furukawa, M.; Katsumata, H.; Suzuki, T.; Kaneco, S. Evaluation of reaction mechanism for photocatalytic degradation of dye with self-sensitized TiO_2 under visible light irradiation. *Open J. Inorg. Non-Met. Mater.* **2017**, *7*, 1–7.
46. Ahmed, A.Z.; Islam, M.M.; Islam, M.M.; Masum, S.M.; Islam, R.; Molla, M.A.I. Fabrication and characterization of B/Sn-doped ZnO nanoparticles via mechanochemical method for photocatalytic degradation of rhodamine B. *Inorg. Nano-Met. Chem.* **2020**. [CrossRef]
47. Asencios, Y.J.O.; Lourenc¸o, V.S.; Carvalho, W.A. Removal of phenol in seawater by heterogeneous photocatalysis using activated carbon materials modified with TiO_2. *Catal. Today* **2020**. [CrossRef]
48. Yao, S.; Li, J.; Shi, Z. Immobilization of TiO_2 nanoparticles on activated carbon fiber and its photodegradation performance for organic pollutants. *Particuology* **2010**, *8*, 272–278. [CrossRef]
49. Argyle, M.D.; Bartholomew, C.H. Heterogeneous catalyst deactivation and regeneration: A review. *Catalysts* **2015**, *5*, 145–269. [CrossRef]

MDPI
St. Alban-Anlage 66
4052 Basel
Switzerland
Tel. +41 61 683 77 34
Fax +41 61 302 89 18
www.mdpi.com

Photochem Editorial Office
E-mail: photochem@mdpi.com
www.mdpi.com/journal/photochem

www.ingramcontent.com/pod-product-compliance
Lightning Source LLC
LaVergne TN
LVHW070906170726
843515LV00005B/714

* 9 7 8 3 0 3 6 5 3 6 4 8 4 *